Contributions to Phenomenology

In Cooperation with The Center for Advanced Research in Phenomenology

Volume 130

Scope

The purpose of the series is to serve as a vehicle for the pursuit of phenomenological research across a broad spectrum, including cross-over developments with other fields of inquiry such as the social sciences and cognitive science. Since its establishment in 1987, Contributions to Phenomenology has published more than 100 titles on diverse themes of phenomenological philosophy. In addition to welcoming monographs and collections of papers in established areas of scholarship,the series encourages original work in phenomenology. The breadth and depth of the Series reflects the rich and varied significance of phenomenological thinking for seminal questions of human inquiry as well as the increasingly international reach of phenomenological research.

All books to be published in this Series will be fully peer-reviewed before final acceptance.

The series is published in cooperation with The Center for Advanced Research in Phenomenology.

Lajos Horváth

The Affective Core Self

The Role of the Unconscious and Retroactivity in Self-Constitution

Springer

Lajos Horváth
Department of Philosophy
University of Debrecen
Debrecen, Hungary

ISSN 0923-9545 ISSN 2215-1915 (electronic)
Contributions to Phenomenology
ISBN 978-3-031-56919-7 ISBN 978-3-031-56920-3 (eBook)
https://doi.org/10.1007/978-3-031-56920-3

This Springer imprint is published by the registered company Springer Nature Switzerland AG
The registered company address is: Gewerbestrasse 11, 6330 Cham, Switzerland

Foundings

This book was supported by the János Bolyai Research Scholarship of the Hungarian Academy of Sciences (BO/00189/21/2) and by the No. K 138745 project of the Hungarian Scientific Research Fund.

Acknowledgment

I would like to thank the editors for their editorial assistance. I am also grateful to Tamás Ullmann, Csaba Szummer, and Bence Péter Marosán for their constructive feedback and suggestions.

Introduction

The general aim of the current work is to examine the role of the self in the contemporary interdisciplinary dialogue between phenomenology and consciousness studies. Recently, there has been a renewed interdisciplinary interest in the problem of the self and self-awareness. Further tasks of the book are to analyze the wide-ranging discussions of the so-called minimal self and to examine the theoretical mutations and utilizations of the concept. The phenomenological psychopathology of schizophrenia will serve a good example to explicate the intricacies of the minimal self. In contemporary phenomenological psychiatry, schizophrenia has been labelled as an anomalous self-experience that clearly shows the intertwined aspects of bodily feelings and world-related experiences on the one hand and sheds light on the fragility of identity on the other hand. In this framework, identity refers to the pre-reflective sense of self including the impressions of persistency, continuity, and coherence. As it will be demonstrated, anomalous self-experiences, for example, hallucinations, delusional states, incoherent bodily feelings, and even accompanying psychotic states, inspire researchers and philosophers to take into account the vulnerability of identity and search for the rules and processes of self-constitution. The logic behind this kind of reasoning is quite simple: if conscious self-experience is prone to fragmentation and disintegration, then there have to be a set of internal and external factors for maintaining stable identity.

This work can be seen as an interdisciplinary endeavor, a kind of hybrid and harmonizing attempt that has many precursors in contemporary philosophy. For instance, Dan Zahavi regularly combines phenomenological insights with the findings of contemporary neuroscience. Regarding the question of memory and identity, Richard Wollheim, Peter Goldie, Thomas Fuchs, and Marya Schechtman have often applied different psychological models of memory to support their arguments. This work also employs the method of hybridization and harmonization by utilizing phenomenological theories including the characterization of the self and the unconscious. Certain results of contemporary memory research and consciousness studies will also be applied. The reason behind these interdisciplinary hybridization tendencies is that certain philosophical issues can be approached like a jigsaw puzzle. This undertaking complicates matters further by using certain insights of psychoanalysis in order to enrich the discussions of the self. This work joins to the family of the multidimensional analyses of the self – as Zahavi has formulated the contemporary interdisciplinary approach to the self.

According to this proposal, the phenomenology of self-constitution and the explanation of anomalous self-experiences cannot be complete without taking into account the explanatory role of the unconscious. The current work attempts to join with those inter- and multidisciplinary undertakings which provide multi-layered as well as multi-dimensional account of self-experience. In order to broaden the scope of the multi-layered investigation, we have to bring into consideration the role of the unconscious in self-constitution, and the problem of the unconscious will directly lead us to the phenomenological problems of time-consciousness and autobiographical memory. Therefore, another important guideline of the work is the phenomenological problem of passivity, since, from phenomenological point of view, the unconscious can be understood as a pre-egoic, pre-intentional process. The crucial task of the book is to explicate the discrepancies between the phenomenological and the psychoanalytical notions of the unconscious. The main theoretical positions of these opposing paradigms will be analyzed in great detail.

Another important task is to examine the role of the minimal self in psychopathological context, more specifically, in contemporary phenomenological research on schizophrenia. The notions of the sense of self and minimal self are closely related and often synonymous. The minimal self refers to the pre-reflective or implicit layer of self-constitution. The question to consider is that is there any way to connect the problem of the unconscious with the minimal self? Consequently, different notions of the phenomenological and psychoanalytical unconscious will be compared and analyzed in detail. In addition, I will try to demonstrate that the cognitive and implicit notions of the unconscious also play a pivotal role in the contemporary interdisciplinary debates on the nature of the self and personal identity. These considerations will lead us to the problems of episodic and autobiographical memory.[1] It is easy to blur the line between episodic and autobiographical memory. For start, it is important to highlight that autobiographical memories come

[1]Autobiographical memory has its own developmental trajectory. There are overlapping features between episodic and autobiographical memories, but the two categories do not refer to the same mental phenomenon. Nelson (2003) states that the feeling that "I was there, I did that" is characteristic feature of both episodic and autobiographical memory. Endel Tulving argued that the main feature of episodic memory is the mental time travel in which we can literally relive the past (p. 11). Of course, reviving or reliving the past is a difficult issue that will be of particular importance. Nelson further highlights the role of *social interaction* and *narrative form* in the development of autobiographical memories: "The narrative form takes a mundane event, gives it a setting of time and place, provides a central action or goal, a motivation, highlights a highpoint of surprise, success, or failure, an emotion, a conclusion, and an evaluation. The verbal form both organizes the experience and provides a rationale for remembering it as significant personally or socially" (Nelson, 2003, p. 13).

with the sentiment that the represented event is happened in our past and they presuppose a personal history into which they can be embedded.[2]

For preliminary purposes, let me point out that the intertwining aspects of autobiographical memory and the self may serve as a convergence zone for attenuating the glaring contextual and perspectival oppositions between the phenomenological and psychoanalytical unconscious. For start, the sense of self cannot be comprehended without the internal dynamics of temporality and environmental determinants. In the past decades, the paradigm of embodied mind and cognition has made enormous efforts to describe the intertwined aspects of the sense of being alive, being in time, and self-awareness. And the notion of the sense of self, which refers to a qualitative state interwoven with the internal experience of time, can be embedded into this dynamic and ecological framework. The sense of self can be investigated from developmental psychological point of view, that is, it is constituted by biological and societal factors. The sense of self is the first-personal manifestation of the *embodied* and ecologically *embedded* self.

The primary aim of the current work is to introduce a *phenomenological theory of psychointegration* that seems, at first sight, a self-refuting undertaking, since the foundational gesture of the Husserlian phenomenology was to keep distance from psychoanalysis and psychology in general due to their inherent naturalistic assumptions. However, as contemporary interpretations have suggested, in the Husserlian corpus the problem of the unconscious has frequently occurred as the limit problem (*Grenzprobleme*) of phenomenology. The underlying assumption of the current undertaking is that the problems of the unconscious and the minimal self can be compared from the viewpoint of *affective identity* that implies the notion of the

[2] Bermúdez (2017) argues that non-autobiographical memories do not possess the belief or sentiment that the episode is part of our history, it happened in our past. An autobiographical memory requires the "awareness of oneself as having a personal history." In this respect, human beings are special kind of narrative creatures who are able to fit their remembered episodes into their ongoing autobiography. A serious consequence of the self-related autobiographical memories is that it is possible that only language users possess autobiographical memories and creatures who experience time in a cyclical fashion are unable to construct a personal identity on episodic memories. In this respect, autobiographical memory is inseparable from narration, but episodic memory does not require autobiographical narrative and self-concern (p. 189). However, the definition of episodic memory is not unequivocal. Definitions of episodic memory often allude to self-relatedness. Nichols (2017) points to the ambiguity with regard to episodic memory: William James and, in contemporary literature, Hayne and colleagues (2011) represent the prevailing view according to which the awareness that an event happened to "me" is a defining characteristic of episodic memory. However, Nichols argues that it is not necessary that the "sense of personal identity" accompanies the retrieval of episodic memories. For example, neuroscientists, who experiment with rats, characterizes episodic memory without the reference to explicit self-representation. Or there is the case study of the patient R. B., who suffered a head trauma, and after the accident he felt alienated from his recollected past episodes. For instance, he reported that he is able to clearly reconstruct the childhood episode of being at the beach in New London with his family, but the scenery seemed like for him as looking at a photo of someone else's vacation. Similarly, he lost of the ownership of a memory from his childhood. He recounted that the episode of studying with his friend at school felt like and imagining. The intellectual comprehension that the episode is integral to his autobiography does not conveyed the feeling of ownership of that memory (pp. 174–175).

affective unconscious. The notions of affective identity and affective unconscious highlight the fact that the recollection of the past and the anticipation of the future are not only elaborative cognitive processes but also affective accomplishments as well. The main question of the current investigation is a classical philosophical and psychoanalytic question: how does the past embedded in the present? Husserl and Freud, despite their totally different frameworks, ruminated over the role of the past in the constitution of the present. The "tyranny of the past"—to use Richard Wollheim's (1984) phrase—can be understood as the presence of the past experiences in the present.

The careful examination of self-constitution takes into account the different aspects of the emerging past in the streaming present. In this respect, I will focus on three more or less interrelated aspects of the past-present relation. First, I will attempt to demonstrate that the recent discussions on the role of body memory in self-constitution may shed some light on the constructive processes in remembering and unravels the implicit presence of the past. Second, the current work will offer a comparative study that focuses on the similarities and differences between the phenomenological notion of retroactivity and the Freudian notion of "afterwardsness" (*Nachträglichkeit*). The Freudian notion of *Nachträglichkeit* will be of paramount importance, and I will attempt to integrate the Freudian insights into the phenomenology of retroactivity. Hopefully these considerations and analyses will establish the comprehensive account of the affective unconscious and affective identity.

The problem of affective identity is not a unique phenomenological issue but rather an encompassing interdisciplinary subject matter in contemporary phenomenology, cognitive neuroscience, developmental psychology, and finally in memory research. It could be argued that the dilemmas of the minimal self and affective identity are combined and closely connected to body memory and narrative identity. It is important to highlight that at the level of lived experience there are no clear-cut distinctions between the minimal self and the person or between affective identity and narrative identity. As Zahavi and Ratcliffe frequently emphasize, the theoretical notions of the self and identity are only useful abstractions, and one's actual lived experience is the unique composition of implicit and explicit elements of self-constitution. As a result, there could be unimaginable individual variations of personality profiles in which narrative self-understanding, involuntarily body memories, autobiographical memories, false memories, fantasies, etc. play their different role in a unique way. To a certain extent, the minimal self poses a challenge to narrative identity and vice versa. By giving a prominent role to *the retroactive constitution of the past*, I will try to demonstrate that there can be an intermediate horizon of inquiry between the narrative and the minimal notions of the self.

The role of affectivity has come to the foreground in the philosophical inquiries on the nature of the self and identity. For Husserl, affectivity plays a crucial role in organizing the impressional sphere of consciousness and establishing the temporal connectedness with the past. For Freud, basic affections (pleasure and pain) have a fundamental role in our well-being, and often unprocessed, unconscious traumatic affections are the impediments of living a harmonious life. In a similar vein, for contemporary embodied mind theories, affectivity plays an important role in establishing the cognitive accomplishments of consciousness. It is not an exaggeration to say that affectivity sheds light on the precognitive layers of intentionality and widens the scope of the phenomenological investigations. Of course, it is easy to fall into the pitfall of essentialist thinking, which, for example, overvalues the role of affectivity in self-constitution and overlooks other important factors, such as the primary role of the lifeworld or narrative and cognitive developmental processes. In order to avoid one-sidedness, the current project aims to take into consideration the intertwining aspects of the affective and narrative identity.

Affective and narrative identity are inseparably connected to the meaning-bestowing process of consciousness. It can be claimed, from psychoanalytic as well as phenomenological point of view, that the process of psychointegration is a crucial aspect of narrative self-understanding that may involve the re-schematization of the affective infrastructure of the self. I will try to show that an *affective core self* can be positioned into the intermediate level between the minimal and the narrative self. On the one hand, the hypothesized affective core self is the subject of a continuous disintegration and reintegration processes (i.e., it is altered owing to the present circumstances and factors), and, on the other hand, it has its own autonomy, since it is able to actuate a pre-reflective meaning-bestowing process. The important question is following: can we imagine or construe a phenomenological theory of psychointegration in order to describe radical and less radical changes in the structure of the self? What can be the phenomenological consequences of the problem of psychointegration? How can we interpret the psychological process of psychointegration in phenomenological terms? To address and specify these general questions, I will turn to phenomenological discussions centered on the problems of *traumatic subjectivity* and *memory research* (including studies of implicit memory and especially body memory). The term "disintegration-reintegration process" refers to the radical changes in personality structure which often engendered by traumatic events. However, a phenomenological psychointegration theory has to take into account the more subtle changes in the self as well. In this respect, the problem of personal identity or the survival of the former self must be investigated.

In this respect, apparently, in contemporary analytic philosophy, a "narrative turn" took place. In the works of Peter Goldie, Maria Schechtman, and many others, the question of the survival of the past self and the reflective and often retrospective access to the past became a crucial topic to discuss. In contrast to Derek Parfit's thought experiments, which discuss the problem of personal identity from a metaphysical point of view, these new approaches highlight the importance of narrative and affective relations to the past. Because of the "phenomenological sensibility" of these approaches, I will try to connect the problem of the survival of the past self with the dilemmas of the retroactive awakening of a traumatic past. With the phenomenology of retroactivity and the Freudian *Nachträglichkeit*, the revision and reconsideration of past experiences can be investigated. In a remarkably similar vein, by means of narrative self-understanding, the relation between the actual and the past self can be examined in various ways. Hopefully, the following interpretations and arguments will show that phenomenology, psychoanalysis, and even the discussion of personal identity can contribute fruitfully and effectively to define the problem of the affective identity.

Due to the complexity of the discussions of the self, I restricted the following investigations to the problem of the minimal self. The problem of the minimal self will serve as a springboard from which several other dilemmas can be examined and interpreted: for example, the relation between the minimal and the narrative self, the disintegration and alienation of the self, the constructive elements in autobiography, the phenomenon of retroactive affective awakening, the survival of the distant past in habitual body memory, and so on.

The current work is not oblivious of the methodological and contextual differences between the different paradigms of the self. Obviously, there are several reasons against the synthesis of psychoanalytic and phenomenological ideas. The thematic convergences on which recent interdisciplinary investigations founded pay attention to the epistemological and methodological dilemmas. For example, the Freud's naturalism is basically incommensurable with the Husserlian transcendental approach. But despite the contextual and epistemological differences, one can extract phenomenologically significant considerations from Freudian oeuvre. To recapitulate, I do not intend to provide the synthesis of the above-mentioned paradigms, rather the current work can be understood as a careful circumspection of thematic convergences between different approaches. The basic ground for the analysis in question is the broad category of *self-constitution* which implies a culturally and theoretically open-ended, incessantly renewable inquiry.

Part I of the work discusses the problem of the self in general and subsequently introduces the notion of the minimal self, that is, the idea of pre-reflective self-awareness. In contemporary discussions of subjectivity and consciousness, the minimal self plays a pivotal role. Moreover, the minimal self serves as an *umbrella term*; it is the common theoretical target of cognitive neuroscience, phenomenology, developmental psychology, and finally of philosophy of mind. The minimal self tries to bridge the gap between philosophy of mind and phenomenology. The aim of the first chapter is to give a detailed account of the self in the context of contemporary consciousness studies. That is, I will delineate the paradigm of the minimal self in the boarder context of philosophy of mind, consciousness studies, and embodiment. Therefore, I will put heavy emphasis on the bodily nature of subjectivity and the minimal self. The convergence between philosophy of mind and phenomenology concerning the problem of the self and personal identity is also an important issue for this chapter. The problem of the self is deeply connected with the issues of the Husserlian inner time-consciousness, self-affection, and pre-reflective self-awareness. As we shall see, these phenomenological ideas are combined in Zahavi's notion of the minimal self. Of course, the terminological divergences and convergences between phenomenology and philosophy of mind must also be addressed. As the contemporary debates demonstrate, the paradigm of minimal self is imbued with semantical and terminological difficulties. In the wider context, skeptical objections can be raised against the trends of consciousness studies and embodiment as well. What is important to see here is that there is no generally agreed definition of the minimal self; therefore, the proliferation, convergence, and intersection of different paradigms and theories will be scrutinized.

The investigation of subjectivity, as Zahavi and others demonstrated, is a mutual theoretical inquiry for philosophers of mind and phenomenologists as well. Zahavi offers a Husserlian and Sartrean solution to the problem of phenomenological consciousness and to the diachronic continuity of the self. However, his notion of the minimal self keeps a distance from the intricate discussions of psychological continuity and personal identity, in general. Nevertheless, Zahavi frequently addresses the problem of the narrative self, but, at the same time, he argues for the fundamental role of the minimal self in self-constitution. Therefore, Part I will give enough room for discussion of the relations between certain narrative theories of selfhood and the minimal self. This train of thought will try to highlight both the contradictory and complementary aspects of the minimal self and the narrative self. Finally, the chapter will clarify the basic theoretical and conceptual considerations with regard to pre-reflective self-awareness.

Part II will examine the problem of the unconscious from both psychoanalytical and phenomenological point of view. The problem of the minimal self addresses not only the mysteries of phenomenal consciousness and affectivity but also the separation between conscious and nonconscious mental states. The customary approach to address this problem is to examine the opposing views of higher order theories of reflective-awareness, and the same order theories of pre-reflective self-awareness. The aim of the part is to turn away from this distinction in favor of the dilemmas of the Freudian unconscious and repression. Repression will be of particular importance in comparing the phenomenological and psychodynamic notions of the unconscious. The analysis of repression in phenomenology and psychoanalysis will highlight the role of affectivity and passivity in the constitution of the self. I will focus on the illuminating works of Rudolf Bernet, Thomas Fuchs, and Anastasia Kozyreva. The crucial differences between the two main alternatives of the phenomenological unconscious, namely the *representational* and *non-representational*, are also to be discussed in Part II. The former comprises the role of remembering and the latter focuses on the implicit layers of the self, that is, body memory and affectivity plays a pivotal role in these discussions. Part II will also look at closely to the problem of the *affective unconscious* and addresses the dilemma of traumatic subjectivity as well. Finally, Part II will foreshadow the conclusion that the constitution of the affective core self is deeply intertwined with the invisible, implicit affective infrastructure of the self.

In Part III, the scope of discussion will be extended further to include psychopathological considerations. In this respect, I will discuss the interconnected issues of self-constitution and self-disorders. In contemporary consciousness studies, the philosophical implications of psychopathology came to the foreground. The notion of *anomalous self-experience* implies that the examination of consciousness is not restricted specifically to the analysis of attention, awareness, or wakefulness; rather, it enlarges the scope of enquiry to various conscious states, including psychiatric illnesses (e.g., depression, schizophrenia, post-traumatic stress disorder, etc.). In this respect, recent developments in phenomenological psychiatry will be rightful addendum to the discussions of the minimal self. The recent interdisciplinary literature is abundant with analyses concerning psychopathological disorders, and the term "self-disorder" is widespread among phenomenologically inclined scholars. In fact, on closer examination, it became clear that the minimal self offers many interpretative and explanatory theories to certain anomalous self-experiences.[3] In Part III, I will scrutinize the ipseity-disturbance model (IDM) of schizophrenia, and in Chap. 6 (and especially in Sect. 6.3), certain aspects of post-traumatic stress disorder (PTSD) will be discussed.[4] Again, I side with the proponents of embodied mind who claim that a mutual enlightenment can be established between phenomenology and clinical practice; however, my approach does not follow the project of naturalizing phenomenology.[5]

The introduction of Part III will briefly consider the methodological problems and internal conflicts of phenomenologically oriented psychiatry. Nevertheless, the analysis of the IDM model may shed light on such anomalous self-experiences which might have serious consequences for the idea of the minimal self. The crucial question to ask is whether there is an endogenous activity that gives rise to alienating experiences (e.g., "unworlding") in schizophrenia, or researchers have to place more emphasis on the social and affective components of the schizophrenic self-disorder. As we will see, currently the phenomenological interpretations of schizophrenia are proliferating and striving for acknowledgment. For start, it is notable that even phenomenologically oriented approaches could not provide an exhaustive account of schizophrenia spectrum disorders. What is philosophically relevant in these approaches is that their explanatory target is the onset of psychosis, the earliest stage of schizophrenia (the prodromal phase). This stage of schizophrenia is not a full-blown psychopathological entity, rather, it can be characterized as anomalous self-experience and world experience.

It is extraordinarily difficult to disentangle the hidden causal factors of such disorders. Nevertheless, the phenomenological investigations of IDM and PTSD might give the opportunity to reconsider the phenomenological account of trauma and

---

[3] For preliminary purposes, it is important to note that the category of "anomalous self-experiences" is not identical with "altered states of consciousness." While each category implies the wide variety of non-ordinary conscious states, the term "anomalous self-experience" derives from phenomenological psychiatry and generally alludes to delusions, hallucinations, altered experience of the lifeworld, i.e., to psychopathological alterations of the self-world relation.

[4] Post-traumatic stress disorder (or PTSD) could be the result of exposure to traumatic events (e.g., rape, war, torture, and accidents, etc.). According to Stocker, the core symptoms are trauma-related nightmares, intrusive flashbacks, and avoidance of reminder cues and activities. Another important feature of this tormenting state is hyperarousal in the form of hypervigilance (cf. Stocker, 2020, p. 322). The phenomenon of PTSD and its special relation to lived time will be covered and discussed in Sect. 6.3.

[5] The debated project of "naturalizing phenomenology" was one of the foundational gestures of embodied mind. In short, the aim of this project is to harmonize neuroscientific findings with the basic phenomenological structure of experience. The paradox of the undertaking lies in the fact that Husserl himself argued for the primacy of phenomenology against the backdrop of natural and other sciences (cf. Moran, 2013).

traumatization per se. As previously mentioned, the role of the unconscious came to the foreground in contemporary phenomenological approaches that also entails the analysis of repression and trauma. In his recent papers, Tamás Ullmann (2015, 2017), following Rudolf Bernet (2000), introduced the notion of "traumatic subjectivity" that not only implies trauma-related self-experiences but also a quasi-autonomic integrative process that is closely related to retroactive self-understanding. Naturally, the notion of trauma plays a crucial role in both psychoanalysis and phenomenology; however, there are crucial differences between the two paradigms. One can find several different notions of trauma from Freud to Ferenczi, from Levinas to Waldenfels; however, there is no generally agreed definition. Bearing in mind the various notions of trauma, the part will focus especially on psychological (and emotional) traumas and their implications for the minimal self. And the above-mentioned psychopathological disorders demonstrate the vulnerability of the self and its inclination for disintegration due to traumatization and other physical and psychical factors. The problem of trauma, as the motive for retrospective revision and even the reconfiguration of the past, will be analyzed in greater detail in Chaps. 6 and 7.

I will synthesize the main arguments of the book in Part IV and further examine the problem of the minimal self. According to the current proposal, one can establish an important distinction between the *minimal self* and the *affective core self*. Growing number of theorists underscore the role of affectivity in self-constitution and self-experience. While the formally defined minimal self remains constant through life, the affective core self can be regarded as a swiftly or slowly changing self-experience, deeply embedded in inner time-consciousness and situated in the lifeworld. Due to traumatic circumstances, the affective core self constantly subjected to disintegration and reintegration processes. Sudden physical or psychical shock or the slower process of alienation can initiate different kinds of maladaptive and adaptive processes. The phenomenological dilemma of the changing self and the Freudian problem of *Nachträglichkeit* is to be addressed in Chap. 6.

The current work proposes that the Husserlian notion of affective relief, as a special kind of unconscious, can illustrate the structural complexity and phenomenal thickness of the above-mentioned affective core self. In this respect, the detailed analysis of the phenomenological unconscious, including the Husserlian notion of *affective relief*, serves complementary purposes. I argue that *retroactive meaning-constitution* is the fundamental function of the affective core self. Furthermore, I will integrate the concept of *implicit body memory* into the affective core self. While the minimal self indicates a formal notion of identity, the affective core self has its own inner complexity. In line with the thinking of Fuchs and Kozyreva, I will argue for the role of affectivity and body memory in the constitution of the affective core self. The affective core self is the implicit infrastructure of the self, including habits, preferences, attitudes, reactive patterns, memories, and emotions of the individuum. However, due to the situatedness and embodied structure of the affective core self, I will also incorporate Ratcliffe's *existential feelings* and Maiese's *affective framing* conceptions as well. Ratcliffe's idea provides the opportunity to shed light on the intersubjective embeddedness of the affective core self. The idea of affective framing draws attention to the pre-cognitive layers of identity and especially highlights the affective schematization processes of the desiderative sublayer of the lived body. Of course, the general discourse around the minimal self also concentrates on the self-other relation and the bodily embeddedness of the self. The affective core self can be understood not only as an implicit habit and memory structure but as a medium that gives emotional and meaningful colorization to the surrounding world. The affective core self not only affects the body due to its drives, dispositions, and habits, but also provides the emotional tonality and colorization of the lifeworld.

There is a general agreement on the bodily foundations of the minimal self despite the disagreements about its ontological and epistemological status. Zahavi (2014) argues that "for-me-ness" (the ownership of our experiences) is embedded into our first-person perspective and it is the defining feature of the minimal self. The problem of ownership is the recurrent theme in contemporary interdisciplinary discussions of the self; therefore, I will examine its role in the constitution of the minimal self. Zahavi proposes that the phenomenological investigation of the self can be posited against the backdrop of the antirealist no-self views. Despite the eliminative attempts of the no-self doctrines, the careful analyses of selfhood will underscore the two main opposing poles of "experiential minimalism" and "social constructivism" in the contemporary literature. The concept of social constructivism, as an umbrella term, encompasses various cognitive and narrative theories of the self. Ratcliffe (2017), contrary to Zahavi, seems to be an advocate of the social constructivist stance, and proposes that the key role of the minimal self is to differentiate between the modalities of intentionality (i.e., it is a different experience being absorbed in daydreaming in contrast to perceiving an object in the surrounding world, and so on).

The conclusion of this undertaking is to incorporate the temporal structure of retroactivity into the discourse of the minimal self. The phenomenon of retroactivity extends and supplements the idea of the minimal self; therefore, the notion of the affective core self will be distinguished from the formal minimal self. Owing to suddenly appearing retroactive awakenings, which are closely related to the reproductive acts of recollection, one can stipulate that the minimal self is not only the constant background feeling of being alive (i.e., the experiential core self) or the simple self-referential aspect of self-awareness (i.e., the minimal self *per se*), but it can also be imagined as a hidden, fluctuating affective relief structure; that is, the dimension of combined affective valances, emotions, dispositions, habits. To put it more radically, the affective core self, as a "less-minimal minimal self," can be interpreted as the dynamic layer of sedimented affective patterns of the (phenomenological) unconscious. Obviously, the integration of the unconscious into the discourse of the minimal self is, at first sight, self-contradictory, since the minimal self is based on the first-personal givenness of experiences and rules out the role of the unconscious in self-constitution. How can we attenuate this seemingly irreconcilable difference between the minimal self and the unconscious? I will try to find solution to this aporia by appealing to the layered, multi-dimensional conception of the self. Hierarchical notions of the self has already been discussed by psychoanalysis and phenomenology; therefore, there is no compelling reason to exclude the problem of the unconscious from the multi-dimensional analysis of the self. I begin my discussion from afar by introducing the problems of consciousness and self-consciousness. After the brief introduction to consciousness studies, I will consider the problem of the minimal self in greater detail.

References

Moran, D. (2013). 'Let's look at it objectively': Why phenomenology cannot be naturalized. *Royal Institute of Philosophy Supplement, 72*, 89–115.

Nelson, K. (2003). Narrative and self, myth and memory: Emergence of the cultural self. In R. Fivush, & C. A. Haden (Eds.), *Autobiographical memory and the construction of a narrative Self: Developmental and cultural perspectives* (pp. 3–28). Lawrence Erlbaum Associates Publishers.

Nichols, S. (2017). Memory and personal identity. In S. Bernecker & K. Michaelian (Eds.), *The Routledge handbook of philosophy of memory* (pp. 169–179). Routledge.

Ratcliffe, M. (2017). *Real hallucinations: Psychiatric illness, intentionality, and the interpersonal world.* MIT Press.

Stocker, K. (2020). Mental perspectives during temporal experience in posttraumatic stress disorder. *Phenomenology and the Cognitive Sciences, 19*(2), 321–334.

Ullmann, T. (2015). A narratív, a traumatikus és az affektív szubjektivitás. In. Bujalos I., Tóth M., & Valastyán T. (Eds.), *Az identitás alakzatai* (pp. 21–37). Kalligram.

Ullmann, T. (2017). Phenomenology of experience and the problem of the unconscious. In M. Gabriel, Cs. Olay, & S. Ostritsch (Eds.), *Welt und Unendlichkeit: Ein deutsch-ungarischer Dialog in memoriam László Tengelyi* (pp. 141–161). Verlag Karl Alber.

Wollheim, R. (1984). *The thread of life*. Yale University Press.

Zahavi, D. (2014). *Self & other: Exploring subjectivity, empathy, and shame*. OUP.

Contents

About the Author

Lajos Horváth is an assistant professor at the institute of philosophy of the University of Debrecen. His current research focuses on the relationship between phenomenology and psychoanalysis. His work and recently published papers are supported by the János Bolyai Scholarship of the Hungarian Academy of Sciences and by the No. K 138745 project of the Hungarian Scientific Research Fund.

Part I
Consciousness and the Self

Chapter 1
Consciousness and the Self

Abstract The first chapter is dedicated to the problem of consciousness and its relationship to the self. The aim of the chapter is to illustrate the shift from the problem of phenomenal consciousness to the self in contemporary consciousness studies. The chapter will also show how the problems of the non-conscious and the unconscious infiltrated the study of consciousness. The section defines the difference between consciousness and self-consciousness and shows that the problem of the self has come to the foreground of consciousness studies. In addition, the analyses of the minimal self established a fruitful dialogue between phenomenology and philosophy of mind. This chapter also discusses the no-self approaches. These approaches deny the importance of the self and regard it as an illusion generated by the brain or mental capacities.

Keywords Consciousness · Self-consciousness · Phenomenal consciousness · Unconscious · Illusory self · Minimal self

The aim of this main chapter is to give a broad outline of the basic ideas of contemporary consciousness studies. Due to the proliferation of the definitions of consciousness and the increasing number of naturalist approaches to conscious phenomena the following survey is doomed to be simplistic, or, to say the least, partial. Admittedly, the following analysis will display the dominance of the phenomenological tradition in searching for definition of the self and especially for the minimal self. However, due to the complicated nature of the self, I made efforts to demonstrate the special "convergence zones" between contemporary consciousness studies (including philosophy of mind) and phenomenology. For start, it is incumbent upon every discussion concerning consciousness to clearly define the main approaches and terminologies. The notorious difficulty lies in the definition of self-consciousness in which the basic characteristics of consciousness and the self are often intertwined and conceptual ambiguity permeates the field of enquiry.

L. Horváth, *The Affective Core Self*, Contributions to Phenomenology 130,
https://doi.org/10.1007/978-3-031-56920-3_1

1.1 From Consciousness to the Self

From the 1990s onward the problem of consciousness came to the foreground in interdisciplinary investigations. Owing to the ground-breaking works of Thomas Nagel (1974), David Chalmers (1995, 1996), Ned Block (1995, 2007) and many others, the problem of consciousness penetrated the fields of philosophy, neuroscience, cognitive psychology and developmental psychology. After the dominant era of behaviorism, the black box of the mind opened up again and scholars began to speak of the perplexity of consciousness that meant especially the mystery of phenomenal consciousness and qualia (i.e. the phenomenal character of our experiences). Blackmore's first introductory remark in her monumental book on consciousness is the following: "If you think you have a solution to the problem of consciousness, you haven't understood the problem" (Blackmore, 2018, p. 1). The perplexity not only refers to the metaphysical dilemmas of Cartesian dualism, but also to the lack of conceptual clarification and multitude of meanings of the term "consciousness". There is no generally accepted definition on consciousness, however the history of the term dates back to early modern philosophy and it has been used frequently in contemporary neuroscience and philosophy as well.

Wilkes (1988) distinguished four kinds of consciousness based on the usage of "conscious" as adjective. First, the adjective "conscious" means wakefulness; somebody is awake and not in the state of dreamless sleep, coma or in any other state which entail the loss of wakefulness. The second meaning revolves around the nature of sensations and their relation to the perceiving subject. Sensations has certain epistemological features such us privacy, i.e., one knows them by directly or immediately, without inference by direct acquaintance. The knowing subject has *privileged access* to her sensations or ideas, and the judgements about them are *incorrigible*. That is, the second meaning summarizes concisely the basic epistemological doctrines of seventeenth century philosophy of mind. The Cartesian transparency of the mind is challenged by contemporary philosophers and studies of cognitive psychology. Recently the principle of "privacy" and the notion of "inalienable" experience (i.e., I cannot transport my pain into another person's mind, may pain is only my pain) were called into question in contemporary mind sciences by the so-called *disownership-symptoms*; limbs, thoughts, emotions and other sensations are not appropriated or owned by patients suffering from neurological or mental illnesses.

Wilkes has shown that the classical "marks of the mental" are unclear and inadequate. That is, the principle of immediacy, the incorrigibility of sensations, or sensations without inference can be called into question when the sphere of the mental is widened to include "tacit knowledge, non-conscious, subdoxastic states" (Wilkes,

1988, p. 175). According to Wilkes, subliminal perception and blindsight[1] clearly bears the mark of the unconscious and in contemporary analytic philosophy the role of implicit knowledge and the Freudian unconscious gained acknowledgement again (Schechtman, 1996).

Subpersonal mechanisms were always examined by philosophers of mind, but what is striking in contemporary interdisciplinary discussions is that phenomenology takes into account the subliminal and the unconscious with increasing enthusiasm as well. According to Wilkes, the fourth category of the mental covers propositional attitudes. She made a careful distinction between *occurrent* and *dispositional attitudes*. In the former case thoughts, words, memories suddenly and inadvertently emerge in the mind, but one can observe and investigate the occurrent mental states with "explicit deliberation" as well. Under the second category Wilkes listed beliefs, desires, memories, emotions, preferences; however, in order to appropriate these mental states, one do not have to be self-aware – Wilkes argues that they can be assigned even to a sleeping man (Wilkes, 1988, p. 179).

The distinction between the occurrent and the dispositional propositional attitudes does not mirror the conscious versus unconscious differentiation. Wilkes embraces psychoanalytical and psychological considerations and believes that "there are countless occurrent thoughts that are not conscious." The phenomenon of dream nicely illustrates the dominance of occurrent but non-conscious thoughts. Another telling example when someone stumbles at the top of a stair and says: "I thought there was another step" (Wilkes, 1988, p. 179). As a result, there are mental states situated between the obscure dimension of the conscious and the unconscious. Wilkes argues that the "disavowed beliefs of the self-deceived" or the thoughts in the Freudian unconscious often wandering between the two sides of the conscious-unconscious spectrum (Wilkes, 1988, p. 181). Wilkes conceptual clarification has led to the realm of the subliminal. Again, consciousness is not a simple on-off stitch of attentional acuity and it cannot be reduced to reflective awareness. Consciousness can be likened to the peak of a "hive mind" in which occurrent thoughts, memories, imaginings, and repressed elements are trying to gain self-manifestation. As we shall see, the realm of the implicit or subliminal will constitute a theoretical obstacle for the idea of the minimal self.

Now, let us turn to the characterization of consciousness per se. Blackmore supposes that studying consciousness entails an existential impact. As she puts it: "Studying consciousness will change your life" (Blackmore, 2018, p. 7). If the aim of philosophy is to discover and analyze the passive layers of day-to-day living, then highlighting the relevance of self-awareness seems reasonable. However, despite the all-encompassing presence of consciousness in daily life, one is rarely puzzled

[1] Blindsight is a paradoxical condition of subliminal perception. Subjects with lesions are often able to correctly guess about visual and auditory stimulus. Other studies demonstrated that blind people successfully employ some form of echolocation. From these intriguing studies Wilkes concluded that "The *esse* of sensory experience is therefore emphatically not *percipi*, nor are we in any important sense privileged about, or incorrigible with respect to, it" (Wilkes, 1988, p. 177). For additional information concerning blindsight see Weiskrantz (2007).

by the very nature or genealogy of self-awareness or self-reflection. Human beings are not only perceiving their environment, but they are also able to reflect on the acts of perception, remembering, or believing and so on. In this sense, consciousness is a kind of meta-awareness that often plagued and astonished intellectual thinkers. In addition, Blackmore gives the impression that consciousness is not only an epistemological or evolutionary conundrum, but also an ontological mystery as well. Nevertheless, concerning the significance of consciousness, a cautionary remark is needed especially in cases of anthropological or ontological assertions. As Wilkes (1988, 1995) demonstrated, different historical periods have placed emphasis on different aspects of our mental life. The Aristotelian philosophy of about the nature of the soul and its modalities is certainly incommensurable with our contemporary idea of consciousness, i.e., with the problem of self-awareness. Contemporary scholars of consciousness studies[2] often trace back the origin of mind sciences to the Cartesian dualism. Philosophy of mind stresses the incommensurability between Descartes's substance dualism and physicalism, and scholars frequently regard Descartes as the philosopher who "impeded the scientific study of consciousness" (cf. Frith & Rees, 2007, p. 10). However, Owen Flanagan (1991) gave a more nuanced view of Descartes when he highlighted the role of the *reflex arc* and the mechanical explanation of bodily processes in Descartes' oeuvre. Nonetheless, the metaphysical chiasm between the mechanical body and non-physical mind remains untouched if we interpret mind-body dualism in ontological terms. According to Flanagan, the crucial question to ask is how can we harmonize certain tenets of Cartesian dualism with scientific psychology? The answer lies, for Flanagan, in abandoning the notion of non-physical mind: instead of the immaterial mind one should entertain the idea of unobservable mind that is far more acceptable tenet from the viewpoints of modern psychology and neuroscience. In short, an epistemological dualism seems compatible with recent scientific worldview. In Flanagan's own words:

> For example, we might reconceptualize the mind body distinction along the lines of electron-table distinction. We have indirect access to the former and direct access to the latter, but both electrons and tables are physical. We could, in effect, trade in metaphysical dualism for a less problematic epistemological dualism: mind and body are not different in kind, but simply differ in accessibility (Flanagan, 1991, p. 22).

Of course, as Flanagan (1991) concluded, this epistemological turn does not mean that the scientific study of conscious phenomena are without challenges, rather the reverse, Descartes demarcates the limits and suitability of mechanical explanations and underscores the irreducible aspects of human nature; namely the "purposeful, cognitive, deliberative, willful, and free aspects of human cognition and action" (p. 22).

---

[2] The term consciousness could be misleading due to the changing contexts. For example, Giampieri-Deutsch (2012b) argued that when neurobiological and cognitive investigations approach to consciousness then the traditional term of "mind" would be more suitable. The term "mind" covers all the conscious, preconscious, and unconscious processes which are the main targets of neurobiological and cognitive approaches (cf. Giampieri-Deutsch, 2012a, p. 94)

According to Wilkes, our language la'cked the word "consciousness" until the seventeenth century and Greek philosophy splendidly operated without it. And not only the distant past of the history of philosophy but also cultural distances and relativism cast doubt on the universal nature of consciousness: the Chinese or Korean term *yi-shi*, Wilkes argues, only partially overlaps with our concept of consciousness as self-awareness (Wilkes, 1988, p. 173, 1995, p. 99). Interestingly, the ancient Chinese *"yi"* is close to our contemporary meaning of tacit, implicit knowledge, but the modern version of it is highly unclear and only partially overlaps with our modern concept of "conscious". The precursor of "conscious" was *cumscire* ("to know together with") until the mid-seventeenth century. After the notion of the *mind* become prevalent in the seventeenth century, as the sum total of mental events, "consciousness" was regarded as a "defining condition for being mental" (Wilkes, 1995, pp. 99–100). Wilkes's approach is not without any precursors; the highly speculative arguments of Julian Jaynes have also revolved around the emergence and history of consciousness. Thiel (2011) points out that the Latin "conscientia", the Greek "syneidesis" with its stoic origin, and the English "consciousness" meant shared knowledge: being conscious of something meant a perception or knowledge of something that one shares with someone else. But in the seventeenth century the notion of "being conscious" received its self-relating sense. The meaning of consciousness was not derived from knowing something with someone else, but rather denoted "knowing something with oneself" (p. 8).

In short, Jaynes provocatively argued that in the period of Homer's Iliad there was no room for consciousness in explaining human behavior and mental phenomena. There were no words for consciousness and mental acts; contrary to the modern concept of mind[3] the antique term "*psuche*"[4] covered different facets of human nature. Jaynes's so-called bicameral mind hypothesis gave rise to heated debates in psychological, anthropological, philological, and philosophical circles; due to its eclecticism and highly speculative nature, it resembles more likely to a science-fiction story than psychological theory. Surprisingly, Jaynes imagined whole civilizations without consciousness, in modern parlance he introduced the hypothesis of "cultural zombies"; adaptive creatures without conscious, second-order reflective processes. He argued that consciousness is not necessary for learning, thinking, and for reason. He placed great significance on the role of automatic inferences and skills. Consciousness is not necessary for making judgment and it is not the seat of reason – he argued. In *The Origin of Consciousness* he concluded that: "…there

---

[3] Wilkes distinguished two main branches or roots of the modern term mind. First, there is the Cartesian notion that comprises the features of immediate awareness, privileged access, incorrigibility, and privacy. Second, the term intentionality ("aboutness") derives from mediaeval scholastics and Brentano (cf. Wilkes, 1995, p. 99)

[4] Wilkes regards the term "psuche" superior to the modern notion of mind. We can imagine the ladder of living beings and their set of capacities, including plants, insects, familiar animals like cats, dogs, or humans. When contemporary scientific psychology (including neuropsychology and neuroscience) ascribes intentional and conscious features and correlates them with brain activities, then, Wilkes thought-provokingly suggests, it basically talks about the "psuche" (Wilkes, 1995, p. 105).

could have existed a race of men who spoke, judged, reasoned, solved problems, indeed did most of the things that we do, but who were not conscious at all" (Jaynes, 2000 [1976], p. 47). Jaynes envisioned the pre-historic man as a "noble automaton" who obeyed the hallucinated voices of the chieftains and gods. To put it drastically, the provocative idea was that ancient civilizations were built on a mass-psychosis in which actions were governed by hallucinated voices. And modern man bequeathed the remnants of "bicameralism" in cases of schizophrenia. Jaynes considered the possibility that schizophrenia might be a relapse into the phase of bicameral mind. However, Jaynes himself admitted that his analogy between the noble automaton and the hallucinating schizophrenic is rudimentary and there is no way to prove it scientifically. Among other things, Jaynes claims that in our modern days religious hallucinations appear in subjects who were educated in orthodox religious circumstances, and, based on cases of paranoia, he draws a parallel between the hallucinated commanding voice of the parents of the schizophrenic and voice of the gods of the "bicameral men" (Jaynes, 2000 [1976], pp. 410–411).[5]

According to Jaynes the mind of the bicameral man is divided into two separate systems of executive and commanding via voices. The two-chamber model collapsed due to the evolving complexities of ancient civilizations and the unitary model of the mind was born. The modern concept of mind is often characterized by its representative capacity described by spatial metaphors and spaciousness per se. But where can we find the seat of the reflective "I" who is the subject of conscious states? In ordinary circumstances, Jaynes contends, we are prone to locate consciousness in the head. Jaynes argues that we intuitively assume a space behind the eyes of a person; however, this classical spatial metaphor of consciousness is misleading. Consciousness is not a hidden reservoir behind the back of our head, but rather an ongoing operation and the bulk of its activity functions unnoticedly. The concept of a central "I" that carefully scrutinizes the inner mental acts is only the result of our imagination based on metaphorical thinking. The "mind-space" is the metaphorical representation of the perceptual world into which we project a so-called analog "I"; there is an imagined self in an imagined inner world in which we make decisions or design our subsequent actions (Jaynes, 2000 [1976], 62–63). The analog "I", or executive ego in more contemporary parlance, is able to narratize in the mind-space and construes an autobiographical narrative.[6] Consciousness is a

[5] Evidently, schizophrenic patients suffer from such a degree of disorganization and disintegration that even automatic inferences and skillful processes are compromised. The lack of agency in action and centralization in mental life eventuates a chaotic, maladaptive mindset. Among other reasons, this is why Jaynes speculative theory is closer to science fiction than to a full-fledged philosophical psychology.

[6] While Jaynes emphasizes that in our mind-space we portray ourselves as a main figure in an ongoing story, his account of narration is more closely related to subconscious Gestalt-perception and filling-in mechanisms. The narrative schematizes the experience; it focuses on important situations that are congruent with the ongoing story. The function of the narration is to selectively perceive situations, to give a causal account of ours and others behavior. In this respect, narratization is the general feature of consciousness: "A child cries in the street and we narratize the event into a mental picture of a lost child and a parent searching for it. A cat is up in the tree and we narratize the

language-driven, culturally mediated phenomenon conjoined with the narrative capacities of the analog "I". On the one hand, Jaynes ideas seem to be radically far-fetched without any chance to corroboration, but, on the other hand, there are elements in his book which are worthy of consideration and inspired major figures of contemporary consciousness studies. Jaynes explicitly states that "spatialization is characteristic of all conscious thought"; this feature enables that the diachronic can be turned into synchronic. If we think of the last hundred years, Jaynes argues, we usually order significant events from left to right. The spatialization of events and pieces of knowledge in our mind-space is governed by rules. The laws of fit, that is, the arrangement of events and knowledge in the mind varies greatly from person to person and from culture to culture (Jaynes, 2000 [1976], pp. 60–61). The constructive theories of memory were also important for Jaynes when he wanted to model the spatialization tendency of consciousness. He argued that consciousness does not copy the experiential content: we are unable to remember the fine details of perception despite the frequent occurrences of them (e.g., we cannot remember the numbers on telephones), moreover, we are prone to reconstruct lived experiences from a bird's-eye view via observer memories (Jaynes, 1986, p. 130). That is, the lived experience with its proprioceptive and other qualitative givennesses and the reconstructed image in our mind-space are not a lifelike representation of the occurred event in time. As we will see in the following sections, Husserlian phenomenology took great pains to develop the notion of time-consciousness in which time cannot be reduced to an abstract container of successive events. The very nature of flowing time resists the attempts of spatialization.

Jaynes imagined the bicameral men without self-awareness and phenomenal consciousness. In this respect, contemporary thought experiments of philosophical and cultural zombies (which demonstrate the epiphenomenalism of consciousness) are indebted to Jaynes' psychological emphasis on automatic processes. Ned Block made a careful distinction between *phenomenal consciousness* and *access consciousness*.[7] In case of phenomenal conscious experiences there is *something it is like to* be in that state, while access consciousness refers to available information for reasoning, rationally guided speech and action (Block, 1995, p. 227). Gary Williams (2011) demonstrated that in Jaynes' system, consciousness is not identical with phenomenal consciousness; the hypothesized nonconscious bicameral mind and the sheer organic reactivity of living beings might presuppose "nonconscious phenomenal experience" without access consciousness (p. 227). Williams argues that "…for most animals, 'what-it-is-like" is merely the experience of acting on autopilot without autobiographical memory and can be explained behaviorally through ecological

event into a picture of a dog chasing it there. Or the facts of mind as we can understand them into a theory of consciousness" (Jaynes, 2000 [1976], p. 64).

[7] More precisely *access consciousness* means: "A state is access-conscious if, in virtue of one's having the state, a representation of its content is (1) inferentially promiscuous, i.e., poised to be used as a premise in reasoning, and (2) poised for [rational] control of action and (3) poised for rational control of speech" (Block, 1995, p. 231). In the sections that follows, the problem of *phenomenal consciousness* will be analyzed in more depth.

psychology" (p. 228). This level of organic activity is characterized by the so-called "floodlight attention" that serves as a cognitive subconscious and includes automatic processes and implicit knowledge. When Jaynes talked about the mind-space with analog "I" he basically referred an introspectable mental workspace with the capacity to narratize (p. 228). In Williams's (2011) reading the real-time operations of the cognitive subconscious bears the mark of phenomenality; that is, in Jaynes' behaviorism there is no room to the additional mystery of phenomenal consciousness (p. 236). Thus, reactive organic behavior and embodied sensory experience has an internal side; the qualitative aspect of "what-it-is-like" to be that organism. However, the idea of phenomenal consciousness has widened the gap between observable behavior and first person experiences. As Levine has formulated, there is an epistemological and explanatory gap between the two modes of access (Levine, 1983). As we shall see, phenomenal consciousness resists all reductive approaches and double-aspect theories of consciousness.

Thomas Nagel, in his ground-breaking paper, called *What is it like to be a bat?*, heavily criticized the contemporary reductionist and functionalist explanations of consciousness and highlighted the pivotal role of first-person perspective and *qualia* (the subjective character of experience) in defining consciousness. Nagel's ground-breaking insight was the following: "Without consciousness the mind-body problem would be much less interesting. With consciousness it seems hopeless" (Nagel, 1974, p. 436). The main idea is that conscious states are inseparable from the first-person perspective of the knowing subject and we also suffer from a cognitive closure;[8] that is, we are unable to subjectively comprehend or experience for ourselves the radically different "way of life" of a bat. We do not know and never will know "what is it like" to be a bat from the inside; we cannot step into the inner "mindset" of a bat duo to the constraints of our physiology and perceptual apparatus. The inner side of the bat's mental life remains a mystery for the third-person observer (Nagel, 1974). The question of the first-person phenomenology is of crucial importance in studies concerning self-consciousness as well.[9] The image of the

[8] Colin McGinn argues that we do not have the required capacities to comprehend what lies beyond the mind-body problem. The core of his arguments is the following: "Conceiving minds come in different kinds, equipped with varying powers and limitations, biases and blindspots, so that properties (or theories) may be accessible to some minds but not to others. What is closed to the mind of a rat may be open to the mind of a monkey, and what is open to us may be closed to the monkey. Representational power is not all or nothing. Minds are biological products like bodies, and like bodies they come in different shapes and sizes, more or less capacious, more or less suited to certain cognitive tasks" (McGinn, 1989, p. 350).

[9] It is important to note that, due to the interdisciplinary nature of contemporary consciousness studies, the meaning of phenomenology varies greatly between Anglophone and Continental philosophers. The simplified meaning of the term often refers to subjective experience and qualia. However, as Zahavi points out, the analysis of subjective experience does not mirror the "technical sense" of traditional phenomenological philosophy: "Although a small number of prominent figures in consciousness research have recently started to take philosophical phenomenology seriously, the vast majority of (Anglophone) philosophers and cognitive scientists are not using the term in its technical sense when they talk of phenomenology, but are still simply referring to a first-person description of what the "what it is like" of experience is really like. In fact, there has

mind as a container or reservoir, filled with mental states, became a misleading idea. In contrast, the idea that mental states directly refer back to an experiencing subject has been the subject of careful investigations in contemporary consciousness studies and phenomenology.

Concerning the problem of the first-person phenomenology, Blackmore reminds us that the meaning of the idiom "what it is like" is ambiguous. Blackmore argues there are at least two different ways to use this idiom: (1) we are comparing things or making analogies with such statements like "the ice cream tastes like rubber" or "his look cut through her like a knife"; (2) the other meaning is about identity and not a comparison. Blackmore illustrates the questions of identity in the following examples: "What is it like to work at McDonald's?" or "What is it like to be a molecule, a microbe, a mosquito, an ant, or an ant colony?" (Blackmore, 2018, p. 33). Blackmore further mentions that, in the same place, that the social psychologist Guy Saunders proposed a similar phrasing to clarify the question of identity (see also Saunders, 2014, p. 146). For him "How it is to be…" or "How it is for you" idioms are more adequately express the conundrum of identity and subjectivity; however, these expressions presuppose the context of the human lifeworld, whereas Nagel's proposal tried to capture the wider context of human and non-human living beings.

Another important step was in the development of phenomenal consciousness David Chalmers' distinction between the easy problems and the hard problem of consciousness. Chalmers claimed that the easy problems can be solved by the methods of cognitive science, they can be explained by computational and neuronal mechanisms. Following in the footsteps of Nagel, Chalmers argues that the hard problem refers to the qualitative nature of experience:

> When we see, for example, we experience visual sensations: the felt quality of redness, the experience of dark and light, the quality of depth in a visual field. Other experiences go along with perception in different modalities: the sound of a clarinet, the smell of mothballs. Then there are bodily sensations, from pains to orgasms; mental images that are conjured up internally; the felt quality of emotion, and the experience of a stream of conscious thought. What unites all of these states is that there is something it is like to be in them. All of them are states of experience (Chalmers, 1995, p. 203).

The problem of *qualia* lies at the center of the hard problem: there is something it is like to live through a sensory experience, or imagine something in front of our mind's eye, or experience an occurrent emotional state spontaneously. The idiom of "what-it-is-like" resurfaced in contemporary phenomenologically oriented philosophical projects as well. In this respect the puzzle of consciousness is, on the one hand refers to the fact that incomprehensible amount of information-processing takes place in the brain without self-awareness or subjectivity, however, on the other hand, from first-person perspective it seems clear that living beings are endowed with phenomenal consciousness and self-awareness. Chalmers defines phenomenal consciousness in the following way: "… an organism is conscious if there is

---

been a widespread tendency to identify phenomenology with some kind of introspectionism" (Zahavi, 2005a, p. 4).

something it is like to be that organism, and a mental state is conscious if there is something it is like to be in that state" (Chalmers, 1995, p. 203). Michael Tye highlights the unique feature of our egocentric perspective: "… some of our mental states cannot *fail* to be conscious. For each mental state, there is a *subjective* perspective that goes along with it. This perspective is conferred upon the subject simply by his or her undergoing the mental state. It is captured in everyday language by talk of 'what it's like'" (Tye, 2007, p. 23). The convergence between phenomenology and philosophy of mind can be illustrated by the mutual relationship of phenomenal consciousness and self-awareness (or self-consciousness). Here mutual relationship means that phenomenal consciousness and self-awareness are indispensable to the characterization of consciousness. As will be discussed in more detail later, recent discussions on the nature of consciousness often explicitly suggest that phenomenal consciousness implies pre-reflective self-awareness and *vice versa*.

Naturally, the problem of phenomenal consciousness cannot be limited to the interrelated problems of the subjective character of experience (i.e., experience or phenomenality per se) and self-awareness. Phenomenal consciousness is also strongly connected to the so-called *creature consciousness*. We presuppose that sentient beings are bearers of qualitatively differentiated mental states, like pains, itches, desires, thoughts, etc. But how far can be the experiential sensitivity extended? Where are the limits of sentience? For example, the different branches of *panpsychism* do not confine the emergence of experiential reality to the upper stages of the phylogenetic ladder. Consequently, panpsychism[10] tries to eliminate the hard problem since it lowers the threshold where sentience comes to sight, but it also sharpens up the distinction between sentience (the pure phenomenality of experience) and reflective self-awareness. That is, not every sentient being is self-aware, but all self-aware beings can be considered sentient. There is no room here, to analyze in great detail the physicalist and panpsychist theories of consciousness, albeit Chalmers himself presented arguments for panpsychism (Chalmers, 1995, 1996). In addition, recent discussions of the so-called minimal mind can also be seen as a point of interest between phenomenology and consciousness studies (Marosán, 2023).[11]

[10] David Skrbina notes that definition of panpsychism relies on ambiguous sub-definitions. When philosophers of mind are trying to provide a comprehensive view of panpsychism, they frequently utilize the terms of "sentience", "experience", or "consciousness". Skrbina observes that Chalmers' approach, in particular, blurs the dividing line between the meaning of mind and consciousness. According to Chalmers, panpsychism means that "everything has a mind" or "everything is conscious". Contrary to the wide variety of ambiguous definitions, Skrbina offers a more restricted functional definition: "All objects, or systems of objects possess a singular inner experience of the world around them" (Skrbina, 2005, p. 16). According to Skrbina, the main theoretical dilemma of panpsychism is that we cannot escape anthropocentric thinking. Terms such as "low-grade awareness", "proto-mentality", or "occasions of experience" often refer back to our own subjective ideas of mentality (cf. Skrbina, 2005, p. 18).

[11] According to Marosán, the term "mind" refers to our entire mental sphere including conscious, subconscious, and unconscious mental acts. The minimal mind is our "essential capacity of having

In summary, while phenomenal consciousness was defined by Nagel, Chalmers, Block, and other figures of philosophy of mind, the term turned out to be an umbrella term that often refers to different approaches to solve the mystery of consciousness. The following chapter will focus on the relationship between phenomenal consciousness and the minimal self.

1.2 Phenomenal Consciousness and the Minimal Self

There is no room here for the detailed analysis of the diverging problems of phenomenal consciousness, however, I would like to list a few problems that are closely related to the nature of phenomenal consciousness and to the minimal self as well. First of all, the problem of *ownership* is a recurrent theme in debates surrounding phenomenal consciousness and self-consciousness. Michael Tye points to the fact that the problem of ownership is closely connected to the above-mentioned *privacy* of mental states. Experiences and feelings, for example, particular pains, after-images, tickles, itches, are private experiences and necessarily owned by the subject. The problem of ownership poses a challenge to the reductionist approaches to consciousness: if mental states are only physical states (i.e., brain states) then why are they require an owner? Pains, thoughts, feelings, and other mental states are always belonging to a specific creature, but physical things in themselves do not require an owner (Tye, 2007, p. 24). The problem of the ownership of mental states has become a central issue in both embodied mind and philosophical psychopathology (cf. Gallagher, 2005; Gallagher & Zahavi, 2012).

In this respect, Stevens and Graham (2007) highlight the fact that the problem of ownership is closely connected to psychopathological phenomena where the line between the self and non-self is blurred, or dissociation ensues. The authors cite William James, who claimed in *The Principles of Psychology* (1983 [1890]) that the elementary psychic fact is not "thought" in itself, but "my thought". We feel an intimate connection with our occurrent thoughts; in normal circumstances we feel that we own our own mental states. And Freud already observed that there are cases where perceptions, thoughts, feelings, or the person's own body may not seem to belong to the ego (Stevens & Graham, 2007, p. 195). The concept of the ownership of experiences is intimately tied to the ambiguous idea of private sensations. As Wilkes (1988) has shown the Cartesian concept of private sensations is associated with ambiguous epistemological features. As previously discussed, the privacy of sensations was defined by "incorrigibility" and philosophers often talekd about "incommunicable" sensations. However, as Wilkes argued, this is plainly false,

---

phenomenally conscious, subjective experiences" (Marosán, 2023, p. 5). Marosán introduces the concept of the "minimal subject", not to be confused with the minimal self. The minimal subject is a Husserlian term that refers to the "minimal set of conscious structures and capabilities which make possible a minimal concrete subjective being-in-the-world". And one of Husserl's biological examples is the jellyfish with its decentralized nervous system (Marosán, 2023, pp. 14–15).

since we can communicate a lot about our mental states and feelings. According to the third interpretation, private mental states are considered "inalienable" experiences which simply point to the fact that our pains are only ours and cannot be transferred to another person (Wilkes, 1988, p. 121).

The notion of *disownership-symptom* in philosophical psychopathology incorporates at least two determinants of privacy in an indirect manner. The ipseity-disturbance model (that examines schizophrenia and depersonalization) often meets with the impulses of "incommunicable" experiences in autobiographical accounts. Moreover, patients often complain that they cannot integrate parts of their bodies into their body image, or they feel disconnected from their own intimate proprioceptive and kinaesthetic experiences. Here, the rationale is that disownership-symptoms indirectly refer back to the "healthy" or "normal" state of ownership of mental states and body parts.

There is a second important characteristic of mental states that is closely connected to the problem of the minimal self and also to the phenomenon of ownership. Nagel convincingly demonstrated that first-person mental states require and presuppose an experiential point of view. Discussions of philosophy of mind mainly focus on the irreconcilable conflict between the physical and the mental. While, in contemporary debates, Cartesian substance-dualism moved to the periphery, arguments of property dualism against physicalism are still prevalent, that is, the physical and mental features are the two aspects of the same universe.[12] Nevertheless, we may consider the idea that the first-person perspective is not immediately given to living beings but rather it is a result of a developmental process. As Tye argues, a child who never experienced pain in his life is unable to perceive or grasp the meaning of pain. Regardless of how much information is available to the child about the physiological mechanism of pain, without the possession of pain-experiences she could never acquire the first-personal access to the phenomenal character of pain. To fully understand the nature of pain, feeling of depression, or visual experiences of red and other qualitative states, we need an *experiential perspective* (Tye, 2007, pp. 25–26). The current proposal holds that the infusion of phenomenological insights into the discussion of mind-body problem may give rise to a more dynamic and fruitful analysis of self and mind. As we shall see, mainly throughout Zahavi's contentions, the problem of self-constitution may shed new light on the above-mentioned problem of "experiential perspective". The genealogy of the self could lead to developmental theories and phenomenological considerations as well.

[12] One classical argument against physicalism is Frank Jackson's knowledge argument that tries to demonstrate the incompleteness of physicalism by a thought-provoking scenario. Consider the case of Mary, the brilliant scientist, who has been confined to a black and white room and she has never experienced colors in her restricted circumstances. Despite of this, she knows everything about the (physical) nature of color perception. His knowledge about the physical nature of perception is unparalleled. Finally, when she leaves the room and looks at a red rose, she learns something new about the world, something qualitatively different aspect of the world is revealed before her eyes. The thought experiment concludes that abstract physical information cannot provide the "what-it-is-like" character of experience (Jackson, 1983).

The third essential problem concerning phenomenal consciousness and the minimal self is the *unity and disunity of consciousness*. How the brain-mind complex can synthesize dispersed perceptual information and discriminate between obtruding Gestalts and background? How do coherent gestalt-perception and object constancy proceed from brain-activity? These questions are not only relevant to contemporary cognitive psychology and neuroscience but also typical philosophical conundrums. For example, Kant and Husserl had a keen interest to explain the experiential unity of our first-person phenomenal field. From the late 1960s onward, *split-brain cases* had a tremendous impact on philosophy of mind and gave rise to the possibility of separated or fragmented minds. In addition, *multiple personality disorder*[13] was equally important for philosophers with empirical inclination. In these strange cases people are prone to attribute their thoughts, feelings, and even dreams to another person who supposedly dwells in the same body (Stevens & Graham, 2007, p. 196). These cases are particularly interesting from the angle of self-consciousness. Stevens and Graham (2007) argue that symptoms in which patients prone to attribute their conscious episodes to another person can be subsumed under the label of *alienated self-consciousness*. The above-mentioned disownership-symptoms and alienated self-consciousness go hand in hand: in diswonership-symptoms patients deny the access to or appropriation of a mental or bodily state, whereas alienated self-consciousness refers to a dislocated or misattributed egocentric perspective. Split-brain and multiple personality cases present their own methodological and theoretical problems. However, they also rise general philosophical questions, such as how to conceive separate minds within one body? What is happening with, or how can we define phenomenal consciousness in these cases? Of course, there is no general agreement on the consequences of split-brain cases and multiple personality disorder (MPD) has been abandoned and replaced with dissociative identity disorder (DID).

Philosophers of mind interpreted the results of split-brain experiments by means of unconscious automatism, disunited access consciousness, or with two separate streams of consciousness or phenomenal consciousness and so on (Tye, 2007, pp. 33–34). In contemporary debates on the nature of the disorders in question empirical and theoretical considerations may constrain the philosophical interpretations. For instance, the split-brain phenomenon cannot be directly interpreted as

[13] Multiple personality disorder (MPD) transmuted into dissociative identity disorder (DID). The subjects of DID are both aware and not aware of the same experience at a given moment, they experience severe amnesia and blackouts, they have "disowned" experiences, and the major provoking symptom is the extreme dissociation that might give rise to quasi-autonomic alter personalities (cf. Maiese, 2016, p. 183.; Wilkes, 1981, 1988). Both Hacking (1995) and Braude (1995) investigates MPD-phenomena from historical and cultural viewpoints. If one tries to take seriously MPD and not to regard it as a therapeutic artefact, then, according to Braude (1995), two main causal determinants can be discerned: (1) capacity to profound dissociation; and (2) a severe or chronic childhood trauma (p. 39). In addition, Braude (1995) mentions the claim that we are all multiples in one way or other: "that is, a colony or synthesis of lower-order functionally distinct selves" (p. 37).

split-mind or divided consciousness.[14] According to Colvin and Gazzaniga (2007), the intriguing result of split-brain cases is the more-or-less intact, unified consciousness despite the corpus callosum being cut through. The authors accentuate that the patients, surprisingly, do not experience themselves as having a divided mind. Instead, consciousness emerges from automatic integration of modular neural processes. As a result, there is a unified subjective experience (i.e., phenomenal consciousness) despite the lack of awareness of missing information from certain parts of the brain (Colvin & Gazzaniga, 2007, p. 191). In summary, we have found four fundamental issues that are closely related to the concept of the minimal self. These were the problems of (1) phenomenal consciousness; (2) ownership and privacy; (3) the problem of perspectival subjectivity; and finally, (4) the problem of the unity and disunity of consciousness. In Sects. 2.3 and 2.4 we will explore how these fundamental problems of consciousness converge and gain new significance in the discourse surrounding the minimal self.

1.3 Self-consciousness and the Unconscious

While consciousness studies remained a fruitful interdisciplinary enterprise, it is not an exaggeration to say that the conundrum of the hard problem denotes an unsolvable mystery and a methodological impasse. While materialist and panpsychist theories of consciousness are flourished, the explanatory gap between physicalist and phenomenological approaches remained untouched (Levine, 1983). That is, naturally, due to the conceptual, methodological, and most importantly to the perspectival difference between first-person and third-person approaches, the subjective character of experience poses a challenge to neuroscience and physicalism. Of course, countless hybrid alternatives were put forward, but the epistemological gap between first-person access to experience and third-person access to physical processes seems to be not only a phenomenological but also a metaphysical (or rather anthropological) fact. Nagel's phenomenal consciousness or Colin McGinn's cognitive closure hypothesis (McGinn, 1989) belong to the *mysterianism* position that argues against reductionist-materialist views of consciousness. At the other end of the spectrum lies the community of *eliminativist* philosophers who deny the significance of phenomenality per se and try to diminish the role of phenomenality from the discourse (cf. Dennett, 1991; Churchland, 1989). These entrenched positions are mirrored in the interpretations of the self.

[14] Naturally, the cautionary remark above does not want to suggest that basic philosophical problems of the mind can be perfectly demystified and solved by empirical investigations. While I inclined to agree with the neurophenomenological approach, that sees a mutually constraining and enlightening relation between neuroscience and philosophy of mind, I accept and adhere to the contention that philosophical problems have their own autonomy and discursive rules (cf. Varela et al., 1991; Varela, 1996).

Wilkes (1988) has already stressed that the pressing problem of the brain and behavioral sciences is to construe an appropriate taxonomy of *explananda.* Moreover, she convincingly argued that consciousness cannot be the *explanandum* of scientific inquiry. Instead, we need to consider the holistic structure of interrelated mental phenomena and subcategories of different mental states. Consciousness refers to heterogeneous psychological phenomena. The term "consciousness" has an adverbial form and it is parasitic on a number of other psychological predicates: "In other words, we presuppose a healthy subset pf a whole slew of psychological ascriptions – to do with perception, motivation, belief and desire, misperception, illusion, recognition, and so on and so forth – when an ascription of consciousness is to make sense" (Wilkes, 1988, p. 194). A useful analogy for Wilkes is "pain". We may try to explain pain in a simplified manner; however, simple behavioral and physiological observations draw attention to the different subcategories and types of pain. Under normal circumstances, we make postural adjustments to unconsciously compensate our "unnoticed discomfort". In addition, in the process of characterizing the meaning of pain we must also consider, for example, the very absence of pain experience in people who suffer from congenital insensitivity. Wilkes underscores that these people do not simply suffer from the absence of *feeling* pain, but rather they are such people who literally do not *have* pain, even in the mildest and unnoticed forms (Wilkes, 1988, p. 186). Therefore, in the case of pain experience, conscious pain experience and behavior is only a fragment of a more complex mental landscape in which "unnoticed pain" or the total absence of pain may also play a crucial role. In turn, take for example a general opposition in the definition of consciousness: consciousness as a waking state is distinctly different from non-conscious states, i.e., dreamless sleep or coma. However, Wilkes meticulously analyses mental anomalies which are hardly fit into the comparative categories of conscious and non-conscious states: epileptic automatism, hypnosis, fugue states, and multiple personality might be problematic in this respect since they cannot be characterized only by the absence of consciousness (cf. Wilkes, 1988, pp. 100–128). Not surprisingly, the folk-psychological, intuitive knowledge about consciousness does not entail to an unambiguous definition, but rather gives way to the description of the spectrum of consciousness. Wilkes does not intend to deny the productivity of the distinction between conscious and non-conscious processing, she only accentuates the plethora of uses of the term "consciousness"; that is, consciousness is a too elusive and flexible concept to be a unique explanandum of a scientific inquiry (Wilkes. 1988, 100). Ultimately, if one aims to specify the meaning of consciousness, it is necessary to circumscribe the relevant phenomena.

Recently, Antti Revonsuo has attempted to firmly define these elusive terms which cut-across the boundaries of philosophical and psychological inquiries. For Revonsuo (2018), phenomenal consciousness is limited to conscious creatures: "A non-conscious creature or object *does not feel or sense its own existence in any way. It does not have a conscious mind*" (p. 12). That is, unconsciousness is devoid of any kind of subjective states. The phenomenally conscious mind is a feeling mind and defined by the presence of felt experiential qualities (p. 13). Furthermore, Revonsuo defines self-awareness based on reflective consciousness and ownership.

Self-awareness is basically a higher level of consciousness that relies on the snapshots of the past and imagined futures. In addition, we arrive at the doctrine of ownership: "When self-aware, we not only undergo experiences, we not only *have* a stream of subjective consciousness; we also become aware of the *owner* of those experiences: These are *my* experiences and the *me* who *owns* them is a person or a *self.* This self is *embodied*; it has a body where its consciousness lives, and it has an *identity*, the self is *someone* with a name and a past and future" (p. 16). It is important to note that Revonsuo distinguished between phenomenal consciousness, reflective consciousness, and self-awareness for the purposes of the scientific investigation of consciousness.

Despite the general agreement on the above-mentioned definitions, the problem the self became a perplexing issue in consciousness studies. Recently, several authors have drawn on insights from phenomenology to redefine the nature of the self. The Husserlian and Merleau-Pontian phenomenology slowly infiltrated the consciousness sciences and rejuvenated the field of philosophy of mind by rediscovering the idea of pre-reflective self-awareness (For detailed discussion of these issues see Sect. 1.3).

Retrospectively, self-consciousness seemed to be an easier phenomenon to explain by means of our inherited mental vocabulary. For Wilkes, self-consciousness meant not only self-awareness (i.e., the ability to reflect on our occurrent mental states) but also the problems of unity and disunity of the self.[15] In my reading, Wilkes implicitly considers the conscious and unconscious factors to be equally important in the constitution of self-consciousness. However, the introduction of the role of the unconscious to the discourse of personal identity could have serious ramifications. The basic notion that comes to our mind with regard to personal identity is the rational concept of the person. Wilkes's claims that the *rational person* is able to reflect on his or her thoughts, desires, beliefs, and strives for consistency between thought and action by ruling out incongruencies. However, the introduction of unconscious processes leads to two serious consequences: (1) the needed unity and continuity of the self are not restricted to the level of consciousness; (2) we must adhere to the notion of the "person" despite the "substantial amount of disunity and discontinuity" (Wilkes, 1988, p. 197). Self-consciousness entails the problem of personal identity, and the concept of the person, not surprisingly, is also an ambiguous phenomenon owing to its vernacular application and holistic meaning. When we use the concept of person, we ascribe an inexhaustible set of physical and mental predicates to ourselves and to others (Wilkes, 1988, p. 97).

On the one hand, there is no unanimous agreement on the definition of consciousness, and, on the other hand, anomalous self-experiences, altered states of consciousness, and psychopathological states have always been a central concern

[15] That is, Wilkes notion of self-consciousness and Revonsuo's self-awareness are only partially overlapping notions. To my mind, Wilkes's approach is more holistic, since she tried to integrate the role of altered states of consciousness and the unconscious into the definition of self-consciousness and personal identity. As I see it, Wilkes's self-consciousness is far more "thicker" and "dense" than Revonsuo's abstract and operational definition of self-awareness.

for this interdisciplinary collaboration – further complicating the field of inquiry. The discussions on the status of the unconscious can also be seen as an important addition to the list. In general, altered or anomalous states of consciousness are inseparable from the problems of the cognitive and psychoanalytical unconscious. According to Wilkes, we must take the Freudian unconscious seriously in order to fully understand the nature of personal identity.

> The picture of the human mind is thus transformed. The mind consists no longer exclusively of the Cartesian self-illuminating consciousness, but becomes a rich and many-layered dynamically structured system which leaves consciousness as no more than the bare tip of an enormous and largely unknown iceberg. [...] Our ordinary beliefs, attitudes, emotions, dispositions, traits, moods, and styles of acting and deliberating can be explained in full only by essential appeal to a huge variety of unconscious or non-conscious thoughts which are always in the background (Wilkes, 1988, p. 82).

Furthermore, Wilkes claims that Freud's alleged discovery of the unconscious can be rephrased as follows: Descartes undiscovered the unconscious by imposing upon philosophy the myth of consciousness (Wilkes, 1988, p. 82). Wilkes argues that the problem of non-conscious and unconscious motivating factors was present in the history of philosophy prior to Descartes and remained significant after the seventeenth century as well.[16] Two serious consequences follow from a willingness to accept the role of the unconscious in the constitution of the rational person. First, Wilkes argues, the "purely rational person" is an abstractum; that is, the person can be described as a certain character who is conditioned by irrational impulses (e.g., desires and drives) and who is endowed with irrational skills like imagination. Second, there is no sharp distinction between the "conscious" and the "unconscious", but rather there are shades and fusions between the different sub-layers of the person (e.g., even the unconscious or pre-conscious elements can emerge into consciousness). And third, the clear line between normality and abnormality is blurred; in Wilkes's words: "Thus there is a continuum between normal and aberrant behavior, a continuum Freud never ceases to emphasize" (Wilkes, 1988, pp. 80–82). The contrast between the Cartesian transparent mind and the Freudian unconscious is undeniable; however, a recent unorthodox thesis might shed new light on this standard opposition.

To put it drastically, the Cartesian mind is barred from the realm of the unconscious. By means of *direct access*, we are able to scrutinize our inner (i.e., conscious) mental states effortlessly and incorrigibly. Through radical skepticism, Descartes develops the idea of self-transparent consciousness and does not differentiate between conscious and unconscious activity. Surprisingly, the concept of the Cartesian unconscious has emerged in recent years. Eshleman (2007) points out that a special kind of (neuro)physiological unconscious can be found in Descartes' speculative, mechanistic explanation of the genesis of affections. The transparency principle and incorrigibility of mental states does not necessarily apply to the passions

---

[16] Among other things, Wilkes cites passages from Heraclitus, Shakespeare, Goethe, and Lichtenberg to demonstrate the all-pervasive presence of the various forms of the unconscious in different periods of western thinking.

of the soul. More precisely, in his own words: "The (causal) 'closeness' of passions to the soul entails that it is impossible not to feel them as they are felt, but their 'closeness' to the body 'renders them confused and obscure' and makes mistake possible" (Eshleman, 2007, p. 177). Eshleman (2007) further argues that there is even a pre-Freudian speculative account of traumatization in Descartes' physiological ruminations (p. 183).[17] According to Eshleman's heterodox thesis, Descartes offered a rudimentary analysis of unconscious determinations prior to Leibniz (p. 169). There are special cases where a "natural defense mechanism" kicks in, and the soul is unable to notice the cause of the traumatic impressions. As a result, afterward, a stimulus could trigger an unusual aversion due to the fact that there was no established connection between the soul and the impression stored in the brain. Descartes observes unusual habits owing to traumatic events in childhood. Certain sense-impressions may have a serious impact on the brain which may consolidate a psycho-physiological habit (Eshleman, 2007, p. 178). In the end, Eshleman (2007) defines the Cartesian unconscious as "impressions stored in the brain without the mediation of the soul" (p. 180). All in all, Eshleman attenuates the impact of the Cartesian mind in the genealogy of subjectivity. Descartes did not eliminate completely the division between the conscious and unconscious, rather in *The Passions of the Soul* and in his letters, he anticipated a neurophysiological unconscious that does not directly contradict to the idea of mental transparency.

This line of reasoning concerning the Cartesian unconscious (i.e., the role of habit structure in identity) will resurface in the discussions of body memory. The problem of the unconscious calls into question two Cartesian epistemological presuppositions: there is no unambiguous evidence for the immediacy and incorrigibility of our mental states given the discovery of tacit knowledge and subdoxastic states. That is "we may be unaware of many of the inferential steps in our thought process" (Wilkes, 1988, p. 175). Firth and Rees (2007) point out in their historical survey that Helmholtz anticipated the contemporary notion of the cognitive unconscious[18] when he discovered that immediate perception is highly dependent on

[17] "Indeed, one might suggest that Freud's neurological account of repression that results from trauma is, at its root, Cartesian. Since, according to Descartes, nerves and animal spirits cause the passions, and sometimes the appropriate passions necessary for memory are not caused, Descartes can easily explain how the nervous system 'gets rid' of overwhelming stimuli. Although the language has shifted from Descartes's 'unusual aversions' to Freud's 'hysteria,' the parallel here is strong" (Eshleman, 2007, p. 183).

[18] The cognitive unconscious means, in a wider sense, unconscious influences without awareness. Kihlstrom does not regard the Freudian theory superfluous, rather he argues equally for Freud's and Helmholtz's role in the discovery of the cognitive unconscious: "Helmholtz concluded that conscious perception was the product of unconscious inferences based on the individual's knowledge of the world and memory of past experiences. Somewhat later, Freud asserted that our conscious mental lives are determined by unconscious ideas, impulses, and emotions, as well as defense mechanisms unconsciously arrayed against them. These nineteenth-century ideas exemplify the notion of the cognitive unconscious-mental structures and processes that, operating outside phenomenal awareness, nevertheless influence conscious experience, thought, and action" (Kihlstrom, 1987, p. 1445). Kihlstrom claims that phenomenal awareness (i.e. consciousness) in not identical with a specific perceptual-cognitive function, rather "consciousness is an experiential quality that may accompany any of these functions" (Kihlstrom, 1987, p. 1450).

"unconscious inferences"; and, as a consequence, he questioned the idea of personal responsibility as well (p. 12). It is an intriguing historical fact that Helmholtz himself was not satisfied with the term and wanted to demarcate his conception from the Schopenhauerian "obscure and unjustified" concept of *will* that closely connected to sexual desire (p. 12).

Recent studies on implicit memory might serve as a new framework for examining the role of the cognitive unconscious in different circumstances. Priming effect is one of the major phenomena demonstrating unconscious influences without awareness. In a very general sense, priming effect means that the processing of one item influences the processing of another. In different experimental settings the influence can facilitate or inhibit the recognition of the target item (for example a meaningful keyword like "doctor") (cf. Kihlstrom et al., 2007, pp. 527–528). On the one hand, there are complicated methodological and theoretical debates on the relationship between explicit and implicit memory, but, on the other hand, researchers seem to agree on that implicit memory represents "the dynamic influence of memory in the absence of conscious recollection" (Kihlstrom et al., 2007, p. 529). According to Daniel L. Schacter, episodic and autobiographical memory can be divided into explicit and implicit systems. The former deals with the conscious recollection of past events, and the latter refers to the influence of an event on a subsequent experience, thought, or action (Kihlstrom et al., 2007, p. 527). Schacter speaks of the "hidden world of implicit memory" that has been unraveled by the unusual achievements of amnesic patients[19] who were able to retain and recognize the influence of the past without conscious recollection or factual knowledge. The simplest definition of implicit memory is the case "when past experiences unconsciously influence our perceptions, thoughts, and actions" or "when people are influenced by a past experience without any awareness that they are remembering" (Schacter, 1996, p. 9, 161). It is a common feat of both amnesic and ordinary people that they are influenced by "recent experiences" without the explicit recall of those influencing events. Kihlstrom and his colleagues (2007) identify implicit memory with "unconscious memory of the past", and they claim that the hidden reason to stick to the term "implicit" rather than to "unconscious" is to keep a distance from Freudian psychoanalysis (Kihlstrom et al., 2007, p. 529). This observation is consistent with Schacter's claim that Freud's and Pierre Janet's patients, who suffered from psychogenic amnesia, most probably retained implicit memories about traumatic events. Thus, the Freudian statement that "hysterics suffer mainly from remi-

---

[19]The priming effect gave an opportunity for memory research to study the subliminal effect of implicit processes: "The first investigators to document this anomaly under controlled conditions were Warrington and Weiskrantz (1968), who found relatively normal retention when the patients were tested with fragments or stems of list items. Thus, after studying a word like *elastic*, amnesics will be unable to recall or recognize it; but when asked to name a word, any word, that starts with *ela*, they are much more apt to say *elastic* than chance alone would predict. This phenomenon, known as a *priming effect*, has been demonstrated many times. When amnesic patients who cannot remember words from a study list are later given an opportunity to use those words in another sort of task, they use previously studied items more frequently than unstudied neutral items—just as neurologically intact controls do in similar tasks" (Kihlstrom et al., 2000, p. 31).

niscences" might refer to the implicit effect of reproduction of past experiences (Schacter, 1996, p. 232). In contemporary memory research the so-called *emotional unconscious* has been introduced as the subtype of the cognitive unconscious. Again, a reformulation of the Freudian paradigm took place when Kihlstrom and his co-authors (2000) argued that Breuer and Freud's repressed memories can be defined by implicit memories: that is, one can identify his or her actual emotional states, but at the same time be unaware of their origin. More precisely, the emotional unconscious is a kind of "emotional processing" that relatively independent from cognitive processing, but its main function is still to analyze stimuli; thus, it is still a cognitive operation (p. 45). To put it simply, there is a layer of emotional appraisal in our mind outside the field of awareness: "People can be aware of their emotional states but unaware of the percepts and memories that evoke these states" (p. 46).

In the Freudian model *defence mechanisms* render the unconscious affections, motivations, and desires inaccessible. In light of implicit processes, the emotional unconscious means that "one may be consciously aware of his or her emotional state yet unaware of its source in current of past experience" (Kihlstrom et al., 2000, p. 37). As a result the "realm of implicit" not only comprises implicit memory, perception, learning, and thought, but emotions as well. One crucial example of the implicit emotional expression is Claparede's experiment with the amnesic patients: Claparede pricked the patient with a pin in his hand and the next visit the patient refused to shake his hand because sometimes people may hide pins in their hand – she contended. Thus, in the amnesic patient there was a dissociation between the conscious awareness of the distress and the conscious recollection of the source of the distressing emotional state (Kihlstrom et al., 2000, p. 38). Another frequently cited example is Pierre Janet's patient Madame D., who suffered from hysterical somnambulism. She apparently forgot the incident when some men brought home her husband, laid him down on her doorstep, and claimed that he was dead. After forgetting the incident, she felt a debilitating terror when she had to be at the front door and often dreamed that her husband was brought back dead. Again, the reasons for the emotional distress and nightmares were outside of phenomenal awareness (Kihlstrom et al., 2000, p. 39; cf. Schacter, 1996, p. 232). The resurgence of psychoanalytical stories in light of implicit processes is only one aspect of contemporary memory research. Kihlstrom and his colleagues emphasize that the dissociation between acquired emotional preferences and explicit memory has been confirmed by experiments. In short, amnesic as well as normal people's preferences can be modified with exposure effects. For example, amnesic patients immediately forgot the biographical information that accompanied the pictures of a "good guy" and a "bad guy", but later showed a strong preference for the good guy without conscious recollection of his introduction and characterization (Kihlstrom et al., 2000, pp. 39–40).

It would be counterproductive to analyze in great detail the cognitive psychological dilemmas of the emotional unconscious. The status of the unconscious is a controversial subject in memory research since it is a recurring thesis that felt emotions cannot be unconscious by definition. However, recent qualitative studies on body memory might, as we shall see, may shed new light on the problem of subliminal emotional appraisal. Nonetheless, as Kihlstrom et al. (2000) argue that there are at least to aspects of the emotional unconscious that is supported by both psychoanalytical case studies and empirical experiments. On the one hand, past events might exert an influence on the actual emotional setting without explicit recollection, and on the other hand, subliminal emotional responses have an impact on current experiences (p. 62).[20] The system approach in question enables dissociation between the emotional and cognitive processing of an actual or past event. In some cases, the authors speculate, the representation of an event is not connected to the self, but the emotional response, including the behavioral and physiological concomitants of that very event, does. The result is affection or emotion without conscious recollection or explicit memory. Dissociation between memory-systems might be the alternative to the Freudian repression.[21] I will focus on the possible scenarios of reintegration in Chap. 3 by analyzing the contemporary debates on body memory and its relation to trauma.

In terms of current research on the phenomenological unconscious, the affective component of implicit memory is of pivotal importance. The affective-emotional impact of forgotten memories may lead us to the phenomenological investigation of affective-schematism.[22] This investigation will lead us to the phenomenological notion of the affective unconscious, which brings about the affective quality of the minimal self. To put it briefly, the affective unconscious, including its habitual and affective impact, may give a qualitative thickness to the minimal self. The brief analyses of the cognitive and the emotional unconscious will also lead us to the layered concept of the self where reflective self-consciousness is only the tip of the

---

[20] The emotional unconscious is defined by Kihlstrom and his colleagues in the following way: "The emotional unconscious, then, has two different aspects. On the one hand, we may be unaware of the percepts, memories, and thoughts that give rise to our emotional feelings. In this case, emotion serves as an implicit expression of perception, memory, learning, or thought. On the other hand, we may be aware of what we are perceiving, remembering, and thinking but unaware of the emotions instigated by these cognitions. In this case, behavioral and physiological changes serve as implicit expressions of emotion" (Kihlstrom et al., 2000, p. 66).

[21] The problem of repression constitutes one of the main axes of my approach. For introductory purposes it is worth pointing out the distinction between repression and dissociation. Freud did not give a totally unequivocal definition of repression. In the contemporary literature we can find the term of "primary repression", that regards repression as an unconscious process at the outset, and "repression proper" that indicates the conscious act of repression. The second notion designates the process of "after-expulsion" that presupposes the conscious processing of the repressed (traumatic) material. Cognitive psychology deals with this kind of mechanism under the label of "motivated forgetting" (Brewin & Andrews, 1998, p. 951). The term dissociation, in the simplest form, means that two mental processes or contents are not integrated by conscious awareness, memory, or identity. Dissociative states are sometimes accompanied by *flashbacks* which may include involuntary mental intrusions (e.g., fragmentary sensorimotor experiences, or vivid sensations of reliving past experiences in the present) (p. 951). The intriguing problem of flashback is to be discussed in greater detail in Sect. 6.3.

[22] Here, I use the term affective-schematism in a broad sense. The problem of schematism alludes to the fact that we never see perceptual objects in their natural, denuded state, rather, owing to the constitutive processes of consciousness, cognitive and affective schematism kicks in that interprets and emotionally colorizes the actual experience.

iceberg. We have to bring into consideration the layered conception of the self, in which dissociation and/or repression occurs, and the ego is not the master in his own house. Before we delve deeply into the intricate problems of the psychoanalytical and phenomenological unconscious, it is worth considering some fundamental notions of the self, developed by prominent figures in philosophy of mind and contemporary phenomenology.

1.4 The Illusory Self

The problem of the *self* attracted a considerable amount of attention in philosophical, psychological and neuroscientific investigations as well. Galen Strawson enumerated 29 different conceptions of the self, including the most basic notion, which defines the self merely as "a subject of experience" (Strawson, 2000, p. 39). The enormous proliferation of the models and concepts of the self may reveal the interdisciplinary nature of the discourse. Contemporary discussions of the self include contributions of philosophy of mind, developmental psychology, phenomenology, and phenomenological psychiatry. Zahavi (2016) has recently shown that there are rudimentary thematic convergences and resonances between philosophy of mind and phenomenology regarding the problem of the self.[23] Furthermore, Zahavi accentuates the ambitious aspect of the investigation of the self, namely, there was a tendency to offer and formulate hypotheses that "empirical science can subsequently corroborate or falsify". However, Zahavi (2016) dismisses this kind of reductionist method and calls for interdisciplinary collaboration that focuses on conceptual clarifications as well (p. 7). For instance, self-disorders (*Ichstörüngen*) have a long tradition in psychopathology, and dementia or schizophrenia highlights the nature of the injured, shattered self, the terms "self" and "self-consciousness" are "left undenied and treated as unequivocal and monolithic notions" (p. 7). Olsen (1998) explicitly stated that the conflicting definitions of the self are the signs of a degenerated debate and it is superfluous to speak of "the problem of the self". Contrary to Olsen, Zahavi tries to preserve the notion of the self and argues that the existence of different traditions of the self does not entail the elimination of the concept per se. Rather, it must be accepted that the self is a multidimensional and complex concept that increasingly appears in current empirical disciplines (e.g., in developmental psychology, social neuroscience, psychiatry, etc.) (cf. Zahavi, 2016, p. 9). Barresi and Martin (2011) claim that the problem of the self means an explicit departure from the mind-body problem. Naturalism dominates the current discourse on the self,

[23] To be more specific, I have to accentuate that the dialogue between continental philosophy and philosophy of mind seems to be largely rudimentary. Strawson, in the introduction of one of his papers, says that he simply means by phenomenology the study of the character of experience; for that reason, he made a clear distinction between his materialist framework and Husserlian phenomenology (Strawson, 2000, p. 40).

since the authors argue that the focus has shifted from the role and attributes of the *rational soul* to the social and embodied aspects of the individual:

> This new question is not the traditional mind-body problem, but one that already presupposes an integration of the experiential and social in an embodied self. The self is taken here as a unified organic individual extended in time, and the question is more about how that individual experiences its selfhood, including identification of the brain and body mechanisms that produce these experiences, insofar as we can figure this out through cognitive science and neuroscience. It also includes how the first-person perspective of self develops in the individual's life history. Important, with respect to this latter issue, is the role that social existence and other individuals play in the development of self-consciousness (Barressi & Martin, 2011, p. 54).

Despite the popularity of the interdisciplinary research on embodied mind there is no unified theory of the self, and in psychology the self appears in many hyphenated roles, such as self-image, self-esteem, self-interest, self-control, self-denial, self-deception, etc., just to mention a few examples. As it stands now, we have to abandon the dream of the unified self: "… the notion of a unified self fell onto hard times of its own. Its demise was gradual, but by the end of the twentieth century the unified self had died the death if not of a thousand qualifications, then of a thousand hyphenations (Barressi & Martin, 2011, p. 51). However, as we will see, the ambiguous concept of the self gives way to the analysis of the layers of the living-embodied individual; that is, the pre-egoic, unconscious structures and processes can be addressed by means of phenomenological and psychanalytical approaches. But before we dive into the subject of the unconscious and its possible relation to the current discourse on the constitution of the minimal self, let us briefly examine some contemporary ideas of the self.

To provide an initial account of the self Zahavi points to the phenomenological presence of the self and he dismisses the widespread idea of the *illusory self*. Another general notion to be discussed is the *narrative self* and its relationship to Zahavi's minimal self. For start, let us address the notion of the illusory self which became quite popular in current philosophy of mind and consciousness studies.

Unusually, Miri Albahari argues in her book *Analytical Buddhism: The Two-Tiered Illusion of Self*, that the Buddhist *witness-consciousness* might be a suitable candidate to explain the conundrum of the self that comprises the dilemmas of phenomenal consciousness and self-awareness as well. The serious consequence of using Buddhist philosophy is the illusory nature of the self. That is, our first-person experience that gives us the sense of self is basically an illusion and not some kind of metaphysical fact or phenomenological evidence. However, to apply Buddhist philosophy in western theoretical context has serious ramifications since the self-other relation in the constitution of the sense of self is mostly irrelevant or illusory from the viewpoint of witness-consciousness, and the embodied nature and the life-world of the self can also be seen as illusory. Despite the cultural and metaphysical differences, the *no-self doctrines* have a serious impact on current interdisciplinary discussions.[24] But, what precisely is this so-called witness-consciousness? First, let us consider the status of the illusory self. The illusory self is essentially constituted by "desire-driven thought and emotion"; and this affectively charged sense of self gives rise to the illusion of "*ontologically unique* or *bounded entity*". More adequately:

> Roughly speaking, the self will, I argue, turn out to be a conscious subject that is a unified, happiness-seeking, unbrokenly persisting, ontologically distinct or *bounded* 'me' who is an owner of experiences, thinker of thoughts and agent of actions. On the proposed theory of 'no-self', then, a self of such a description will not actually exist, although most of us will, through our very mode of living, be reflexively *assuming* that we are such a self. In assuming that we are such a self – a self that turns out to not actually exist – we will therefore be in the grip of a *deep-seated illusion* (Albhari, 2006, p. 2).

Witness-consciousness is another layer of the self. Or, to put it another way, the *self* as an autonomous agent can be distinguished from a more primordial region of *subjectivity*, namely the ability of awareness or *witness*. The sense of self and pure subjectivity are two "tiers" of the self. The contrast between the sense of self and witness-consciousness can be illustrated by the contradictory nature of a sound in the dream and the real sound in the physical environment.[25] Witness-consciousness is basically a point of view, a locus of apprehension. The crucial question is for Albahari: how could this kind of awareness be transformed into a person? The answer lies partly in the problem of *ownership*,[26] that is, in the mechanism of identification and personalization. Inspired by Indian and Buddhist traditions, Albahari argues that the subject's modus operandi is to observe or witness objects through perceptual or cognitive modalities. The embodied viewpoint, i.e., the locus of the first-person perspective, delineates the concept of *perspectival ownership*. But,

---

[24] There is no room here to analyze the complexity of the dialogue between Buddhist philosophy and philosophy of mind (including neuroscientific approaches). For example, Davidson and Harrington (2002) provided illustrative examples of metaphysical, methodological, and terminological incommensurabilities between Buddhism and neuroscience.

[25] "The illusory self as depicted in Western theories, by contrast, is more akin to a dreamt-of voice whose shrill quality is entirely dreamt up, with no recognisable input from the world outside of the dream" (Albahari, 2006, p. 3). Further examples demonstrate the illusory status of the empirical self: "…someone on hallucinogenic drugs who 'hears' a chorus of angels as purporting to originate from heaven (as opposed to from their mind) will not, should she know this to be the effect of the drugs, believe that she really is hearing a heavenly chorus. The heavenly chorus (the content of a hallucination) will nevertheless be illusory, purporting to originate from heaven – and hence independently of her perception – when it does not actually exist in this manner (it exists only 'in her mind'). […] The schizophrenic who hears voices in his head and is ignorant of his condition is likely to completely buy into the objective reality of these voices, believing, for example, that the voice is of God telling him to go on a mission. When beliefs as major as this are drawn into the illusion, the phenomenon tends to be classed as a delusion" (Albahari, 2006, pp. 123–124).

[26] The dilemma of *ownership* will frequently resurface in this book. The gist of the phrase occurs in the works of William James and Pierre Janet. The William James in *The Principles of Psychology* claimed that the possession or appropriation of our thoughts, desires, belief etc. is the undeniable mark of the presence of a personal self: "The universal conscious fact is not 'feelings and thoughts exist,' but 'I think' and 'I feel.' 1 No psychology, at any rate, can question the *existence* of personal selves" (James, 1983 [1890], p. 221; cf. Kihlstrom et al., 2000, p. 64).

Albahari claims, the focal point of awareness cannot be subjected to awareness; metaphorically speaking, the observing eye cannot draw attention to its own attentional processes. Albahari further argues that, the subject or the sense of self has a subtle phenomenal character – which can be enhanced during meditation – but it is characterized by a kind of *elusiveness*.[27] That is, to put it in a phenomenological context, because consciousness is oriented to its own intentional states the very nature of awareness remains unreflected. Moreover, the subject is not only the perspectival owner of occurrent experiences, but its identity is also thickened or crystallized by *personal ownership*. Everything that occurs (regardless of their perceptual or emotive aspects) in the phenomenal field of the subject is appropriated or owned by the subject. To use the phrase of Kihlstrom and his colleagues: we "take possession" of our mental and behavioral states or "inject ourselves into them" (Kihlstrom et al. 2000, p. 64).

Personal ownership constitutes a *me* beyond the impersonal awareness of the egocentric point of view. The constitution of *me* entails not only a perspectival and a personal, but also a *possessive ownership* as well. The latter designates our relation to the material world; we possess cars and houses and other kind of assets. Our elusive sense of self is widened or thickened owing to our possessions and social status. In short, objects in the world or mental states are "warmly infused with the sense of mineness" and crystallize and amplify the sense of *me-ness* (Albahari, 2013, pp. 84–85). To put it more powerfully, the personal self is constituted by different levels of infusions into the realm of phenomena. Personal ownership means the identification with mental and behavioral states, possessive ownership means the clinging to material objects. We do not have a clear, reified idea about ourselves, but it seems obvious that we are explicitly entangled in everyday concerns, identify with others, and take possession of and invest in material objects. What we lack, according to the no-self doctrine, is the metaphysical pure witness-consciousness deprived of any form of ownership. The crucial question to ask is how can we make a strict demarcation between perspectival and personal ownership? Albahari argues that perspectival ownership is basically our spatiotemporal-bodily locus, but, at the same time, the thought of "my body" is personally owned as well. Moreover, certain ideas (gender, race, character traits, etc.) ingrained into the subject's first-person perspective as filters for perception and reflection, and for her, this implicit knowledge also implies personal ownership. However, from phenomenological point of view, the distinction evaporates and dissolves into the passive infrastructure of the ego. As we will see in the following chapters, the implicit realm of experiential reality will encapsulate the seemingly different branches of ownership. From a Husserlian point of view the zero point of orientation, the kinesthetic experiences of

[27] In this sense, for Albahari, it is impossible to grasp the minimal self as a reified notion or something that directly occurs in front of our attentive processes: "Of note is that the subject (and the wider self) cannot appear directly to itself in the focal manner of a perspectivally owned object, so although seeming to have a subtle phenomenal character (and hence a sense of itself), which can be enhanced during meditational practice, the subject is perpetually elusive to its own focally attentive purview" (Albahari, 2013, p. 84).

"I can", and the convictions and beliefs of the "natural attitude" are conjoined aspects of the empirical person (The Husserlian layered conception of the (empirical or personal) self will be analyzed in more detail in Chap. 2).

Albahari (2006) takes pains to try to demonstrate the illusory nature of the *personal self.* In one of her examples, she presents a thought experiment in which Ben wins a race and then feels proud because of it. He naturally believes that his personal self remembers the act of winning and the feeling of being proud is intimately linked to his self. He is an agent, who won the race, and he is the personal owner of the joy. Nevertheless, for Albahari (2006) what is the true nature of the self is not the embodied-personal self with its intentional-affective states but the non-personal witness-consciousness (pp. 190–191).[28] The counterintuitive witness-consciousness runs counter to the basic tenets of the embodied mind which tries to specify the source of the self in bodily skills and desiderative nature of the bodily self (cf. Maiese, 2011, 2016). And it also stands in marked opposition to the social constructive alternatives of the self as well, since from the unworldly perspective of witness-consciousness the self-other relation is also regarded as illusory. In short, every kind of identification – that is, the appropriation of a felt emotion or a societal act, etc. – produces the false illusion of a *bounded sense of self.*

A practical alternative for Albhari is meditation, which gradually erodes the sense of self and leads to the recognition of ownerless consciousness (Albahari, 2006, p. 192). The bounded sense of self is the result of the intentional arc. There is a reciprocal relation between "me", the personal owner of experiences, and some item (which can be internal or external). As a result, "me" the subject experience the item as "mine", and a core self or a sense of self ensues (Albahari, 2006, p. 173). The diminishment of the sense of me, through eastern practices, leads to the state of *nirvana*, i.e., to ownerless consciousness. Albahari tries to interpret the state of *nirvana* as a psychological state. She suggests that certain psychopathologies may shed some light on this counterintuitive experience of pure awareness. In her book, *Analytical Buddhism*, first, she considers the possibility that depersonalization syndrome can be characterized with the absence of personal ownership. However, she readily observes in Damasio's and others case studies, that depersonalization is a distressing symptom, and patients frequently complain about stress and constant emotional turmoil resulting from the experience of detachment from and de-realization of the world. Patients suffering from depersonalization might be separated from the world and others but she at the same time is the subject of negative emotions which implies the sense of (bounded) self and personal ownership (p. 174). Furthermore, she cites Damasio's (1999) case studies concerning epileptic automatism, akinetic mutism, and advanced stages of Alzheimer's disease. In these debilitating illnesses there is no trace of an autobiographical self or of the capacity of reflection. In Albahari's own words, there is no sign of "reflexive sense of bounded

[28] "There is only, at any one time, a perspectival witnessing presence with a bundle of thoughts and emotions that, through the mechanism of appropriation, collectively comprise the sense of a self that owns them. There is no actual personal self who has these emotions; that is what it means to say the self is an illusion" (Albahari, 2006, p. 191).

self" in these pathologies (Albahari, 2006, pp. 175–176). Albahari (2006) also suggests that awareness exists without the "reflexive sense of a bounded self" in newborns and primitive organisms (p. 177). As a result, the concept of witness-consciousness boils down to *Shankara's monism* in which: "The wise man ignores the jar, water and the sun's reflection in it, and sees the self-illuminating sun itself which gives light to all three but is independent of them" (citation in Albahari, 2006, p. 194). In other words, witness-consciousness is the nirvana-state of the Buddhist "Arahant" (or *arhat*), who attained enlightenment and liberated himself from the sense of self that implies the "attendant feelings of 'me' and 'mine'" (p. 79). Later, in her paper (Albahari, 2013) she admits that the comparison of nirvana with case studies of psychopathology was premature. What is the reason behind that the state of nirvana does not seem to be a "debilitating pathology" compared to, for example, depersonalization? Ultimately, she claims that the clue may lie in the cultivated attention during meditative practice. Whereas the quality of attention is abnormally low in cases where personal ownership is suspended, for example in akinetic mutism or Alzheimer's disease, cultivated meditative practice can achieve a high quality of attention without negative effects (Albahari, 2013, p. 112).

For Albahari, the concept of witness-consciousness cannot be reduced solely to the mechanism of attention. She argues that subliminal impressions of unattended noises fall into the category of "unattended witnessing". Light or luminescence are recurrent metaphors of witness-consciousness; witnessing can be likened to light in general, and attention can be likened to the directed beam of a flashlight.[29] Another intriguing feature of witness-consciousness is that it has an intrinsic or generic phenomenal character. The "what-it-is-like-ness" of experiential reality is tied to this concealed awareness. Albahari argues that there is "something it is like to be aware simpliciter", and the fundamental phenomenal character of awareness is distributed like the colors of a beam of white light filtered through a prism:

> There is nevertheless, on the proposed concept, something it is like to be aware simpliciter and this quality of awareness (or knowing) imparts its own generic 'flavour' to all conscious states. A useful analogy may be that of a beam of white light being diffracted by a prism into a spectrum of colour. While each colour (sensory-experience) will differ from the others, all will share the unifying and generic character of luminosity (awareness) (Albahari, 2006, p. 143).

At first sight, Albahari's "Analytical Buddhism" seems to be an eccentric move in contemporary consciousness studies. However, the interdisciplinary dialogue of

[29] "A useful analogy for the relation between witnessing and attention is that of a flashlight, whose beam can be concentrated and focused on various objects, while nevertheless casting a dimmer pool of light on objects not focused on. Witnessing can be likened to the light in general, while attention can be compared to the focusing or directing of the beam on this object and that. Even with the beam focused, there is a surrounding, dimmer circle of light, illuminating objects not focused on, analogous to the inattentive witnessing" (Albahari, 2006, p. 143).

embodied mind has always reckoned with the methodological and phenomenological significance of Buddhist philosophy.[30]

Dreyfus (2013), in his critical remarks, points out that Albahari's witness-consciousness is not necessarily representative of *the* Buddhist position. Far from being homogenous, the Buddhist tradition encompasses a wide variety of views. Nevertheless, Dreyfus highlights, Albhari's idea of the transcendent self is supported by certain views within the Mahāyāna tradition, though it is not universally accepted by all movements. And there is a conspicuous discrepancy between the idea of subjectivity as a transcendent awareness and as an experiential flow. Dreyfus resist to the metaphysical conception of a "transcendent and static presence" and rather defines self-awareness as an "always renewed background awareness". Contrary to Albhari's witness-consciousness, Dreyfus places heavy emphasis on the embodied nature of consciousness, and based on Hanna and Maiese's (2009) work, suggests a subliminal, baseline consciousness that is generated by the living body and gives rise to a nuanced phenomenal background feeling that remains constant even under the state of dreamless sleep (Dreyfus, 2013, pp. 141–142). Dreyfus rejects Albahari's metaphor of light in explaining the most elementary model of phenomenal consciousness. Hanna and Maiese's Deep Consciousness thesis is the consequence of their Essential Embodiment Theory, which traces the origin of first-person perspective, and thus minimal self-awareness, back to the primitive bodily awareness of minded animals.[31] The notion of deep consciousness or occurrent consciousness suggests that information processing in the nervous system of embodied beings involves a non-focal, pre-reflective consciousness. The latter conceptualization suggests that consciousness is not an all-or-nothing phenomenon, but rather there is a wide spectrum between the most primitive sensorimotor awareness and extremely attentive, focused, or conceptually mediated consciousness.[32] Of course, according to their framework, consciousness is inseparably tied to embodied-sentient beings (endowed with agency and desiderative-affective characteristics).

---

[30] See for example Varela et al. (1991), Varela (1996), Varela and Shear (2002), Thompson (2007).

[31] The authors reflect upon the problem of mind-body dualism and its Nagelian challenge when they claim: "…we hold, this ability of minded animals to have a single point of view is grounded in egocentrically-centered and spatially oriented essential embodiment, and a primitive bodily awareness that includes proprioception (including kinaesthesia, and the sense of orientation and balance), bodily pleasures and pains, tickles and itches, the feeling of pressure, the feeling of temperature, and the feelings of vitality or lethargy" (Hanna & Maiese, 2009, p. 32).

[32] Maiese claims that all higher, self-reflective consciousness presupposes the pre-reflective sensorimotor subjectivity. A deep consciousness almost invisibly permeates the conscious life of a subject: "…necessarily, whenever a creature with a consciousness like ours is in any sort of mental state, then it is also occurrently conscious. Of course, being occurrently conscious is not the same thing as being maximally conscious. States such as dreaming sleep, dreamless sleep, automatism, trances, reflex action, divided attention, subliminal awareness, and cognitive priming are all examples of states that do not involve full conscious awareness. And yet even these relatively non-conscious states still involve some subjective experience and a point of view. Indeed, all access conscious states whose contents are poised for use in thought and action are occurrently phenomenally conscious, at least in a minimal sense" (Hanna & Maiese, 2009, p. 23).

Despite the unwavering popularity of Buddhist philosophy in current mind sciences, I agree with Zahavi's critical remarks. First, Buddhist views are strongly motivated by metaphysical and soteriological concerns, and their tenets often incommensurable with basic phenomenological ideas. Second, if one accepts – with or without hesitation – that Buddhist no-self doctrine is in par with contemporary views of self-sceptics (e.g., Dennett (1991) and Metzinger (2003, 2009), then the notion of an illusory-self renders the social reality fictional as well. And this latter line of thinking stands in stark contrast with the phenomenological attempt to examine the lifeworld (*Lebenswelt*) (Zahavi, 2013, p. 70). Moreover, as Zahavi has shown, there is an implicit self-contradiction in Albahari's notion: on the one hand, she is an advocate of the no-self doctrine (while the personal I is an illusion), but on the other hand, there is an ontologically independent witnessing presence that implies a "robust notion of selfhood" (Zahavi, 2013, p. 45).

While the supposedly unconstrued, metaphysical reality of pure awareness may seem at least counterintuitive, if not completely alien, to Western thought, analytic Buddhism holds its position in contemporary philosophy of mind. In fact, one can find several alternatives to the no-self theory within naturalist frameworks as well. In line with Albahari, Thomas Metzinger, who is also the advocate of the no-self doctrine, accentuates the role of attentional control. However, contrary to Albahari, he argues that in cases of severe drunkenness or senile dementia one may lose the ability to control attention and, as a result, feel that the self is falling apart. For Metzinger the self is not some a sort of metaphysical entity, but rather an agent generated by the self-representing systems of the brain-mind complex. On his account, the self has its own developmental and phylogenetic origin which makes it vulnerable to disorders. Other examples of lost "attentional agency" are the mind of the infant and dream state. By paraphrasing Husserl, he insists that the loss of control of the "ray of attention" entails to the disintegration of the self (Metzinger, 2009, p. 120). Metzinger's evolutionary perspective or "unrestrictive scientism" – as Zahavi (2005b) has noted – leaves no room for a hidden self-awareness that lurking behind the sight of focal attention.

Metzinger's self-model theory of subjectivity also tries to specify the essential nature of the self (and its holistic phenomenal character), but ends up joining the eliminativist approaches of Churchland (1989) and Dennett (1991). Metzinger in his monumental synthetic approach developed the "phenomenal self-model" (PSM). His idea is reminiscent of the representationalism of Locke and Descartes, and it had an enormous impact in mind sciences. Nevertheless, Metzinger was frequently accused of scientism and unreflective materialism as well. Based on certain psychopathological case studies and altered states of consciousness, Metzinger came up with the idea that the brain generates a virtual model of the environment and the body. The Ego, he further argues, is the content of this phenomenal self-model generated by the brain. Based on neuroscientific interpretations of the rubber hand illusion[33] and out-of-body experiences (OBEs) Metzinger believes that the brain generates a "reality tunnel", a simplified model of the world that is similar to the virtual reality of advanced video games, and an "Ego Tunnel", defined as the simplified model of the subject's interiority. The ego tunnel is, of course, not qualifies as the realistic representation of the environment and the body, rather it also comprises "unconscious filter mechanisms" (Metzinger, 2009, pp. 8–9). As a result, the subject is unable to establish a direct link to "reality" as such:

> The idea of an Ego Tunnel is based on an older notion that has been around for quite some time now. It is the concept of a "reality tunnel," which can be found in research on virtual reality and the programming of advanced video games, or in the popular work of non-academic philosophers such as Robert Anton Wilson and Timothy Leary. The general idea is this: Yes, there is an outside world, and yes, there is an objective reality, but in moving through this world, we constantly apply unconscious filter mechanisms, and in doing so, we unknowingly construct our own individual world, which is our "reality tunnel." We are never directly in touch with reality as such, because these filters prevent us from seeing the world as it is. The filtering mechanisms are our sensory systems and our brain, the architecture of which we inherited from our biological ancestors, as well as our prior beliefs and implicit assumptions. The construction process is largely invisible; in the end, we see only what our reality tunnel allows us to see, and most of us are completely unaware of this fact (Metzinger, 2009, p. 9).

While the concept of Ego Tunnel seems appealing from the angle of indirect realism, it is often burdened with contradictions. On the one hand, Metzinger does not deny access to the world and says that the improvements in scientific communities are the clear signs of leaving behind the interiority of the mind, on the other hand, in defining the nature of consciousness he frequently employs simulation and brain-in-the-vat hypotheses. For him, the problem of consciousness is "all about subjective experience, about the structure of our inner life, and not about knowledge of the outer world" (Metzinger, 2009, p. 11). The problem of qualia or phenomenal content also received a simplified explanation: the mental representations of the PSM have *ineffable* phenomenal contents (e.g., "amber-ness" and "sandalwood-ness"). In the case of emotions, the phenomenal content is the feeling in itself without reference to intentional objects. In conclusion, he reduces the origin of qualitative states to "internal properties of brain states" that can be activated by receptor cells in case of perceptual experiences or by psychoactive substances in cases of emotional experiences. Based on the progress of neuroscience, he supposes that complex experiences may be reactivated by their "minimal neural correlate" (Metzinger, 2009, pp. 10–11). The only reason why we are no living in a dream, or in a psychotic

[33] "In 1998, University of Pittsburgh psychiatrists Matthew Botvinick and Jonathan Cohen conducted a now-classic experiment in which healthy subjects experienced an artificial limb as part of their own body. The subjects observed a rubber hand lying on the desk in front of them, with their own corresponding hand concealed from their view by a screen. The visible rubber hand and the subject's unseen hand were then synchronously stroked with a probe. The experiment is easy to replicate: After a certain time (60–90 s, in my case), the famous rubber-hand illusion emerges. Suddenly, you experience the rubber hand as your own, and you feel the repeated strokes in this rubber hand. Moreover, you feel a full-blown 'virtual arm' – that is, a connection from your shoulder to the fake hand on the table in front of you" (Metzinger, 2009, p. 3).

delusion that there is an adaptive coupling between the PSM and environment. But, in the end, we only experience the "internal shadow" of an infinitely complex environment and the self is basically the internal, perspectival reference point of the world-model. Due to the imaginable fast information-processing on the level of first-order representations we are oblivious to the fact that we have only direct contact with the content of the PSM. We can move closer to a television screen and enlarge the on-off switching pixels to reveal the mechanism of movie broadcasting, but the brain is unable to reflect upon its own unbelievably swift information-processing mechanisms. According to Metzinger's physicalist framework, we are *naïve realists* due to our evolutionary development, and the PSM is *transparent*, which means we are looking directly at the outer world through filters and images generated by the brain (Metzinger, 2009, pp. 41–42).

The PSM is not only the medium of internal and external perception, but also relies heavily on the process of meta-representation. The PSM of the *homo sapiens*, Metzinger claims, invented the conscious Ego that enables us to consciously perceive ourselves and others. The content of the PSM is identical with the content of the Ego, which is imbued with the feeling of "mineness" or, in other words, the "sense of ownership". However, we not only "own" of our body parts, feeling, and other mental states, but we also have a "global ownership", a general mode of identification with our body. The rubber hand illusion or full-body illusions (including OBEs)[34] demonstrate that ownership is generated by an inner *body image*, which is also the content of the PSM, and, in certain circumstances, it is prone to disintegration or duplicates itself (Metzinger, 2009, p. 5). Metzinger argues as follows: "Subjectivity is not a *thing*, but a property of complex representational processes unfolding in certain physical systems" (Metzinger, 2003, p. 577). According to Metzinger, scientific psychology and cognitive neuroscience demonstrate that subjectivity is reducible to the phenomenal model of intentional relation (p. 577).

Zahavi (2005b) accuses Metzinger of unreflective reductionism and calls for competing definitions of the self that are missing from Metzinger's model. Zahavi also criticizes Metzinger for denouncing phenomenology as an inadequate method and for not attempting to consider different phenomenological movements. Despite his rejection of "phenomenology", he uses the term in at least in three different ways: (1) it means only the introspectively accessible experiential domain; (2) he often speaks of "folk phenomenology"; (3) and infrequently refers to the phenomenological tradition (Zahavi, 2005b, p. 11). Metzinger accepts most of Zahavi's criticism, and is also intrigued by the "poetic metaphors" of the minimal self (as an "invariant dimension of first-personal givenness"), but, at the same time, he rejects

[34] In one of his papers, Metzinger tries to show that the concept of soul and subtle body can be explained by OBE experiences: "Centuries of phenomenological reports describing it as subtle body pointed in the right direction, and now we begin to see how it actually is a purely informational structure modeling bodily self-experience in cases of absent or disintegrated somatosensory/vestibular input" (Metzinger, 2005, p. 81). While the heuristic value of altered states of consciousness seems promising in mind sciences, Metzinger's analogy is overly simplistic and idiosyncratic.

the alleged charge of scientism and instead adheres to scientific realism (Metzinger, 2006). Albahari and Metzinger presented two alternatives of the no-self doctrine. Albahari draws heavily on Indian and Buddhist philosophy and Metzinger is the advocate of scientific realism and regards the phenomenal self as an all-encompassing concept that will empirically demystify ancient philosophical problems as well.

As Zahavi pointed out, there is no unequivocal definition of the self; there is no classical, commonsensical, or standard notion. Moreover, diverse notions of the self can be found not only in philosophy but also in empirical research. The common denominator might be an overly simplistic notion of the self, for example, an *internal homunculus* or a *principle of identity*. Zahavi defines the latter notion as follows: "invariant *principle of identity* that stands apart from and above the stream of changing experiences; something that remains unchanging from birth to death; something that remains entirely unaffected by language acquisition, social relationships, major life events, personal commitments, projects, and values, something that cannot develop and flourish nor be disturbed or shattered" (Zahavi, 2013, p. 66). Zahavi's approach is based on the denial of the distinction between *subjectivity* and *selfhood*. The analysis of subjectivity (and, consequently, the minimal self) is closely connected to the problem of time-consciousness and passivity. By contrast, selfhood suggest a more rigid, substantial notion of an enduring I (Further elaboration ensues in Chap. 2). Before delving into the issues of subjectivity and pre-reflective self-awareness, it is worth considering certain notions of the narrative self. The narrative self can be contrasted with the minimal self; therefore, it is necessary to briefly examine the current discourse around it. I will highlight the problems of fictional elements and constructiveness in the constitution of the narrative self.

References

Albahari, M. (2006). *Analytical Buddhism: The two-tiered illusion of self.* Palgrave Macmillan.

Albahari, M. (2013). Nirvana and ownerless consciousness. In M. Siderits, E. Thompson, & D. Zahavi (Eds.), *Self, no self?: Perspectives from analytical, phenomenological, & Indian traditions* (pp. 79–113). OUP.

Barresi, J., & Martin, R. (2011). History as prologue: Western theories of the self. In S. Gallagher (Ed.), *The Oxford handbook of the self* (pp. 33–56). OUP.

Blackmore, S. (2018). *Consciousness: An introduction.* OUP.

Block, N. (1995). On confusion about the function of consciousness. *Behavioral and Brain Sciences, 18*(2), 227–247.

Block, N. (2007). Consciousness, accessibility, and the mesh between psychology and neuroscience. *Behavioral and Brain Sciences, 30*(5–6), 481–499.

Braude, S. E. (1995). *Multiple personality and the philosophy of mind.* Rowman & Littlefield Publishers.

Brewin, C. R., & Andrews, B. (1998). Recovered memories of trauma: Phenomenology and cognitive mechanism. *Clinical Psychology Review, 18*(8), 949–970.

Chalmers, D. (1995). Facing up to the problem of consciousness. *Journal of Consciousness Studies, 2*(3), 200–219. Retrieved from: http://consc.net/papers/facing.pdf

Chalmers, D. (1996). *The conscious mind: In search of a fundamental theory.* OUP.

Churchland, P. S. (1989). *Neurophilosophy: Toward a unified science of the mind-brain.* MIT Press.

Colvin, M. K., & Gazzaniga, M. (2007). Split-brain cases. In M. Velmans & S. Schneider (Eds.), *The Blackwell companion to consciousness* (pp. 182–208). Wiley-Blackwell.

Damasio, A. (1999). *The feeling of what happens: Body, emotion and the making of consciousness*. Vintage Book.

Davidson, R., & Harrington, A. (2002). *Visions of compassion: Western scientists and Tibetan Buddhists examine human nature*. OUP.

Dennett, D. C. (1991). *Consciousness explained*. Penguin Books.

Dreyfus, G. (2013). Self and subjectivity: A middle way approach. In M. Siderits, E. Thompson, & D. Zahavi (Eds.), *Self, no self?: Perspectives from analytical, phenomenological, and Indian traditions* (pp. 114–156). OUP.

Eshleman, M. C. (2007). The Cartesian unconscious. *History of Philosophy Quarterly, 24*(2), 169–187.

Firth, C. D., & Rees, G. (2007). A brief history of the scientific approach to the study of consciousness. In M. Velmans & S. Schneider (Eds.), *The Blackwell companion to consciousness* (pp. 9–22). Wiley-Blackwell.

Flanagan, O. (1991). *The science of the mind* (2nd ed.). The MIT Press.

Gallagher, S. (2005). *How the body shapes the mind*. OUP.

Gallagher, S., & Zahavi, D. (2012). *The phenomenological mind*. Routledge.

Giampieri-Deutsch, P. (2012a). Psychoanalysis: Philosophy and/or science of subjectivity? Prospects for a dialogue between phenomenology, philosophy of mind, and psychoanalysis. In D. Lohmar & D. Brudzińska (Eds.), *Founding psychoanalysis phenomenologically* (pp. 83–103). Springer.

Giampieri-Deutsch, P. (2012b). Perception, conscious and unconscious processes. In F. Barth, P. Giampieri-Deutsch, & H.-D. Klein (Eds.), *Sensory perception. Mind and matter* (pp. 245–264). Springer.

Hacking, J. (1995). *Rewriting the soul: Multiple personality and the science of memory*. Princeton University Press.

Hanna, R., & Maiese, M. (2009). *Embodied minds in action*. OUP.

Jackson, F. (1983). Epiphenomenal qualia. *Philosophical Quarterly, 32*(127), 127–136.

James, W. (1983 [1890]). *The principles of psychology*. HUP.

Jaynes, J. (1986). Consciousness and the voices of the mind. *Canadian Psychology/Psychologie Canadienne, 27*(2), 128–148.

Jaynes, J. (2000 [1976]). *The origin of consciousness in the breakdown of the bicameral mind*. Mariner Books.

Kihlstrom, J. F. (1987). The cognitive unconscious. *Science, 237*(4821), 1445–1452.

Kihlstrom, J. F., Shelagh, M., Tobias, B., & Tobis, I. (2000). The emotional unconscious. In E. Eich, J. Kihlstrom, G. Bower, J. Forgas, & P. Niedenthal (Eds.), *Cognition and emotion* (pp. 30–86). OUP.

Kihlstrom, J. F., Dorfman, J., & Park, L. (2007). Implicit and explicit memory and learning. In M. Velmans & S. Schneider (Eds.), *The Blackwell companion to consciousness* (pp. 525–539). Wiley-Blackwell.

Levine, J. (1983). Materialism and qualia: The explanatory gap. *Pacific Philosophical Quarterly, 64*, 354–61.

Maiese, M. (2011). *Embodiment, emotion, and cognition*. Palgrave Macmillan.

Maiese, M. (2016). *Embodied selves and divided minds*. OUP.

Marosán, B. P. (2023). The genesis of the minimal mind: Elements of a phenomenological and functional account. *Phenomenology and the Cognitive Sciences*. https://doi.org/10.1007/s11097-023-09946-7

McGinn, C. (1989). Can we solve the mind-body problem? *Mind, 98*, 349–366.

Metzinger, T. (2003). *Being no one: The self-model theory of subjectivity*. MIT Press.

Metzinger, T. (2005). Out-of-body experiences as the origin of the concept of a 'soul'. *Mind and Matter, 3*(1), 57–84.

Metzinger, T. (2006). Reply to Zahavi: The value of historical scholarship. *Psyche, 12*(2), 1–7.

Metzinger, T. (2009). *The ego tunnel: The science of the mind and the myth of the self*. Basic Books.

Nagel, T. (1974). What is it like to be a bat? *The Philosophical Review, 83*(4), 435–450.

Olsen, E. T. (1998). There is no problem of the self. *Journal of Consciousness Studies, 5*(5–6), 645–657.
Revonsuo, A. (2018). *Foundations of consciousness*. Routledge.
Saunders, G. (2014). *Acts of consciousness: A social psychology standpoint*. CUP.
Schacter, D. L. (1996). *Searching for memory: The brain, the mind, and the past*. Basic Books.
Schechtman, M. (1996). *The constitution of Selves*. Cornell University Press.
Skrbina, D. (2005). *Panpsychism in the West*. The MIT Press.
Stevens, G. L., & Graham, G. (2007). Philosophical psychopathology and self-consciousness. In M. Velmans & S. Schneider (Eds.), *The Blackwell companion to consciousness* (pp. 195–208). Wiley-Blackwell.
Strawson, G. (2000). The phenomenology and ontology of the self. In D. Zahavi (Ed.), *Exploring the self: Philosophical and psychopathological perspectives on self-experience* (pp. 39–54). John Benjamins.
Thiel, U. (2011). *The early modern subject: Self-consciousness and personal identity from Descartes to Hume*. OUP.
Thompson, E. (2007). *Mind in life: Biology, phenomenology, and the sciences of mind*. HUP.
Tye, M. (2007). Philosophical problems of consciousness. In M. Velmans & S. Schneider (Eds.), *The Blackwell companion to consciousness* (pp. 23–36). Wiley-Blackwell.
Varela, F. (1996). Neurophenomenology: A methodological remedy for the hard problem. *Journal of Consciousness Studies, 3*(4), 330–349.
Varela, F., & Shear, J. (Eds.). (2002). *The view from within: First-person approaches to the study of consciousness*. Imprint Academic.
Varela, F., Thompson, E., & Rosch, E. (1991). *The embodied mind: Cognitive science and human experience*. The MIT Press.
Warrington, E. K., & Weiskrantz, L. (1968). A study of learning and retention in amnesic patients. *Neuropsychologia, 6*(3), 283–291.
Weiskrantz, L. (2007). The case of blindsight. In M. Velmans & S. Schneider (Eds.), *The Blackwell companion to consciousness* (pp. 175–180). Wiley-Blackwell.
Wilkes, K. V. (1981). Multiple personality and personal identity. *The British Journal for the Philosophy of Science, 32*(4), 331–348.
Wilkes, K. V. (1988). *Real people: Personal identity without thought experiments*. OUP.
Wilkes, K. V. (1995). Losing consciousness. In T. Metzinger (Ed.), *Conscious experience*. Schöningh: Imprint Academic.
Williams, G. (2011). What is it like to be nonconscious? A defense of Julian Jaynes. *Phenomenology and the Cognitive Sciences, 10*(2), 217–239.
Zahavi, D. (2005a). *Subjectivity and selfhood: Investigating the first-person perspective*. MIT Press.
Zahavi, D. (2005b). Being someone. *Psyche, 11*(5), 1–20. Retrieved from: https://cfs.ku.dk/staff/zahavi-publications/metzinger.pdf
Zahavi, D. (2013). The experiential self: Objections and clarifications. In M. Siderits, E. Thompson, & D. Zahavi (Eds.), *Self, no self?: Perspectives from analytical, phenomenological, and Indian traditions* (pp. 56–78). OUP.
Zahavi, D. (2016). Analytic and continental philosophy: From duality through plurality to (some kind of) unity. In S. Rinofner-Kreidl & H. A. Wiltsche (Eds.), *Analytic and continental philosophy: Methods and perspectives. Proceedings of the 37th International Wittgenstein Symposium* (pp. 79–94). Walter De Gruyter.

Chapter 2
The Narrative Self and the Minimal Self

Abstract The aim of the chapter is to provide a detailed introduction to the multidimensional analysis of the self. The chapter examines the contrast between the narrative self and the minimal self and compares Peter Goldie's and Maria Schechtman's views on the constitution and the nature of the narrative self. The chapter also examines the developmental perspectives on the emergence of the narrative self. The chapter shows that the concept of implicit narrative reflects the influence of psychoanalysis in the characterization of the narrative self. Finally, following this line of argument, the problems of the affective unconscious and the affective core self are introduced and compared to the minimal self.

Keywords Narrative self · Personal identity · Minimal self · Self-awareness · Memory · Autobiographical memory · Affective identity

2.1 The Narrative Self and Personal Identity

It would be impossible to review all the different notions of the narrative self in an introductory chapter. In this section, I will focus on contemporary approaches that bears heavily on the issues of the minimal self. The idea of the narrative self proposes that the self is constituted by stories, it is the protagonist of an ongoing life history. The concept of the narrative self also implies the crucial role autobiographical memory; that is, the self is fundamentally an extended self that includes social and developmental factors, and, of course, time-consciousness.

Peter Goldie distinguished between *narrativist* and *narrative skeptic* positions. At one pole, there are the *narrativists* who claim that our lives are composed of lived narratives and we are both the authors and the protagonists of our narrated life. Goldie highlights that, according to the narrativists argument, our lives are similar to or the same as the lives of characters in literature. Goldie explains: "Our having the right kind of narrative of four lives is, in some sense, integral to or constitutive

L. Horváth, *The Affective Core Self*, Contributions to Phenomenology 130,
https://doi.org/10.1007/978-3-031-56920-3_2

of our being the persons that we are. Our very survival depends on our having such a narrative" (Goldie, 2012, p. 1). By contrast, on the other pole, *narrative skeptics* do not acknowledge the leading role of narratives in the understanding of our lives. The "proper home" of narratives is in literature and not in our lives, since the constructed narratives are perspectival and distort reality, truth, and objectivity. As a result, they are unable to give an authentic picture of "what it is to be a person" (p. 1).

At first glance, it may seem a provocative question to ask: What is it like to be a person who narrates? Are there any significant qualitative differences between a person who construes meaningful and coherent narratives about herself and someone who is unable to reconstruct her life through an all-encompassing narrative framework? Both memory research and philosophical insights indicate that there are conceivable qualitative differences. Here we can see a connection between the studies of phenomenal consciousness and the self, i.e., the narrative self. Marya Schechtman argues as follows:

> The phenomenological life of an individual who narrates her life in the way we narrate ours is much different from that of an individual who does not. The pulling together of events and experiences occurring at different time into a single narrative does more than simply change the quality of conscious experience at a given moment, it changes the quality of conscious experience overall – narrators have a consciousness that extends over time. A new psychological and behavioral repertoire comes into being when and individual forms an identity-constituting self-conception and is lost when the narrative thread is lost (Schechtman, 1996, p. 148).

Contrary to Schechtman, Goldie denies that the subtle presence of a sense of self, i.e., a narrative self, could be constituted through the narrative reconstruction of significant life events. On Goldie's account, the "narrative sense of self" does not require a stable self whose identity or survival could be threatened by the breakdown of narrative capacities. For Goldie, the appearance of the stable self is the result of the use of the word "I" in autobiographical narratives. The first-person pronoun is sufficient "for literal life and death, for legal and moral responsibility, and for backward- and forward-looking reactive attitudes and emotions of self-assessment…" (Goldie, 2012, p. 141). Both Schechtman and Goldie discuss the subject's relationship to his or her past (Later I will discuss the problem of relating to the past in greater detail). However, they disagree about the status of the narrative self. Goldie defines the "I" of the narratives as a linguistic construction. Schechtman, on the other hand, argues for the phenomenal thickness of the autobiographical or narrative self. In other words, Schechtman comes closer to the problem of phenomenal consciousness than Goldie.

What happens when a person loses his or her ability to tell a meaningful story? Owen Flanagan in his book, *Consciousness Reconsidered*, introduces the problem of narrative identity through Luria's case study. For the veteran called Zazetsky the integrity of his identity and his personal world shattered because of a battle injury. Flanagan highlights Zazetsky's efforts to regain the wholeness and meaningfulness of his life through automatic writing; inspired by this method, he produced a lengthy, three-thousand-page autobiography. Despite the heroic attempt to reproduce the objective story of his life, Zazetsky is unable to find and identify with the protagonist of this incessant story. According to Flanagan, the moral of the case study is that: "No one can grasp the whole narrative of his or her life at once. Indeed, there is always more to our identity than what we can narratively capture or convey" (Flanagan, 1992, p. 197). We can assume that patients suffering from memory problems (i.e., from different types of amnesia) may have a different sense of self in comparison to subjects with intact autobiographical memory. The synthesis of various meaningful events into a single, coherent narrative might give rise to a sense of self that differs from the shattered and disoriented identity of patients with severe memory impairments. Flanagan's observation support Schechtman's claim that the construction of a narrative could change the *quality* of conscious experience.

Hutto (2017) argues that the main role of autobiographical memory is not to faithfully represent the past, but rather to find a suitable narrative to describe our present purposes and circumstances. Autobiographical memory has a forward-looking aspect, and it also plays a crucial role in shaping interpersonal experiences. Hutto cites several studies that indicate that the pivotal role of remembering is "forging and maintaining" intimate relationships and preserving mental health (p. 200). It is striking to discover that narrative storytelling with *causal explanatory language* and *emotional language* lowers the level of anxiety and depression, and even higher immune system functioning may ensue.[1] In this context, autobiographical memory can be seen as a coping skill for dealing with stressful events, and, by remembering, it prepares the subject for the upcoming stressful events. Hutto points to the fact that the overemphasis of the representational fidelity of memory thwarted the recognition of the affective dimension and the social benefits of autobiographical memory (Hutto, 2017, p. 200).

Let us now turn back to the contrasting views of Schechtman and Goldie. I suggest that the contradiction between Goldie and Schechtman can be mitigated if we endow the narratively generated sense of self with a subtle phenomenal character and do not consider it as a reified, observable mental entity. Schechtman accentuates the relationship between narration and time: the narration of an autobiographical element brings about a special, extended phenomenal consciousness. However, every speculation about the nature of narratively generated self-experience and personal identity has serious ramifications. Flanagan surprisingly asserts that we are all like Zazetsky, constantly striving for wholeness, but we are unable to grasp our identity at once. Because of our finitude, we are unable to capture or recapture our self in its fullest form (Flanagan, 1992, p. 198). Flanagan concludes that memory and the ability to project ourselves to the future are indispensable for the constitution of the narrative self. On the one hand, he emphasizes the "streamlike" structure of life and consciousness; however, on the other hand, he argues that remembering, the prospective capacities of consciousness, and the fact that we are social beings make us "inveterate story tellers" (p. 198).

Occasionally, self-transformation takes place in life and we try to give reasons for the radical or the subtle changes. And this is where the gap between the inner

[1] Hutto quoted the study of Fiviush and MacDermott-Sales (2006) (cf. Hutto, 2017, p. 200).

and the outer perspective opens up. Let us examine Flanagan's example of *conversion* from hedonism to ascetic Buddhism. From the third-person perspective one may see fission in the fabric of the subject life history that significantly reconfigures the structure of habits, dispositions, beliefs, motivations, and emotional preferences. While we see a radical self-transformation in the radical change of the subject's way of life, the subject may feel the same and sees nothing extraordinary in the act of conversion (cf. Flanagan, 1992, p. 198).

How can we preserve or even characterize the diachronic identity of the self? Flanagan seems to suggest that the stream of consciousness may remain the same while narrative identity is intentionally or unintentionally reconfigured by life events. The difference between first-person and third-person perspectives further complicates matters. We can mitigate the impact of these issues by adhering to the multidimensional or multilayered conception of the self. Let us discuss some other theories of narrative identity; the following list is not exhaustive, but serves to highlight the contrast between the narrative and the minimal selves.

Gallagher. in his review article, gives a rudimentary definition of the narrative self: "A more or less coherent self (or self-image) that is constituted with a past and a future in the various stories that we and others tell about ourselves." (Gallagher, 2000, p. 15). It is important to note that, the constitution of the narrative self is inevitably linked to language. Human beings are remarkable creatures with the ability to recollect past-experiences and imagine future possibilities. Furthermore, through storytelling we confer meaning on a fundamentally meaningless physical word and, as a result, we exceed our basic biological boundaries: "We use words to tell stories, and in these stories we create what we call ourselves. We extend our biological boundaries to encompass a life of meaningful experience" (Gallagher, 2000, p. 19). Another frequently highlighted feature of narrative storytelling is the contribution to and even participation in others' narrative. Krueger (2013) summarizes the main tenet of the narrative accounts of the self in the following way: "Narratives help us organize and interpret our own experiential histories, share these histories with others, and meaningfully participate in the lives and experiences of others by entering into their ongoing narratives" (Krueger, 2013, p. 34). He adds that narratives not only convey a "dramatic texture to our lives" but we can also tell stories about *how we reason*. Narrative constructions may open up the space of self-reflection and self-knowledge, and certain naturalistic conceptions claim that our brain is basically hardwired to "spin narratives" and constitute the illusion of an enduring self. Developmental and evolutionary psychology examines the basic constituents and conditions of possibility of our basic narrative competencies. As Schechtman has shown, proponents of the narrative approach differ significantly in characterizing the *nature of the narratives*. For instance, Dennett (1992) puts a lot of emphasis on the role of an all-encompassing, single life-narrative, while Velleman (2006) argues that self-narration takes place in small unites (Schechtman, 2011, p. 401).

Dennett defined the self as a "center of narrative gravity" which is a useful fiction. To illustrate these claims, he uses psychical and computational analogies. In physics the center of gravity is a useful "fictional object" or "abstractum" for predicting the behavior of objects. However, the terms of fictional or abstract are not meant to be pejorative, since physical science can legitimize the use of the center of gravity. In a rather convenient way, he obfuscates the differences between physical and "soul-sciences" and claims that both of these paradigms attempt to "interpret" their target phenomena:

> The physicist does an *interpretation*, if you like, of the chair and its behavior, and comes up with the theoretical abstraction of a center of gravity [...] The hermeneuticist or phenomenologist – or anthropologist – sees some rather more complicated things moving about in the world – human beings and animals – and is faced with a similar problem of interpretation. It turns out to be theoretically perspicuous to organize the interpretation around a central abstraction: Each person has a *self* (in addition to a center of gravity). In fact we have to posit selves for *ourselves* as well (Dennett, 1992, p. 105).

It is unlikely that human behavior – embedded in complex psychosocial relations – and psychical particles or more complex objects could be explained on the same horizon. However, the crux of the argument is not to dissolve the borders between the different explanatory horizons of natural and human sciences. Dennett tries to demonstrate that the reified notion of the self is a theoretical (albeit folk psychological) creation, similar to the construction of theoretical entities in physics (such as elementary particles). His intuitively more appealing example is the "novel-writing machine", which creates the fictional character of Gilbert out of nothing (more precisely, it is a dumb machine with the ability of write a novel). Furthermore, we can embed the writing-machine into the body of a robot, but his change does not make any difference. The imagined robot will not know that he creates a fictional character, and similarly, our brain is also oblivious to the fact that it creates a (fictional) character at the intersection of automatically generated narrative threads. Therefore, from the perspective of a "materialist philosopher or neuroscientist", it is a *category mistake* to look for the self in the brain. Our brains are narrative-generating machines and ourselves are the protagonists of the generated narratives.[2] Despite the overt reductionism, Dennett emphasizes the role of narratives in the process of self-constitution:

> And just as spiders don't have to think, consciously and deliberately, about how to spin their webs, and just as beavers, unlike professional human engineers, do not consciously and deliberately plan the structures they build, we (unlike professional human storytellers) do not consciously and deliberately figure out what narratives to tell and how to tell them. Our tales are spun, but for the most part we don't spin them; they spin us. Our human consciousness, and our narrative selfhood, is their product, not their source (Dennett, 1991, p. 35).

Dennett's proposal neglects the intricacy of human actions and the complexity of self-other relations.[3] The narrative understanding of life cannot be reduced to a

---

[2] Dennett enthusiastically claims that multiple personality disorder exemplifies the curious situation in which a single brain generates multiple but divided narratives. Each narrative gives rise to a unique form of identification. Sybil, one member of multiples, developed different personalities due to the suggestions of the therapist. Dennett claims that Sybil was a "living novel", a real "novel-writing machine" who lived a life with alternate personalities (Dennett, 1992, p. 111).

[3] Zahavi, on the grounds of the phenomenology of the lifeworld, explicitly rejects the fictional self and dismisses Dennett's scientism: "To declare everything peculiar to human life fictitious simply

brain-generated evolutionary capacity, but it also reveals the interwoven aspects of affectivity, subjectively lived time, meaning-bestowing processes, and the lifeworld. In contrast to Dennett, Goldie and Schechtman brings into consideration the *emotional import* of the living present in autobiographical narratives. In their reading, the human mind is not identical with complex, subpersonal information-processing systems, but rather our mind strives for coherence and meaningfulness by accumulating narrative self-knowledge. The *embodied situatedness* of agents also plays a significant role in the constitution of narrative selves; different horizons of meaning might have different impact on the ongoing narrative understanding. Moreover, the contemporary proposal of *embodied narratives* poses a challenge to Dennett's proposal. In this respect, Menary made a significant distinction between "abstract narrative accounts" and "embodied narrative accounts" of the self (Menary, 2008, p. 63).[4]

Menary argues against the center of narrative gravity (CNG) thesis that the protagonist of the narratives is both a narrator and embodied consciousness. On Menary's reading, Dennett does not pay significant attention to the role of the body; for him, there is no room for "minimal embodied self", the self cannot be identified with the *acting* and *living* body. The self is rather an illusory image, the representation of the embodied agent that emerges from the interpretation of its own behavior (Menary, 2008, pp. 66–67). He points out a glaring contradiction in Dennett's account of CNG. On the one hand, the CNG plays an important role in the "cognitive economy of a human organism", on the other hand, Dennett considers the self as a fictional object (Menary, 2008, p. 69). From the perspective of embodied mind, Menary claims that the main dilemma of the "abstract narrativist" is that he or she neglects the self-referential character of affective and perceptual experiences. From this point of view, the (embodied) subject of the narrative is the prerequisite and not the result of the spinning narrative scripts. Menary tries to close the gap between the narrator of events and the embodied subject of those very events:

> So when I tell the story of how the cricket ball that hit me on the left fore-arm last Saturday 'bloody well hurt!' I am ascribing the pain in the forearm to a collection of narratives. This sounds wrong. I feel pain after being struck on the arm by a hard cricket ball propelled at me at 85 miles per hour. That is what the narrative is about, the narrative is about a subject who feels pain, and that subject who feels is me. [...] Whilst we want to distinguish between a sense of the self as an embodied agent and a sense of the self as a narrator, in practice we do not want to distinguish between them, because the embodied self is also the narrator. The pain I still feel after being hit on the arm is the pre-narrative fodder for the narrator who wishes to discursively tell an interlocutor about the previous Saturday's events on the cricket ground (Menary, 2008, p. 73).

---

because it cannot be naturalized, because it cannot be grasped by a certain mode of scientific comprehension, merely reveals one's prior commitment to a naive scientism, according to which (natural) science is the sole arbiter of what there is" (Zahavi, 2005, p. 112).

[4] As we shall see, Zahavi often characterizes the status of the minimal self in comparison to alternatives of the narrative self. In line with his considerations, I will focus on those accounts of the narrative self that point to the role of the pre-narrative embodied and affective dimensions.

Menary concludes that the self is not simply a narrative construct, but rather it is the "entire human organism", and that the embodied, feeling self is more fundamental than the narrative self. Based on developmental studies Menary tries to show that the temporal structure of experience (i.e. the felt connection with past experiences and expectations toward the future), and the ability to create primitive or more complex narratives are grounded in a *minimal (embodied) self* (Menary, 2008, p. 75). Rather than moving from one extreme to another, Menary tries to demonstrate the interrelated aspects of language acquisition and embodied situatedness. The interiorization of exterior linguistic communication in children is the result of a long and complicated developmental process; in the end, when children have finally mastered narratives, the "narrative sense of self" is born. The higher-order narrative self is able to complement and interpret the embodied and affective processes of self-awareness.[5] And, conversely, speech is not only the representation of action but also the dialogical relation between the self and others provides "cognitive structuring to the embodied perceptions"; as a result, the self as an agent is born (Menary, 2008, p. 82).[6] Of course, the narrative self does not emerge from emptiness; it has a specific developmental structure and relates to the gradually evolving self-consciousness.

There is no room here to review all the rival theories on the development of theory of mind and narrative understanding, but Nelson (2003) provides a general explanation. Experimental evidence has shown that children around one-year can remember the brief sequences of events and verbalize familiar scripts at age three. These rudimentary representations of personal experiences (i.e., canonical events or event sequences) serve as the ground for narratives. These unreflected canonical events form the basic layer of the "landscape of action" (Nelson, 2003, p. 26). What is phenomenologically most relevant in these findings is the development of the sense of temporality. In the case of a two-year-old child, it would be difficult to find the general temporal structure of past-present-future. As Nelson puts it: "…for the child of 2 years the only temporal differentiation is that of the present activity and everything else that has been entered into memory as a representation of the

[5] The developmental considerations put heavy emphasis on the role of our theory of mind. Understanding others, in a sense, is incorporated into the narratives of our lifeworld. Katherine Nelson (2003) addresses the close connection between the development of consciousness and the acquisition of narrative understanding in the following way: "…it is a developmental view that sets narrative not aside as some kind of special individual human gift but as part and parcel of the wide-ranging developments that take place during the critical years when the child can enter fully into the linguistic world but is not yet a participant in formal schooling. These developments include awareness of self and other, of the wider world beyond self, of past and future; in traditional cognitive developmental terms, they include theory of mind, perspective taking, autobiographical memory, and self-concept. This view then suggests the close connection between narrative and the emergence of a specifically human level of consciousness. It is possible, I believe, to draw these connections more tightly" (p. 22).

[6] There is a subtle intertwining between embodied action and narration of action. Menary, thorough citations from Vigotsky, tries to show that speech and action are part of the same psychological process. One of the striking discoveries of Vigotsky was that speech (i.e. narration of action) is so significant that young children cannot accomplish a given task without it (Menary, 2008, p. 82).

experienced world. These 'temporally unfixed' representations may consist of sequences of events, of people and their routines, or of places and their associated objects" (Nelson, 2003, p. 27). The development of the "verbal-narrative representation" makes it possible for the child to leave the confined level of "direct-experience-activity". The general idea is the following: "shared experiential narratives are the symbolic vehicles, available only to humans, through which such insights are gained, including the important insight that the other has a past and a present that differ from one's own, as well as the accompanying insight that one's own past is unique to oneself" (Nelson, 2003, p. 29). The recollection of the remembered past is structured by narratives, but children of 2–4 years often conflate their own experiences with others. Several studies on false memory demonstrated that children often appropriate other children's stories as their own, implying the lack of clear differentiation between self and other. Therefore, the capacity of verbal recapturing of an event does not necessarily allude to a fully developed autobiographical memory and to a differentiated self-other relation (Nelson, 2003, p. 31). According to Menary's strong hypothesis, language and especially narrative is a vehicle through which a second awareness, the "self-and-other awareness", emerges that also comprises the broader horizons of the social world and the awareness of past and possible future (Menary, 2008, p. 33). From the developmental perspective the narrative self is not a kind of automatism that is randomly generated by the brain. The narrative self emerges through the interaction between the child and caregiver; it is basically a socially mediated and facilitated ongoing process.

2.2 Empathic Access and Free Indirect Style

As previously discussed, we can approach the problem of the narrative self from the standpoint of information processing or from developmental and social perspectives. What is central to the present argument is the relationship between the narrative self and autobiography. Goldie (2012) underscores the role of narratives[7] in autobiographical thinking. Goldie's main idea is that there is a close analogy between literature and real life. Imagine the scenario when one is enthusiastically absorbed in a literary work and inadvertently identifies with several different perspectives. There is the perspective of the reading subject, who can identify with the main protagonist, with the narrator of novel, with the protagonist's past self, and

[7] Goldie defines narrative in the following way: "My claim is that a narrative, *whether or not narrated*, requires an external perspective. According to me, a narrative can be a sequence of thoughts with the three characteristic features of narrative: coherence, meaningfulness, and evaluative and emotional import. And the thinker is the person who occupies the external perspective of narrator – he is the external narrator just in the sense that it is he who thinks through the narrative" (Goldie, 2012, p. 40).

with several other characters.[8] What is really matters here is the "ironic gap" between different, divergent first-person perspectives. A specific literary style, namely, the so-called "free indirect style",[9] gives the reader the opportunity to open up and simultaneously bridge the gap between the different perspectives. To put it simply, we can effortlessly keep in mind of the protagonist's perspective and our own perspective on a particular action in the story. The free indirect style has a serious impact on the nature of narration: what is expressed in a narrative is more than the original intention of the narrator. The surplus of meaning and "unconscious efforts" can also be observed in autobiographical narratives. Goldie's illuminating example is when one tries to elicit sympathy by telling the story of one's father's cruelty, while the hidden reason behind the narrative was the guilt for not paying enough attention to him as a person (Goldie, 2012, p. 33). It is easy to see the previously mentioned example as a close analogy of the Sartrean "bad faith", or the Freudian "rationalization". In addition, Goldie argues that the intricate workings of autobiographical memory can be reconsidered in terms of free indirect style. By reconfiguring or reconstructing a past event through a novel viewpoint or because of an emotional import, an ironic gap opens up in the subject. In cases of recollection, the past self serves as the protagonist and the present self, who has an evaluative stance, is compared to a narrator or to the audience of a theater.

The narrative self is incomprehensible without the helping hand of autobiographical memory. From the perspective of the broad category of personal identity there can be a connection between *psychological continuity* and the *narrative constitution* of the self. But there can also be significant differences. As Schechtman summarizes: "Whereas psychological continuity theories define continuity in terms of gradual change, the Narrative Self-Constitution View looks at it in terms of organized development. What is story-like about lives is that they unfold according to a certain kind of trajectory" (Schechtman, 2014, p. 107). In this respect, the narrative

[8] Goldie contends that literary works can be understood as "training grounds" for accepting and dealing with discrepancies between divergent perspectives in real life: "All kinds of narrative, when concerned with people, have this special explanatory, revelatory, and expressive power, which can remind us of, and throw light on, the subtle and complex ways in which perspectives can diverge, and it is for this reason that good literary works can be a kind of training ground for appreciating and dealing with dramatic irony and other kinds of diverging perspective in the so-called 'real world', and for evaluating and responding to those perspectives that diverge from our own: the therapist's perspective diverges from our own qua patient, the bank manager's from our own qua customer, the house-buyer's from our own qua house-seller, the child's from our own qua parent, the colleague's from our own qua colleague, the politician's from our own qua voter, and so on" (Goldie, 2012, p. 32).

[9] Goldie introduces free indirect style thorough James Wood's work (*How Fiction Works*). The gist of the concept is that we "see thorough" the character's eye, but at the same time we can gain more information than the character has. In Kafka's *Metamorphosis* we can immerse ourselves into the inner perspective of Samsa as a "monstrous verminous bug", but also can know him from an external, observer perspective. The role of free indirect style is to close the gap between the two perspective from the vantage point of omniscience (Goldie, 2012, pp. 34–36)

view serves as a critique of Lockean and neo-Lockean views on personal identity.[10] Schechtman (2014) argues that identity cannot be defined in terms of "relations between time-scales", or by reidentification of an object, rather, identity can be analyzed by *characterization*. Characterization means: "the question of which actions, experiences, and traits are rightly attributable to a person" (p. 100). Goldie tries to widen the notion of narration[11] by including *narrative thinking*; that is, he tries to "bring narrative back inside the mind". Goldie also takes into account the intricate nature of autobiographical narratives, but he envisions narrative thinking from an external perspective in which the narrator gives *coherence*, *meaningfulness*, and *emotional import* to a series of events (Goldie, 2003, p. 307). The significant point here is that the dilemmas of *gradual* and *radical change* are common phenomenological bedrocks for both the psychological continuity thesis and the narrative-constitution thesis. Moreover, emotion and affectivity, in general, also play an important role in these approaches. Goldie points to the role of traumatic experience in narrative understanding by referring to post-traumatic stress disorder. Schechtman, as we shall see, is primarily interested in the gradual change of personality and discusses thought experiences of alienation and dissociation. For start, let us look at Schechtman's illustrations.

The affective relation to the past and traumatic experiences may shed new light on the nature of the narrative self. It is important to bear in mind the distinction between "identity-threatening change" and gradual "personal development" (Schechtman, 2001, p. 108). In either case, the past can be seen as the constellation of memories which can determine the present. The notion of network or constellation of memories, however, is not sufficient to fully represent the intricacies of the distant past. The gradual and sudden (traumatic) changes in the personality structure are related to an ongoing process of *affective schematization*; that is, certain past events become affectively relevant, while others sink into the distant past. The problem of traumatic subjectivity will be analyzed in detail in Chap. 4, but here I focus on the issue of gradual change. In the following passages, the access to the distant past will be discussed in light of the question of personal survival.

---

[10] These views are trying to capture personal identity by observing or finding a "chain of sufficient numbers of psychological connections between person time-slices stretching from one to the other" (Schechtman, 2016, p. 18). Modern psychological continuity theories focus on psychological relations. The main inspiration for these theories stems from Locke, who believed that the "sameness of consciousness", i.e. the conscious recollection of autobiographical memories, is what makes a person one and the same in different timescales. However, Schechtman shows that the memory theory is vulnerable to criticism; for instance, it entails to circularity for there is no satisfactory criteria to clearly distinguish "genuine memory" from "delusional memory-like experience". As a result, personal identity hangs on the presupposed genuine memory and vice versa. Another obstacle attributed to Thomas Reid. It is plausible to suppose that a fifty-year-old man cannot remember the memories of his ten-year-old self, but it would be a counterintuitive inference to deny the continuity of the person on these grounds (cf. Schechtman, 2009, p. 637)

[11] For him "the notion of narrative is traditionally taken to be a notion of a kind of text or discourse – of something that involves written, or spoken, or signed, language" (Goldie, 2003, p. 301).

Schechtman (2001) defines *emphatic access* as a special kind of past-relation that often moves on implicitly. Emphatic access is a crucial ingredient that complements the Narrative Self-Constitution View (NSCW) of the self. She invites us to consider the lifestyle of a party girl from the perspective of her older self. The thought experiment branches into different scenarios that can be understood as illustrations of emphatic access. First, the so-called *Serious Matron* distances herself from her lecherous lifestyle, she literally alienated from her former passions and desires and almost unable to accept her former choices. The former choices and behavior of the party girl are comprehensible for her, but she is significantly alienated from the desires, choices, and passions of her former self. Second, for the *Less-Serious Matron* the distant past retained its "phenomenology", its affective significance. While she has reconsidered the lifestyle of the party girl from the perspective of new circumstances, she may still occasionally feel a wishful nostalgia towards her former habits, or on other occasions she may remember the parties as boring and empty. When her daughter makes similar choices, she can easily comprehend her motivations and drives, even if she disapproves them because of her mature way of thinking and changed perspective. The crux of the second scenario is not an affective alienation from the past, but rather the "recognition of changed circumstances" (Schechtman, 2001, pp. 101–102). The second scenario exemplifies a gradual change, a maturation that relies not on the replacement of beliefs, but rather on the expansion of them.[12] Another matron, the so-called *Mortified matron* demonstrates a radical change between the actual and the former self. In this latter case an affective revision takes place: the mortified matron vividly remembers the experiences of the young party girl, but at the same time, she feels shame and disgust.[13] The mortified matron literally suffers from the recollections, however, Schechtman argues she has lost the emphatic access to the past (Schechtman, 2001, p. 105). Schechtman concludes that only the Less-Serious-Matron retains the emphatic access to the past, and that kind of access allows the survival of her former

[12]The affective significance of the past may present itself in the somewhat-less-serious matron's behavior, actions, current decisions, or values: "The relation that the not-so-serious matron has to her past is more than just cognitive recollection; the passions that belonged to the party girl are still there. She experiences them and they are represented in the decisions she makes. It is for this reason that this woman's change seems like ordinary maturation and development rather than a loss of identity. The alterations in lifestyle and outlook may be just as pronounced as those in the case of the serious matron, but these alterations are the result of an expansion of beliefs, values, desires and goals rather than a replacement. New decisions are informed by a recognition of the nature and pull of past characteristics" (Schechtman, 2001, p. 102). However, Schechtman (2001) admits that the indicators of the living past in the present way of life are quite murky, we often turn to "tell-tale signs" (p. 109).

[13]The key feature of empathic access in not the authentic recollection or reconstruction of the past, but rather a kind of affective-emotional endorsement of the past: "The mortified matron has access to the feelings and thoughts of her past, what she lacks is the *empathy* – she is totally unsympathetic to the psychological life of the party girl. What is needed for empathic access is thus not an exact recreation of past emotions, thoughts and feelings, nor just some sort of ability to call them up from a first-person perspective. What is needed is this ability plus a fundamental sympathy for the states which are recalled in this way" (Schechtman, 2001, p. 106).

identity in the present (cf. Schechtman, 2016, p. 19). However, Schechtman admits that the mechanism of the affective appropriation of the past is shrouded in mystery: "What we do not yet know is how much change with respect to the perception of earlier thoughts and emotions (and in their behavioral implications) is permissible before empathic access is lost" (Schechtman, 2001, p. 106). In her recent paper, Schechtman (2016) argus that what is at stake here is not the loss of identity in the metaphysical but rather in the *figurative sense*. The loss of self implies the loss of contact with a former point of view (p. 18). The most radical change in a person's first-person perspective is based on the fact that he or she is unable to remember "what it was like to be" before the radical change. Schechtman argues that traumatized persons (e.g., an addict, an abused spouse, a depressed or chronically ill patient, victims of PTSD) are unable to inhabit their former life and first-person perspective.[14]

While empathic access in its original formulation was concerned with the endorsement and evaluation of the past, Schechtman in response to Goldie's criticism, reformulated empathic access as a kind of "phenomenological access" to the past. In cases of gradual changes in the matrons, personal growth was conceivable due to the preserved phenomenological access, which could involve the positive or negative evaluation of the past self. However, in the case of the traumatized subject, there is no way to recreate or restore the former self's first-person perspective on the world (Schechtman, 2016, pp. 20–22). But how do all of these considerations relate to the problem of narrative identity? Schechtman (2016) claims that there is a "largely implicit autobiographical narrative which serves as the lens through which we experience our lives" (p. 28). The ongoing narrative has a psychointegrative function; that is, the past and the anticipated future event coalesce into a coherent, meaningful whole in the present through an implicitly operating narrative context. The result is the *diachronic unity* of the subject. In this respect, empathic access constitutes a "phenomenological relation" to the past: the term covers the intuitively appealing observation that there are several (affectively) vivid and meaningful events in our past embedded in our actual narrative context (pp. 28–29). Schechtman (2007) illustrates the affective impact of the past on the present in the following way:

> Temporally remote actions and experiences that are appropriated into one's *self* narrative must impact the present in a more fundamental sense than just constraining options or having caused one's current situation and outlook. These events must condition the quality of

[14]The first-person perspective of an individual is not reducible to the pervasive presence of moods or to experiences of embodiment altered by illnesses. The crucial aspects for Schechtman are not bodily or psychological states but the way of being in and interacting with the world: "…the difference in perspective here is not best understood as a difference in values, beliefs, desires, commitments, or memories, but rather as a different way of being in, and interacting with, the world. The general point is that a first-person perspective is something more pervasive than individual states that colours the totality of experience and cannot be reduced to the details of what is experienced" (Schechtman, 2016, p. 17). What is most relevant and sympathetic in Schachman's (2016) approach is that she directly concedes that the nature of "first-person points of view" is almost unfathomable and it cannot be imagined as a pair of glasses or as a record and replication of a previous life-phase (p. 17).

> present experience in the strongest sense, unifying consciousness over time through affective connections and identification. [...] ... one must implicitly organize one's experience according to a narrative that recognizes past and future experiences as one's own in the sense that one sees the past as having implications for one's present situation and choices, and the present as having similar implications for the future (Schechtman, 2007, p. 171).

It is important to note, however, that Schechtman does not presuppose a concrete unconscious process behind narrative self-understanding. Instead, the narrative framework functions as an implicit organizing principle that appropriates and endorses past experiences. In this regard, the interrelated aspects of experiential time, autobiographical memory, and narrative came to the forefront. Schechtman bolsters her idea of empathic access with philosophical-psychological reflections on the nature of memory. She argues that the memories that give rise to emphatic access to the past are basically iconic states of event-memories, or more precisely, "centered event-memories"[15] which can be described from the first-person point of view as "a sort of theatrical presentations to ourselves" (Schechtman, 2001, p. 103). We can recapture the past event through centered event-memory and by the *sympathy for* the psychological makeup of our former self. Unlike centered event-memory, the "acentered event-memories" do not involve a particular point of view. Another crucial aspect of event-memories that they are imbued with *affect*. The third relevant feature of event-memory is its relative stability. Based on Wollheim's considerations, Schechtman (2001) argues that reconstructive processes occur in event-memories due to the changing psychological makeup of the person, but the cogency of the memory remains relatively stable (p. 104, cf. Wollheim, 1984, p. 103).

Goldie has risen objections to the role of Schechtman's emphatic access in the constitution of narrative identity. For him, the key to understanding the distant past is not the affective resurrection of a latent self, rather he highlights the role of the external perspective in narrative self-understanding. However, the free indirect style, as the main instrument of the narrative self, does not exclude the role of emotional responses in self-understanding. Goldie openly acknowledges the role backward-looking and forward-looking emotional appraisals with the caveat that there is no need for neither a stable self (composed of latent, dispositional traits), nor for empathic access. Experiences of fear or grief, he argues, are basically *processes* and not mental states; they unfold in time and manifest themselves in action, bodily responses, feelings, facial expressions, appraisals, and so on. Goldie claims that narrative accounts have "powerful explanatory, revelatory, and expressive powers". That is, in case of narrative understanding the expression of an emotion (e.g. the lengthy process of grief) is more significant than the *causal ordering* of events (Goldie, 2012, p. 69). Goldie offers a distinction between an "indifferent act of memory" and "the art of recollection". Meaningful recollection has an evaluative

[15] Schechtman relies heavily on Wollheim's (1984) distinction between centered and acentered event memories. In the instable acentered memories the subject is "edited out" of the scene, whereas in the case of centered event-memory the subject remembers an event from his own point of view as he lived it at that time. Ownig to this phenomenon, Wollheim regards every centered event-memory as *experiential memory* (Wollheim, 1984, p. 105, see also McCarroll, 2018, p. 9).

aspect that is missing in certain cases of trauma-processing where one feels locked in or trapped in the past. What is needed for the integration of the traumatic past is *narrative* and *emotional* closure. In real life, there is no way to a achieve narrative closure, to construct a fully developed, satisfactory story of our lives, but in trauma-processing the innate tendency toward wholeness, to tie together lose ends, is exaggerated.

The second possibility to deal with incomprehensible events is to find a way to emotional closure. For Goldie, this means the looking back to the past "in the right way" from the present, from an external perspective. Making sense of past events is not restricted to offering a causal explanation, but also requires finding the appropriate emotional responses for that event. The talking cure of psychoanalysis finds its place here: the right emotional closure and evaluation often brought about by the act of telling. The act of telling has its advantages and disadvantages. Goldie mentions harrowing autobiographical stories to demonstrate the limitations of public narration. A woman may refer to the ineffable qualities of the experience of finding her husband dead in bed. Goldie recounts the horrifying story of Leon Greenman, an Auschwitz survivor who saw his wife being taken away on a track, but back then he did not realize that she was being transported to the gas chamber. Goldie (2012) claims that ineffability is not the same as incommunicability (pp. 74–75). As we have already seen, Schechtman also considered radical – traumatic changes – in life; she stated that patients with severe depression or PTSD cannot feel or look at the world as they did before. Empathic access has been lost, they cannot inhabit their former perspective even if they feel the urge to identify with their former healthy and vigorous self (Schechtman, 2016, p. 20). For Goldie the framework of empathic access gives rise to the illusion of a distinct, stable selves. According to Goldie, the affective connection to the past can be established without empathic access. For example, in shame or any other kind of backward or forward-looking emotion we are "riveted" to the past. We do not need to imagine our former self as a condensation of latent dispositions, all we need for personal identity is the use of the first-person pronoun in autobiographical narratives (Goldie, 2012, p. 141).

The narrative understanding of life gives us the ability to consider different, often divergent, perspectives, and the emotional and evaluative appraisal emerges from the interaction of these perspectives. According to Goldie, this process is the crux of free indirect style. Let us consider the distance between the past self and the present self through time-consciousness. It seems that the linear stream of time-consciousness can be suspended by free indirect style. The multiplication of perspectives can be seen as a spatial network in which significant life events light up or fade out respectively before the eyes of the conscious narrator. From an external or distant perspective, we can even observe our multiple earlier perspectives about something simultaneously. I can observe a large number of previous perspectives on the same matter. The internal (earlier) and external (later) perspectives are constantly expanding over time, or, to put it more simply, our perspectives are nested within each other, forming a kind of hierarchy over time. In contrast to free indirect style, the arrow of time is far more significant for Schechtman's empathic access. When we speak of our lost selves in everyday life, we unwittingly place the pearls of quasi-stable selves on the string of time. Another contrast arises with regard to the role of affectivity in autobiographical memory.

In order to make clear the main difference in Goldie's "narrative sense of self" and Schechtman's "stable self" conception, we need to examine the role of memory in self-constitution. For Schechtman, as we saw previously, empathic access relies heavily on centered event-memories in which the point of view of the subject is the same as the point of view of the original experience. That is, centered event-memory preserves the original field perspective, and the recollected memory is filled with emotion and feeling (McCarroll, 2018, p. 261). In a recent study, Stocker (2020) characterized the field perspective as observing a future or past image "from within one's body", or "looking out from within one's own body". He also cites studies that demonstrate that PTSD patients regard their trauma memories as more emotional when they recall them through a field perspective. When patients were instructed to use an observer perspective, the result was a decreased physiological emotional reactivity (studies referred to decreased heart rate and skin conductance) (pp. 321–323). However, Goldie and McCarroll have raised objections against the view that one has to resurrect memory by means of field perspective in order to reexperience its affective impact. McCarroll complements Schechtman's model with the role of the observer perspective. The crux of the argument is that, not only memories with first-person (or field) perspective can preserve embodied (e.g. kinaesthetic states) and emotional contents, but that the observer perspective can also be infused with affection. McCarroll refers to empirical studies that demonstrate the oscillation between internal and external perspectives:

> The problem lies in the fact that Schechtman seems committed to the view that all the internal perspectival modalities are inevitably bound together. In order to remember one's past emotions, thoughts, or even kinaesthetic states one must adopt an internal visuospatial perspective – a field perspective. Yet there is evidence from film, sports psychology, and studies on autobiographical memory that internal and external perspectives can come apart or converge in interesting ways (McCarroll, 2018, p. 266)

The dynamics of the divergence and the convergence between the present and past perspectives alludes to a temporally extended self, or as McCarroll claims: "We are diachronically structured units, simultaneously inhabiting the past and the present…" (McCarroll, 2018, p. 275). McCarroll (2018) remarks that Schechtman in one of his former papers alluded to the dynamic nature of memory: memory is often condensed, generalized, or schematized. In her earlier work Schechtman admits that event-memory is only one type of autobiographical memory. In this respect, McCarrol argues, the strict adherence to field perspective memories in explaining personal identity is puzzling, and an unrealistic criterion for continuity (p. 265).

In a very similar vein, Goldie argues that Schechtman underestimates the role of observer memories in empathic access. According to Goldie (2003) narrative thinking does indeed make room for perception-like remembering. However, narrative thinking does not only revive our previous perspectives, but enables the multiplication and simultaneity of perspectives:

> One perceptually imagines oneself from an external perspective: that is, one looks back in perceptual imagination on what happened, from the perspective of no person within the imagined scene, so that oneself, as one then was, appears as part of the content of what one remembers. In remembering this way, from an external perspective, one's later self can evaluate and respond emotionally to what one then did; and one can do this in a way that one could not if the remembering were just of the stream of consciousness variety. One sees oneself as another, and in just this respect, on is at the same time both actor and spectator, both agent and judge (Goldie, 2003, pp. 311–312).

Thus, according to Goldie, emotional or other kinds of evaluation does not necessitate the occurrence of event-centered memories. Emotional and evaluative appraisal can also be facilitated by observer memories. Observer memories, neglected by Schechtman, does not deprive the remembering self of affectivity. Goldie (2012) argues against Schechtman that we can feel sympathy for events occurring in observer memories. The subject sees himself or herself from an external perspective in observer memories, and Goldie raises the possibility that the seemingly empty semantic memories are also capable of generating sentiments and emotional appraisals of the past. There is no reason to suppose that observer and "autobiographical semantic memories" are deprived of qualitative and affective content (pp. 138–39).

What is autobiographical semantic memory? Is it only the faint echo of erstwhile vivid event memory, or rather the condensation of repeated events into factual information? Goldie emphasizes the transformation and the synthesis of different types of memories. He draws attention to the complexity of autobiographical memory, which contains memories of *lifetime periods*, recurring *general events*, and *event-specific knowledge*.[16] Contemporary memory research has shown that experiential memory can transform into semantic memory. However, it is impossible to determine at which point an event-specific memory becomes semantic memory. Furthermore, lifetime periods and general events are also possess experiential character.[17] Not only particular episodes, but also repeated episodes can give rise to certain affective allure or feeling tone. Goldie shows that how the above-mentioned

[16] The typology is based on Conway's work on autobiographical memory (Conway, 1992; Conway & Rubin, 1993; Conway & Playdell-Pearce, 2000). Goldie uses a simplified but plausible model to illustrate the main components of autobiographical memory. Of course, contemporary memory research refined the above-mentioned categories. For example, event-specific memory or knowledge seems to be a too restrictive idea: "In the current model, the term 'episodic memory' has replaced the original term of 'event-specific knowledge'; this was considered misleading as it simply referred to the content of episodic memories and did not encompass the sensory-perceptual-conceptual-affective processing which is summarised in episodic memories" (Williams et al., 2008, p. 39).

[17] "First, as we have already seen, an experiential memory of a particular episode might fade into a semantic memory, as I might now merely remember *that* my father fell down drunk at my twenty-first birthday party, where the point at which it becomes merely semantic is indeterminate. Secondly, and more interestingly, memories of lifetime periods and of general events can be experiential (Thus, it would be more accurate to reserve the term 'episodic memory' for experiential memories of particular episodes, allowing the term 'experiential memory' to include more than just memories of particular episodes)" (Goldie, 2012, p. 45).

layers of memory can be integrated into a mini-narrative: "When I was in my early teens I would go to bed early. One night I was sitting on the edge of the bed with a knife in my hand and my father burst in." – in this scenario the levels of lifetime period (early teens), the general event (going to bed early), and an event-specific memory come together in a seamless mini-narrative (Goldie, 2012, p. 47).

The primary function of narratives is to connect the different levels of specificity. There are diverse ways of recollection, and our memories are the result of a certain synthetic activity that can lead to autobiographical semantic memories. Goldie denies the possibility of "unalloyed" episodic memory and adheres to the constructive memory concept. While Schechtman's empathic access presupposed the two-state model of remembering, Goldie is satisfied with the one-state model. There is no need to presuppose a quasi-stable target state (a lifetime period or a stable self), to which we gain (empathic) access from the present during remembering. The recollected memories are always infected or colored by the actual self's "reactive attitude toward the past" (Goldie, 2012, p. 134). From the narrative perspective, the past state without interpretation is counterintuitive. In a debate with Schechtman, Goldie (2012) argues that the mortified matron does not recollect a shameful past in itself, but rather she construes the shameful and disgusting past by the explicit narrative of "I shamefully remember the disgusting desires and passions of my youth" (p. 134). The theory of constructive memory retrieval leads to two intriguing consequences: First, the psychological correlation of the ironic gap can be found in autobiographical narratives. Second, if emotional and other kinds of evaluation are initiated in the present, then the distinctiveness of event-centered memory is highly questionable, again. Goldie highlights the role of *observer memory* that was neglected by Wollheim and Schechtman (p. 139).

According to Goldie remembering can be understood as an oscillation between field memories and observer memories. For Goldie the deployment of observer memories is based on two interrelated factors: the age of the memory and the ironic distance between the remembered event and the time when it was remembered. The dialectic between field and observer perspective boils down to a special affective relation: when one focuses on one's actual emotional situation, one is likely to adopt the observer perspective, on the other hand, when one tries recollect a distant memory along with its affective allure, one is very likely to adopt the field perspective. In Goldie's own word:

> My suggestion is that the contrast in emotion between field and observer memory should not be between more emotionality and less emotionality, but rather between recall of the emotions that were experienced at the time remembered, and the emotions that were experienced at the time of remembering. Thus if one were trying to 'orient' to one's feelings at the time one would naturally adopt the field perspective, whereas if one were trying to orient to one's feelings now, one would naturally adopt the observer perspective (Goldie, 2012, p. 51).

Goldie's observations are consistent with the findings of contemporary memory research. The constructive concept of memory emphasizes the role of the subject's actual knowledge and emotional state in the process of recollection. Actual beliefs, desires, feelings, and concerns contribute to the recollection of remembered episodes, regardless of the observer or field perspective of the recollected memories. In Goldie's description, the irony, that is, the divergence between the *now* and *then* perspectives, is infused into field and observer memories.

From a phenomenological perspective, we can reconsider Goldie's observations as the process of *affective schematization*: he explicitly argues that feeling can be invested into the *content* of memory and into the *way* of remembering that memory (Goldie, 2012, p. 53). Goldie (2012) also argues that the Freudian *Nachträglichkeit* (afterwardness) clearly demonstrates the mechanism of the ironic gap, where the past is reinterpreted in light of new, actual knowledge. For example, what was once only a frightening act, becomes a traumatic sexual abuse because of new understanding and the onset of puberty.[18] One can reach the status of "omniscience" by reinterpreting the past event in the different light of the present knowledge; the "triply ironic gap" manifests itself in the poles of the lived event, the present knowledge, and the observing I (cf. p. 53). Goldie suggests that remembering is not *excavation*, but rather a *construction* that inevitably leads to the *infusion* of current feeling and knowledge into the past. From the viewpoint of narrative understanding, the type of autobiographical memory (i.e., episodic, or semantic) becomes irrelevant. The essential feature of narrative self-understanding is that constructive processes are at work in different kinds of memory.

From the considerations above, it is clear that Schechtman highlights the role of event-centered field memories, while Goldie's approach focuses on the role of observer memories. Schechtman tries to reach back into the past to find a condensed, quasi-stabile self, in which the backward-looking empathic access can find its intentional object and evaluate it from the present perspective. The past can be excavated more or less effortlessly. For Goldie, the memories of the distant past and self that is supposedly situated in our episodic memories are reconstructed in the present.

According to the current proposal, the opposition between Schechtman and Goldie can be mitigated by the phenomenology of retroactivity and, more precisely, by retroactive self-understanding. In a curious way, Schechtman and Goldie have indirectly shown the two poles of *nachträglich* or *retroactive* self-understanding.[19] On the one hand, traumatic events may obtrude into consciousness in a decontextualized way, and, on the other hand, the cognitive and affective evaluative processes of narrative self-understanding colorize our important episodic memories.

Goldie takes seriously the constructive memory processes in autobiographical narratives, but dismisses the idea of empathic access. He argues that the cognitive

[18] The problem of afterwardsness is a crucial topic in this book, and it will be examined in greater detail in Chap. 4.

[19] The similarities and differences between the Freudian *Nachträglichkeit* and Husserlian *retroactivity* are central to Sect. 6.4. Here, I just wanted to emphasize the problem of reconstructing past experiences in light of the evaluative effect of the present knowledge. In the following chapters, I will attempt to demonstrate that retroactive self-understanding processes are not only restricted to autobiographical narratives, but they can also be traced back to implicit constructive mechanisms. See for example the notion of *body memory* in Chap. 4.

and affective evaluation of the past and semantic autobiographical memories pose difficulties for empathic access. According to Goldie, empathic access or the lack of empathic access to past episodes is not required to establish the narrative sense of self. Goldie invites us to imagine an aging rock star who has no explicit autobiographical knowledge of the past, yet his sense of being a continuous self (i.e., his diachronic continuity) is preserved owning to a stable defining trait.

> Consider the rock star of the 1960s for whom living a life of sex, drugs, and rock'n'roll was then, and is still, fifty years on, as much a part of his identity as it was when he was 20. He remembers that he lived the life of a rock star in those days in the 1960s, but, not surprisingly given how stoned he was, there are *no memories from the inside* of those times, and *all phenomenology is gone*. So the rock star has a stable defining trait and yet no memories from the inside (Goldie, 2012, p. 138, *emphasis added*).

According to Goldie, our actual emotional horizon is far more important in the act of recollecting than the type of the recollected memory. The actual feelings of shame or grief enable us to feel the direct contact with the past. Epistemic and even affective access to the past often takes place in terms of diffuse memory states, and, in addition, constructive processes always schematize the recollection of the past.

In conclusion, what is phenomenologically significant in the previous debate is the contention that "the whole gamut of backward- and forward-looking emotions of self-assessment goes with the basic sense of survival" (Goldie, 2012, p. 141). The narrating 'I' and the 'I' of the narrative refers to the same human being; therefore, alienation from past experiences and traits does not threaten personal identity. For Goldie (2012), the presupposition of a stable self and emphatic access are unnecessary for the sense of narrative self. The ideal of the stable self is erroneously hypostasized. Of course, Goldie accepts that there are subversive, traumatic experiences, but the sense of self can be preserved despite the changes of the person's character traits and dispositions (pp. 145–146). According to Schechtman, the narrative self has a kind of phenomenological thickness, but Goldie rejects this kind of stable narrative self. According to him, the self is a linguistic construction that holds together the multiple perspectives that were opened up by the free indirect style. The disagreement between Schechtman and Goldie can be mitigated by the layered or stratified conception of the self. The peculiar concept of empathic access is open to debate but, Schechtman's view of the narrative self gives room for the consideration of implicit processes.

2.3 Implicit Elements and the Narrative Self

It is an unexpected development in contemporary debates on the nature of the narrative self that unconscious, or more precisely, implicit processes have been highlighted by several thinkers. Of course, it is generally agreed today that the traditional notions of narrative identity are unable to tackle the problem of the unconscious. That is, narrative self-understanding, the elucidation of autobiographical narratives are basically conscious deliberations. By contrast, Zahavi came up with an intriguing question: Is it possible that we learn something about ourselves when we act "out of character"? There are certain cases in which we surpass our seemingly unified narratives (Zahavi, 2005, p. 112). In the same way, Ullmann argued that the concept of narrative identity has to bring into consideration the unreflective actions of the subject, the "blind spots" of consciousness which are the result of the implicit meaning-structure of an *affective unconscious* (Ullmann, 2017, pp. 158–159). Kozyreva (2017) introduced the notions of affective subjectivity and affective identity in order to demonstrate the "tacit influence" of the past in the constitution of subjective experiences (pp. 134–138).[20] Furthermore, in recent years developmental studies and psychopathological cases infiltrated into the discussions on the narrative self. Prior to these inquiries, the standard conceptions of narrative identity, elaborated by for example MacIntyre, Ricouer, and Taylor, were closely connected to issues of moral philosophy and personal identity. Schechtman (2011) recognizes two conjoined elements in the hermeneutical conceptions of the narrative self. The first is that the *sense* of self must be the result of a narrative, the more radical second claim is that the very *lives* of selves are also exhibiting a narrative structure. She goes on to add that the two elements are not distinct but rather two sides of the same coin (p. 395).

The main insight of the hermeneutically inspired views is that selves are fundamental agents and self-interpreting beings. The criterion to be an agent is the intelligibility of our actions to ourselves and others. The important role of language and social institutions is apparent when one tries to identify any seemingly simple human behavior: for example, gardening or writing a book. In MacIntyre's own words: "Intentions thus need to be ordered both causally and temporally and both orderings will make references to settings, references already made obliquely by such elementary terms as 'gardening', 'wife', 'book', and 'tenure'." (MacIntyre, 2007, p. 206). MacIntyre shows that the simple activity of digging is underdetermined by interpretation: one can speculate from the sight of a man who is digging that he is exercising, preparing for winter, or pleasing his wife, etc. The temporal and causal ordering of actions presuppose narrative frameworks over and above human intentions and actions. For MacIntyre, the unity of life is established by means of a narrative quest, and there is an inherent *telos* in the intelligible human actions. Someone who complains that his or her life is meaningless or is tempted to commit suicide has lost his or her intelligible narrative framework (MacIntyre, 2007, p. 2017). I do not wish to address the ethical issues of narrative identity; the main concern here is the contrast between *lived* and *constructed* narratives.

According to MacIntyre, every human transaction, including conversation, battle, chess, philosophy seminar, negotiation and so on, belongs to a genre just as plays and novels do. Furthermore, every human transaction has a beginning, a middle, and an end. Conversations in particular and human actions in general are "enacted narratives". In his own words: "Narrative is not the work of poets,

[20] From phenomenological point of view certain traumatic experiences may give to rise different ways of self-constitution. Part IV will elaborate further these considerations and considers the possibility of the reintegration processes of the traumatized minimal self.

dramatists and novelists reflecting upon events which had no narrative order before one was imposed by the singer or the writer; narrative form is neither disguise nor decoration" (MacIntyre, 2007, p. 211). To put is simply: we live out narratives and we understand our lives in terms of narratives. Meaning cannot be understood without appropriate context; that is, the human action is embedded in a broader narrative structure, or, conversely, a general narrative gives appropriate meaning to a particular action.

Of course, MacIntyre's thesis, according to which "we all live out narratives in our lives" is not without criticism (MacIntyre, 2007, p. 212). MacIntyre takes issue with the problem of *retrospective* narrative understanding that plays significant role in Louis O Mink's claim and Sartre's philosophy. MacIntyre mentions that Mink casts doubt on the inherent narrative structure of life and draws attention to the retrospective or retroactive understanding of human affairs:

> The comprehension at which narratives aim is a primary act of mind, although it is a capacity which can be indefinitely developed in range, clarity, and subtlety. But to say that the qualities of narrative are transferred to art from life seems a *hysteron proteron*. Stories are not lived but told. Life has no beginnings, middles, or ends; there are meetings, but the start of an affair belongs to the story we tell ourselves later, and there are partings, but final partings only in the story (Mink, 1970 p. 557)

Contrary to Mink, MacIntyre insists on the close connection between life and narrative and argues that the genre of life tragedy. Nevertheless, he acknowledges that there is an *uncertainty* in both history and life. The beginning and the end of the narrative cannot be fixed and there are many events in life which are both endings and beginnings. It is only in retrospect that we can speak of unfulfilled hopes and decisive battles. However, retrospective understanding is present in both life and art. Human life represents the tragedies of the Homeric characters (MacIntyre, 2007, p. 212).

MacIntyre shows that Sartre's protagonist, Antoine Roquentin, in *La Nausée* exemplifies a more radical qualm against the possibility of lived narratives. In this context human actions are "unintelligible occurrences" or "discrete actions which lead nowhere". While a storyteller imposes order on human actions, the absurdity of life shows even more powerfully the retrospective rearrangement of life and even demonstrates that there are *no true stories* at all. That is, the role of retrospective understanding is a central epistemological and hermeneutical question in the discourse of narrative self-understanding. To put it in another way, the retrospective ordering of past events and looking forward to the future are the structural components of narrative self-understanding. The narrative structure of life is threatened by fictionality or absurdity.

One possible way out of this epistemological impasse is to consider the pivotal role of lived time. MacIntyre argues that there is no self-contradiction in conceiving of human life as the coexistence of unpredictability and teleology:

> Unpredictability and teleology therefore coexist as part of our lives; like characters in a fictional narrative we do not know what will happen next, but nonetheless our lives have a certain form which projects itself towards our future. Thus the narratives which we live out have both an unpredictable and a partially teleological character. If the narrative of our

> individual and social lives is to continue intelligibly-and either type of narrative may lapse into unintelligibility – it is always both the case that there are constraints on how the story can continue and *that* within those constraints there are indefinitely many ways that it can continue (MacIntyre, 2007, p. 216).

Ricoeur, in contrast to MacIntyre, emphasizes the discontinuities between life and literature (Schechtman, 2011, pp. 396–97). According to Ricoeur, McIntyre avoids the "thorny problem" that life is reconfigured through fiction. Ricoeur contends that literature is "an immense laboratory for thought experiments" where the connection between the agent's action and the literary fiction can be examined by endless imaginative variations. According to Ricoeur, we can imagine ourselves as author, narrator, and character at the same time in autobiographical narratives. However, in contrast of MacIntyre, he places more emphasis on the role of co-authorship. We are not the authors of our life in a strict sense. Ricoeur's other important critical remark concerns the contrast between literary closure and the unpredictability of life. One can find a narrative beginning and end even in Proust's *Remembering of Things Past*, on the other hand, there is no narrative unity (i.e., a fixed beginning and end) in life. (Ricoeur, 1992, pp. 159–160).[21]

According to Ullmann (2015), the tension between stories and real life is always the inevitable consequence of narrative self-understanding. Another tension can be seen between the concrete experiences of the living person and the protagonist of the life history. The lived through experiences become objectified due to the application of collective social codes (p. 29). As MacIntyre and contemporary memory research has demonstrated: we are not the sole author but rather the *co-author* of our life-history. Ullmann identifies three main challenges of the narrative self: (1) the inner coherence of a story does not guarantee its truthfulness, it has to be harmonized with the phenomena of the "outer world"; (2) our story can transform into a rigid and unchangeable fixation, the seemingly correct narrative could turn into a self-depiction of a false-self; (3) the most important question is the status of the narrative self: is it possible to presuppose a self under the narrative structure, or, *vice versa*, the self is the result or the intersection of narrative threads (as Dennett has similarly suggested above) (p. 29).

Narrative self-understanding is inextricably linked to question of gaining self-knowledge through introspection or intersubjective feedback. Jopling (2000) came up with four criteria which are "singly necessary and jointly sufficient" for the narrativist theory of self-knowledge. The first is *internal coherence*, the second is *external coherence* that relies on others' narrative, but of course, coherence is not the same as truthfulness, both first- and third-person descriptions are fallible.[22]

---

[21] Tengelyi highlighted that Ricoeur's thought experiments with time complement the phenomenology of time. Ricoeur's analysis of personal and intersubjective time bridges the gap between the phenomenological micro-analysis of time-consciousness and life history (including the problem of Dilthey's *Lebensgeschichte* and narrative identity) (Cf. Tengelyi, 2007b, p. 310).

[22] The criterion of coherence is one of the major features of narratives. Goldie argues that causal explanations are significant in constructing a narrative through the "process of emplotment" of related events. However, narrative explanations are based on our *interests*, that is, we often high-

Furthermore, he does not rule out the possibility that hermits or non-conformists may be able to achieve certain self-knowledge by keeping distance from social norms. Third, *applicability* means that self-narrative is not only the interpretation of past events, but also has a forward-looking aspect. And the fourth substantive criterion is *empirical adequacy*: the self-narrative must incorporate empirical content. Empirical building blocks are the roots of the "possible permutations of the narratives" (Jopling, 2000, p. 50; see also Zahavi, 2005, p. 110). Despite the above-mentioned set of criteria for self-knowledge, Jopling unsurprisingly claims that the narrative is not the one-to-one representation of the self. The narratively constructed self is likened to an artistic expression (e.g. a painting), but, evidently, the selective and constructive elements in narration do not transform the self into a fictional entity. Instead of fictionalism, Jopling focuses on the creative and coherence-preserving processes of narration:

> …the creation of a self-narrative involves the selection, simplification, and abstraction of narratively relevant materials, directed to the goal of creating a unique synthetic whole. An indefinitely large amount of information about the self may be left out of the narrative in order to bring to the fore certain similarities and differences in the patterns of phenomena that capture the essential truth of the whole self. As such, self-narratives enjoy a degree of interpretive flexibility. If enough coherence-preserving revisions are made, then certain factual errors, false memories, and psychologically incorrect observations about the self can be accommodated into the ongoing fabric of the narrative without its being rendered false (Jopling, 2000, p. 51).

According to Jopling (2000), narrative self-understanding mainly focuses on the *meaning* of historical and psychological facts in virtue of "interpretative statements" (pp. 46–47). Nevertheless, "interpretative flexibility" rises epistemological concerns. First, it is difficult to distinguish coherent and accurate self-narratives from coherent but inaccurate ones. Second, it is not an easy task to distinguish "the pre-narrative self from the self that is overlaid with adventitious narrative-generated artifacts" (p. 51). A narrative self-understanding might be historically and psychologically false and yet adaptive. The upshot is that, while narrative emplotment is an essential form of self-knowledge, there is a tension between veridical self-knowledge and self-deception.

Jopling's (2000) main concern is the status of self-knowledge in the process of narrative self-understanding. In this respect, he highlights two processes that "make narrative self-understanding unsuitable as a model for self-knowledge". The first are the *artefacts of crystallization* and the second are *temporal artefacts* (p. 54). In short, narrative accounts interpret and distort the incipient desires and the memories

---

light the most significant events in the nexus of causally related events. When we narratively explain why we missed a train or describe a car accident, we are looking for a salient, determinant event. Another significant feature of narrative emplotment is *coincidence*. Unexpected events can be seamlessly integrated into narrative descriptions (cf. Goldie, 2012, pp. 14–15). The problem of traumatic subjectivity is concerned with cases where this kind of narrative integration is thwarted by the affective impact of the event and the responsiveness of the subject. See Chap. 6 for more detailes.

of the distant past. The temporal artefacts might be the result of the retrospective fabrication that lies at the heart of narrative self-constitution. Jopling observes that:

> Narrative streamlining and filling in does not adequately capture the phenomenology of the temporal grain and the temporal ambiguity of experience, because it confuses the prospective perspective with the retrospective perspective. Events in the present moment are not normally experienced narratively in terms of plot developments linking together beginnings, middles, and ends. These are determinations that are read into events after the fact… […] Time's arrow is crossed backward and forward at will, allowing events to be interpreted in light of results that were not knowable at the time of their occurrence. When captured in a narrative, for example, small details that were considered insignificant during the time of the event take on new meaning, as signs portending a determinate development. The narration of events distorts the phenomenology of the experience of temporal anisotropy with a layer of temporal artifacts, because the realized future is read back illicitly into the description of the past event, with the achieved outcome of a series of events being used as the key to their meaning (Jopling, 2000, pp. 53–54).

The contradiction between the fictional and the real narrative self is the result of the reconstruction of the past by means of the temporal structure of a plot. In the quote above, Jopling (2000) claims that "the realized future is read back illicitly into the description of the past event" (p. 54). It is not clear that the present, as realized future, gives rise to "temporal artefacts" without any fragments of appropriate self-knowledge. The re-interpretation of past events might generate false memories due to source amnesia, but the reverse is also possible: in certain cases of reconstructions, the past not only appears in a new light, but also gives new insights and knowledge to the ruminating subject.

Narration may indeed distort lived experiences through of emplotment, but "narrative streamlining and filling in" does not lead to a total alienation from the lived experience. On the contrary, the integration of the previously insignificant, unnoticed details can also be regarded as an enrichment or improvement of narrative self-understanding. In the process of narrative self-understanding we do not necessarily distort or fictionalize the past, but rather "after the fact determinants" – to use Joplings wordings – might be the testimonies of meaning-bestowing processes facilitated by the primary experience of change. Ullmann (2010) argues that we are under the mercy of constant change, which is an elusive, almost unconscious phenomenon in lived time. The process of change and the genesis of new meaning is often unnoticeable in lived time, but the retroactive understanding of the past might shed light on the nature of the experience of change (pp. 284–285). These considerations will be explored in more detail in Part IV.

As was mentioned previously, it is not only the lived time that is the subjected to crystallization processes, but also desires. One of many functions of narratives is to crystallize diffuse, inchoate states into a "unified morphology"; in this respect narration has a crucial role in the economy of desire as well. In Jopling's own words:

> Prior to exposure to narration, some desires (for example) have neither narrative structure nor determinate unity; they are plastic, and subject to alternative, even incompatible, crystallizations. The narrative confers upon them a unified morphology, in such a manner that what appears to be discovery is in fact an artifact of the narration that would not have been encountered independently (Jopling, 2000. p. 52).

This does not mean that the expression of a desire in a narrative form counts as the representation of the desire in minute detail. Jopling (2000) introduces the term of "artifacts of desire" that mirrors the possible divergencies in the process of interpreting and organizing incipient desires. The result of the crystallization process might indirectly show the indeterminate nature of the initial desires and drives. In the end, it is difficult to realize what was the original grain of the desire and what is "artifactually desired" (p. 53). Jopling came to the conclusion that different self-narratives will schematize the incipient desires in different ways, or, in a broader sense, there is no universally stable self under the overlay of narrative constructions. As he puts it: "…self-narratives transform the self so that the self comes to fit them, thereby manufacturing some of the very facts about the self that they appear to uncover" (Jopling, 2000, p. 54). Jopling also argues that the somatic sense of self and the narrative self cannot be completely separated. The somatic sense of self (i.e., the proprioceptive, vestibular, or kinaesthetic systems) is related to self-knowledge. In early development periods, the child acquires certain somatic determinations that can later be associated with core values and expressed in bodily states and behavior.[23] According to Jopling (2000), core beliefs and values can be expressed psychosomatically and, in turn, the psychosomatic sense of self (for example a negative body image) can interfere with self-understanding (p. 56). Phenomenologically speaking, an underlying affective schematism is closely connected to narrative self-understanding and it might have a constitutive role that often works hidden from explicit awareness. To conclude, despite the epistemological concerns about self-knowledge and the denial of a hidden self, Jopling does not rule out the "constitutional passivity" of bodily dispositions (i.e., somatic determinations). Jopling's considerations narrow the gap between the advocates of the narrative self and the embodied mind.

Ullmann (2015) argues that the self of the narration is basically a *rational self* that tries to analyze its own life and strives for transparency; in Freudian terms, the Ego is the master in its own house. However, the problem of *traumatic subjectivity* may shed some light on the limits of the narrative self. Ullmann's radical claim is that the self could be better represented by the untold, unarticulated stories in contrast to explicit storytelling (Ullmann, 2017, p. 158). Ullmann distinguishes explicit self-understanding from implicit-affective schematism. Narrative understanding is always accompanied by the gaps or blind spots of self-knowledge, which can be

[23] The prehistory of dispositions and attitudes can be traced back to early childhood: "For example, the sense of oneself as calm or nervous, active or passive, or guarded or open begins in infancy at the hands of one's caregivers, and develops across childhood as basic somatic determinations that come to be expressed in terms of posture, vocalization, and movement. At later stages of development they may come to be associated with core values and ideals, exerting pressure on the subject's self-understanding in ways that go unrecognized. An infant who experiences himself as passive, for example, and who later in infancy and early childhood fails to develop a strong subjective sense of agency, may at later stages develop attitudes and beliefs that express this constitutional passivity: for instance, self-serving beliefs about the value of the idle life, attitudes of resentment or indifference, or beliefs about the relative unimportance of conforming to conventional social norms of physical beauty" (Jopling, 2000, p. 56).

understood as the indirect signs of an underlying affective schematism. This general consideration is one of the main concerns of the current investigation and in Chap. 2 and especially in Part IV I will widen the scope of the analysis of affective schematism.

Zahavi has reservations not only about the anti-realist accounts of the self but also about the narrative self. He claims that some narrative approaches, for instance Dennett's NGC, explicitly state that the self is only a linguistic or social invention (Zahavi, 2005, p. 110). According to Zahavi, the claim that the self is exclusively a socio-cultural construct is "social reductionism". Zahavi cannot accept the "exclusivity claim" that regards the self solely as a "narratively structured entity" and denies that we could only gain knowledge about the self through narratives (Zahavi, 2007, p. 184).[24] In my view, rather than social reductionism, the term *social expansionism* would be more appropriate to capture the essential features of the narrative approaches. Not only the self is subjected to multifactorial analyses, but also the narrative accounts involve different frameworks, such as developmental, hermeneutical, phenomenological, ethical, or judicial contexts. In order to articulate his main ideas, Zahavi sharpens up the distinction between the minimalist and narrativist notions of the self. But, as we shall see, Zahavi often alludes to Sartre's analysis of shame or Husserl's "original reciprocal coexistence", which also underscores the social embeddedness of the minimal self (Zahavi, 2005, pp. 171–173).

It is important to note that the minimal and narrative notions of the self are theoretical abstractions and there is no reason to deny their intermingled, interrelated aspects in defining personal identity. For Zahavi, the minimal experiential self can be seen as a necessary precondition of the narrative self. The idea is illustrated by the distinction between the a formal and a "more tangible" kind of individuation that lies behind the demarcation of the person and the self:

> The fact that the person (i.e., the narrative self) presupposes the experiential self (but not vice versa) does not diminish the significance of the former. Due to the first-personal givenness of experience, our experiential life might be inherently individuated. It remains, however, a purely formal kind of individuation. A description of my experiential self will not differ in any significant way from a description of your experiential self, except, of course, in so far as the first is a description of me, the second a description of you. By contrast, a more tangible kind of individuality manifests itself in my personal history, in my convictions and decisions. It is through such acts that I define who I am, thereby distinguishing myself from others; they have a character-shaping effect (Zahavi, 2007, 193).

Zahavi adds that the minimal self is not a mere abstraction or a reified self, but rather an "experiential core self" that plays a fundamental role in self-constitution and can be harmonized with empirical sciences. More precisely, Zahavi states that "pathological limit situations", especially schizophrenia, reveal the fundamental nature of the minimal self. (Part III will analyze in detail this kind of attempt to harmonize philosophy and psychopathology.) Regarding the minimal self, there is a

---

[24] Of course, Zahavi admits the diversity of the narrative approaches. The opposition between Schechtman (2001) and Goldie (2012) was discussed in Sect. 2.2. Schechtman argued for an enduring narrative self and Goldie argued against the "ideal of a stable self".

dichotomy in the paper cited above. On the one hand, the minimal self it is defined as a *formal* kind of individuation based on the primordial difference between *my* experiential reality and that of *others*; on the other hand, it refers to an "experiential core self" that is closely related to Damasio's core self. In fact, Zahavi tries to bolster the idea of the minimal self, against the backdrop of the narrative approaches, with Damasio's neuroscientific findings:

> We would be forced to concede that there must be more to being a self than what is addressed by the narrative account. This is in fact the conclusion drawn by Damasio, who explicitly argues that neuropathology provides empirical evidence in support of the distinction between core self and autobiographical self. Neuropathology reveals that core consciousness can remain intact even when extended consciousness is severely impaired or completely absent, whereas a loss of core consciousness will cause extended consciousness to collapse as well (Zahavi, 2005, p. 139).

Zahavi also refers to Damasio's patient, David, whose memory is limited due to a severe brain injury. David is unable to learn new facts or create a coherent narrative about himself; his experiential reality is limited to a very short window of time (less than a minute). His short-term memory remained intact, but he was unable to recognize Damasio after three minutes of their meeting (Damasio, 1999, p. 115). The implication of the case study is that a core self[25] remains intact in cases of severe amnesia in which autobiographical memory is absent. Zahavi emphasizes the stability and the continuity of the core self during our lifetime. However, the core self cannot be completely identical with the minimal self; in fact, Damasio suggests that implicit narrative mechanisms are embedded in the representational structure of the brain. Damasio, claims that verbal and wordless narrative is a natural representational mechanism of the brain. That is, the nonverbal core consciousness (which constitutes the core self) is not without narrative structure. When we encounter objects in our environment or recall memories, these experiences immediately enter core consciousness; that is, there is the level of a fundamental, self-referential loop in embodied perception and cognition. According to Damasio, narrative storytelling is the process of "creating a nonlanguaged map of logically related events", which can be compared to a movie; however, Damasio admits that the movie is not the most appropriate analogy, but it can also be called as a "second-order nonverbal narrative" that can be converted into language (Damasio, 1999, p. 189). In case of humans, he further argues, there is an "uninhibitable verbal translation"; that is, the nonverbal narrative of interacting with the objects of mental states is automatically

[25] Damasio argues that the core self undergoes minimal changes during a lifetime and its main function is to register the happenings within and outside the brain-body complex. The core self is also the locus of consciousness, the result of the self-referential loop of perception. In Damasio's words: "The transient protagonist of consciousness, generated for any object that provokes the core-consciousness mechanism. Because of the permanent availability of provoking objects, it is continuously generated and thus appears continuous in time" (Damasio, 1999, p. 175). Consciousness arises via a feeling component of the representation of a more basic, nonconscious proto-self (which is a structure that regulates and represents the internal bodily states). The core self can be placed in the hierarchy of selves: it lies between the autobiographical self and the nonconscious proto-self.

translated to a third-order verbal narrative. As a result, the core self, which is pulsatively generated by intentional directedness, can also be verbally presented in our mind (Damasio, 1999, pp. 188–189). Damasio's notion of core consciousness stands in stark contrast to the language-dependency hypothesis that underlies the ideas of Julian Jaynes, Dennett, and Varela. It seems reasonable to conclude that Damasio's core self is only partially analogous to Zahavi's experiential core self, since Damasio assumes preverbal narrative sequences in the brain/mind.

Zahavi does not endorse ideas that accentuate the role of narratives beyond the boundaries of language. Referring to the notion of "implicit lived narratives", which implies the elemental narrative configuration of lived experiences, he says: "... by severing the link between language and narrative, it threatens to make the latter notion too all inclusive and consequently too diluted" (Zahavi, 2014, p. 58). Despite Zahavi's concerns, the notions of micro and implicit narratives (which mainly imply segments of experience) are prevalent in developmental and in narrative studies. It seems undeniable that film-like representations of action-sequences or behavioral dispositions are far from the usual ideas of narrative identity. The implicit elements of pre-narrative understanding may or may not play an identity-constituting role. However, as the following sections will show, discussions of body memory and affectivity may provide further support for this kind of "pre-reflective narrative understanding". For preliminary purposes, let us conclude that an automatic meaning-bestowing process may take place in certain circumstances.[26]

As previously discussed, both Zahavi and Ullmann emphasize the importance of those behavioral and affective responses that are alien to the ego. It is not clear whether narrative self-understanding could integrate the problem of the unconscious. Menary (2008) criticizes Schechtman for locating narratives in the obscure realm of the unconscious (p. 71). Schechtman's "implicit narratives" are motivating factors in action, but they are not available to conscious awareness. The notion of implicit narrative can be approached through cases of "self blindness", in which the explicit and implicit self-narrative diverge.[27] Self-deception, self-destruction, surprising emotional outburst, alien motivations, the leaving of a partner and other disappointments tend to endanger the intelligibility of our lives and allude to unspoken and unintegrated elements of the psyche (Schechtman, 1996, p. 115). Nonetheless, the articulation constraint of narratives "does not allow a person's self-narrative to remain entirely subterranean" and the narrator should be able to articulate the narrative (Schechtman, 1996, p. 114). Schechtman (1996) provocatively

[26] Take, for example, the cases of psychoneurosis. Freud attributed hidden symbolic meaning to neurotic symptoms (i.e., repetition, obsessions, psychogenic paralysis etc.). It is important to highlight the difficulties of determining the layers or levels of identity constitution. It is also necessary to reflect on the presuppositions of certain archaeological studies (e.g., psychoanalysis) that explore the implicit meaning-bestowing and identity-constituting processes.

[27] "The implicit narrative is understood as the psychological organization from which his experience and actions are actually flowing. It may not be obvious why I want to call this unarticulated (and sometimes inarticulable) psychological organization a self-narrative at all. The person explicitly denies the features we are attributing to her, and so it may seem perverse to say that they are part of her self-conception-even her implicit self-conception" (Schechtman, 1996, p. 115).

states that "people actually *narrate* very little their lives in any self-conscious way. Instead, they allow a general set of background assumptions about themselves and their lives to guide the unfolding of experience" (p. 116). There are unstated elements that are readily available and not readily available for narrative integration. Schechtman (1996) considers a whole spectrum of denied, partially acknowledged, and totally appropriated elements in one's personality. Often the repressed and unconscious elements of an implicit narrative play different roles before and after articulation (p. 119). She also makes explicit reference to Freud and cognitive psychology to point out the limits of articulation and intelligibility of actions:

> It is not uncommon for us to be at a loss to explain some particular action or affective response. Sometimes emotions or impulses seem to come out of the blue, subverting the order and intelligibility of our lives – we often cannot explain our self-destructive behavior, or why we are leaving our partners, or where we find the strength to stand up for our rights, or why we are so undone by trivial disappointments. There are cases where we simply cannot tell some parts of our life stories in a way that makes them intelligible. Moreover, even when we do have ready accounts of our actions, it is sometimes all too obvious that our stories are inaccurate – a commonsense observation given formal expression, in different ways, by both Freudian and Cognitive psychology (Schechtman, 1996, p. 115)

Schechtman pays particular attention to the case of self-deception in which a man fails to recognize his own hostility toward his brother; he unwittingly serves him his least favorite food or inadvertently humiliates him. Schechtman contends that not only unarticulated or unstated elements but also unrecognized, *repressed* elements are part and parcel of a person's self-conception. The repudiated hostility of the brother is not part of his explicit self-conception, but it plays a significant role "in shaping his experience, actions, and emotions". Schechtman (1996) incorporates the basic psychoanalytic tenet according to which we attribute unconscious affects and motivations to ambiguous emotions and actions in order to make them intelligible (p. 116).

Schechtman's "self blindness" and Ullmann's "blind spots" might be based on similar mechanism, although Schechtman (1996) does not fully endorses the intuition that unconscious desires and emotions are "*more* definitive" than conscious reasoning (p. 117). However, the emphasis on the constitutive role of alien, blind elements in self-narratives provides a common ground. Schechtman states that "the repressed or unconscious elements of a person's implicit narrative are less fully her own than those of which she is aware" (p. 118). Schechtman alludes to Freud's observations of the rituals and automatisms of obsessive and hypnotized patients. The subjects are alienated from their unconscious impulses. The greater the realm of incomprehensible actions and experiences the bigger the chance for fragmentation of personality. Schechtman draws attention to the range of intelligibility: on the one hand, a restricted, "local intelligibility" is an integral part of personal identity – there might be unassimilated elements in our general narrative understanding, but, on the other hand, if we are unable to comprehend a large proportion of our actions and emotions, then the danger of fragmentation and incoherence is greater. According to this scenario, one might be under the influence of unknown (alien) forces, and, as a result, cannot function as an *autonomous subject* (Schechtman,

1996, p. 119). However, despite the influence of the implicit elements, we are not fully alienated subjects. Autonomy can be reclaimed by acknowledging the unreflected traits and habits. That is, when the hateful subject in the above example admits his own self-deceptive behavior and liberated from the restraints of automatisms, he is able to *fill in* the missing pieces of his explicit self-narrative. Therefore, Schechtman does not deny the defining character of the unconscious "blind spots", she simply places greater emphasis on the process of acknowledgement that is the prerequisite for conscious integration:

> The features of our narratives that are below the surface *are* revealing of who we are, because they represent the missing elements of our explicit life stories – they fill in the pieces that make the incomprehensible elements of our explicit stories intelligible. In essence, then, they tell us what aspects must be incorporated into an explicit narrative for a given person to develop fully as a person. Nonetheless, until and unless they are seen and acknowledged by the subject herself, they play a different sort of role in her life than articulated aspects of her narrative, and so are less fully hers (Schechtman, 1996, p. 119).

That is, integrated unconscious elements can change their characteristics and be incorporated into an explicit narration. However, a curious contradiction comes to surface. Namely, the unconscious is filled with rigid, inflexible elements which are insensitive to alternating contexts.[28] Moreover, the reciprocity between consciousness and the unconscious is absent: the nonconscious psychical elements influence consciousness, but not vice versa. In short, their effect is incomprehensible, and they cannot be integrated into explicit narrative of the subject. As we can see, the unconscious elements are contradictory in their nature. Certain implicit element can be revelatory for the subject, as the quotation above suggests, but other elements might resist to conscious appropriation and acknowledgement. Schechtman accepts the dynamic nature of the unconscious that means the unconscious material finds different forms of expression. Schechtman also endorses the view that the flexibility and responsiveness of conscious states are not all or nothing phenomena; that is, the unconscious elements could also exhibit these characteristics to a greater or lesser extent (Schechtman, 1996, p. 141). Consequently, there is no need to assume an unconscious reservoir in which the most rigid implicit elements reside. On the contrary, Schechtman divides consciousness into to the two sections of center and periphery:

> At any given time some elements are the forefront of consciousness – the center of attention – and are having the most powerful influence on the overall character of experience. Other aspects are at the periphery – there and accessible, but pushed to the corner; others are not conscious but can easily be brought to consciousness; other cannot be made conscious without immense effort; and some may never come to consciousness at all. The view that consciousness is not univocal state is increasingly less controversial and has the support of psychologists and philosophers of mind alike (Schechtman, 1996, p. 142).

Schechtman is not alone in accentuating the role of the unconscious in narrative self-constitution. Flanagan also laid heavy weight not only on the role of *others* in

[28] The Freudian inspirations of these considerations are explicit in Schechtman's book (cf. Schechtman, 1996, p. 118).

the constitution of the narrative self, but also on the role of the *unconscious*: "I produce in thought and words a narrative self. I do this largely, but not completely, unconsciously. My narrative self is not its own source, although, of course, past self-representing has a causal influence on subsequent self-representing. My narrative self is a product of complex social interaction and much mental processing, both conscious and unconscious on my part" (Flanagan, 1992, p. 204). Flanagan (1992) also echoes the implicit aspects of identity when he speaks of productive, generative forces of identity formation that are causally effective but concealed from first-person access (p. 202).

All in all, Zahavi's and Menary's critique of implicit narratives seems to be misguided: Schechtman does not speak of unconscious storytelling; she has merely widened the scope of narrative self-constitution to address the issue of unconscious influences in explicit self-narratives. Implicit narrative is not the other side of explicit narrative, but rather a hidden constitutive process in the constitution of the person. Surprisingly, there are common tendencies in Zahavi's and Schechtman's approaches. Both seek to expand the scope of consciousness to include the implicit elements that constitute the synchronic and diachronic unity of the self.[29] As we shall see, for Zahavi the implicit layer of experience is pre-reflective self-awareness, while for Schechtman the unity of the self is the result of conscious self-understanding that continuously affected by a dynamic organizing principle (i.e., implicit narrative). Zahavi argues that the minimal self could serve as a foundation for the higher levels of self-constitution, since the alternatives of the narrative self are unable to specify the solid ground of the narrative self; it is characterized as abstract, fictional, transitional, or remains undetermined. Schechtman is concerned with the impact of the past on the present and possible future; the Narrative Self-Constitution View presupposes a temporally extended subjectivity. The main characteristic of the self, besides the blind influence of the unconscious, is the ability to project itself backwards and forwards in time. The normal human adult's self is differentiated from the child or sufferers of dementia who are unable to think retrospectively or imagine a possible future. Therefore, Schechtman concludes, narrative self-understanding gives rise to a *different phenomenological life* (Schechtman, 1996, p. 148). In contrast, Zahavi painstakingly emphasizes the all-encompassing presence of the minimal self under the level of narrative self-constitution. In my view, there is no irreconcilable contradiction between the two ideas. It is equally reasonable to consider what-is-it-like to have a narrative and a minimal self as well.[30]

[29] As Schechtman argues: "The moments of conscious awareness in a single person's life are not distinct entities that are somehow strung together, but rather a dynamic interactive system that integrates to produce a subjectivity that extends over time" (Schechtman, 1996, p. 143).

[30] In Part IV, I will try to substantiate the claim that the experiential (i.e., the *affective*) core self consists of an invisible network of implicit habits, dispositions, and beliefs that is prone to disintegration and reintegration under traumatic circumstances. Within this constantly evolving affective core self, there is a passive, bodily meaning-bestowing process that often runs in parallel with narrative self-understanding and is embedded in time-consciousness.

Ullmann (2015) argued that the theory of narrative identity, in general, suggest that the "Ego is master in its own house"; that is the ego is able to overview its own life, governing rules, and intersubjective relations. Furthermore, the subjectivity that was suggested by narrative identity is rational, conscious, and enlightened, and relates to itself in a harmonious way (p. 31). As we have previously seen, Schechtman's considerations on implicit narratives call into question the image of rational subjectivity at the core of narrative identity. Schechtman convincingly demonstrated that self-deception and implicit influences of the periphery of consciousness play significant role in self-constitution. However, the influence does not mean that the unconscious can be embedded seamlessly into consciousness; let us remind that Schechtman highlighted the incomprehensible aspect of the unconscious and the obscure process of articulation. In short, articulation transforms the unconscious affects and motivations. Ullmann also underscores the gap between unconscious blind spots and conscious self-understanding. That is, there is *alterity* in our narrative understanding and this kind of internal otherness-experience obtrudes into awareness in traumatic experiences (Ullmann, 2017, p. 160).

I follow Ullmann's argument according to which the experiences of *trauma*, and *change* per se, can shed light on the peripheral processes of consciousness; that is, on the schematization process of the affective unconscious. This does not mean that the underlying "affective horizon" of the self may consist of repressed or dissociated unconscious memories. This affective horizon is broader than the Freudian repressed unconscious. I propose that this kind of implicit-affective horizon can be designated as an *affective core self*.[31] Metaphorically speaking, the affective core self – as a special form of experiential core self – is the condensation of the affective and behavioral infrastructure of the self. It can be described as a version of Zahavi's "(less) minimal self". A more detailed discussion of the topic ensues in the following chapters.

2.4 Pre-reflective Self-Awareness and the Minimal Self

The dual aspect of the minimal self, i.e., the *formal* and the *experiential*, is of pivotal importance for the problem of psychointegration that lies at the heart of the experience of personal survival and in the elusive experience of change. On the one hand, the notion of the minimal self refers to a formal pole of identity, to the brute fact that the occurrent experiences are *mine* and do not belong to others. On the other hand, the idea of the *core self* illustrates the more dynamic and layered notion of the self, including the passive-unconscious processes and the problems of radical and

[31] The idea of the affective core self relies heavily on Ullmann's notion of affective unconscious: "In phenomenological terms: the affective unconscious is an emotional or affective horizon (*Spielraum der Möglichkeiten*) or a prefigured implicit sense-structure. [...] Unconscious affectivity is not the counterpart of our rational subject but its hidden structure, its invisible foundation" (Ullmann, 2017, p. 160).

gradual change of personality. Schechtman's counterargument against the pure minimal self is based on the appealing hypothesis that the development of the narrative self gives rise to a new kind of consciousness that enriches the "brute first-personal experience". In consequence, the basic character of first-personal experience may be different for self-narrators. In Schechtman's own words:

> There is, however, a much more interesting line of response. A narrative theorist might argue that the kind of phenomenological self-consciousness that makes a self is a qualitatively different kind of consciousness than the brute first-personal awareness we presumably share with many animals, and that this different kind of consciousness requires narrative. The idea would be that the character (and not just the content) of first-personal experience is different for self-narrators than for non-narrators (Schechtman, 2011, p. 410).

The dilemma of the phenomenal character of self-awareness will be discussed in greater detail in the following chapter, but, for preliminary purposes, let me mention another example that could alter the very nature of the first-personal givenness of experiences. In my view, Schechtman's claim, the idea that first-personal experience can be altered by narrations, can be extended in order to include the problem of traumatic subjectivity. In short, the psychoanalytic aspect of narration suggests that subversive, traumatic experiences and therapeutic interventions (including narrative self-understanding) may change the very core of the experiencing self. More specifically, not only the *content* of recollection might be altered (by means of bringing the unconscious into awareness), but also the *way* or the *character* of remembering might undergo hardly noticeable but still significant alterations. Zahavi accepts the possibility of the first scenario but denies the second. His basic tenet is that our autobiographical narrative is only the enrichment of the more basic, already existing pre-reflective self-awareness. For him, the *mineness* of experiences is independent from the character-shaping effects of language acquisition and psychopathological alterations. In my view, both narrative theories and psychoanalysis challenge the formal notion of the minimal self and, at the same time, contribute to the introduction of the (less minimal) affective core self. These considerations will lead us to the intricate problems of time-consciousness in phenomenology and *Nachträglichkeit* in psychoanalysis. The nuanced differences between the minimal self and the core self will be analyzed in more depth in Chap. 7.

The formal notion of the minimal self is also too empty from the viewpoint of developmental psychology. Menary and Jopling accentuated the role of the *embodied subject* in the constitution of the self. Nelson (2003) also demonstrates the interlacing aspects of embodiment, language, and culture. She analyzed meticulously the evolutionary and developmental stages of symbolization and language acquisition and concluded that narrative capacities emerge in community; through intersubjective feedback loops that shape the individuality of children. As a result, a new level of self-consciousness emerges that comprises the sense of temporality and the awareness of other minds (Nelson, 2003, p. 33). For Zahavi, as we have seen, the minimal self is defined as a formal notion that, paradoxically, comprises the experiential flow of time-consciousness and self-other relations – those characteristic aspects that are the result of complex developmental processes for Menary.[32]

The crucial question to ask is: can we imagine the minimal self as the locus of an integration process in time-consciousness? My contention is that there is an unconscious disintegration-reintegration process at the level of the *experiential core self* that can be contrasted with the narrative construction of the self. For this reason, Part II will analyze the possible role of passivity and unconscious in self-constitution. My second consideration revolves around the possibility that traumatic experiences not only shatter the narrative self (i.e., instigate the reconstructive-integrative process of self-understanding) but also disintegrate the core self at least on its affective-bodily (psychosomatic) basis. My approach will deviate from the formal definition of the minimal self and introduces the more complex and flexible notion of affective core self.

In order to clarify the above-mentioned theses cautionary remarks are needed. There is an ongoing heated debate in the literature about the status of the minimal self. Before proceeding to the problem of self-awareness, it is important to address the key elements of the debates around the minimal self. As I see it, the misunderstandings are resulting from ambiguous terms and from the interdisciplinary nature of the discussion. Zahavi attempted to bridge the gap between continental philosophy and philosophy of mind. The result of his extraordinary undertaking was the characterization of the minimal self at the intersection of phenomenology and philosophy of mind. The interdisciplinary nature of the project further complicates matters: phenomenology, developmental psychology, psychopathology, neuroscience, and psychoanalysis have already introduced their own versions of the self. For start, it is advisable to focus on the phenomenological roots of the minimal self. Zahavi's main concern is the notion of pre-reflective self-awareness, which can be traced back through the contributions of Sartre and Merleau-Ponty to Husserl and even to Brentano and other thinkers (cf. Zahavi, 1998). What is intriguing in Zahavi's work is that he uncovers not only the origin of pre-reflective self-awareness, but also tries to explain the mystery of phenomenal consciousness *per se* (i.e., "phenomenality as such").[33] As a result, his undertaking shows a hybridization tendency:

[32] For instance, in his book, *Self and Other*, Zahavi states: "It is difficult to see how language acquisition should change and transform the very basic structure of pre-reflective self-consciousness. It is consequently important not to lose sight of how formal a notion of selfhood the minimal notion is. To stick to the dye and water analogy, the fact that the water becomes coloured shouldn't make us overlook the fact that the coloured water remains water and retains its liquid properties" (Zahavi, 2014, p. 62).

[33] In a similar vein, Kozyreva (2017) gives a phenomenological answer to the mystery of phenomenal consciousness. Following Husserl, she claims that it is a "naturalistic fallacy" to speak of "what-it-is-like" qualia even in a holistic manner. Phenomenal consciousness cannot be treated as natural objects or laws of nature. It might be that transcendental-phenomenological reduction leads to a new horizon of understanding in which the traditional opposition between the subjective and the objective can be bracketed. It is only a "naturalistic attributional bias" to place phenomenal consciousness into the realm of natural objects, i.e., to place it among sense-qualities (cf. Kozyreva, 2017, p. 93).

the mystery of phenomenal consciousness is addressed by phenomenological considerations. However, as we shall see, developmental issues and psychopathology broadened the domain of investigation, and conceptual analysis is needed to make clear the similarities and differences between the proliferating versions of the minimal self.

The critical views of Zahavi's approach revolve around the problem of the intersubjective constitution of the minimal self; critiques accentuate that the self-other relation is underrepresented in Zahavi's undertaking. However, Zahavi clearly demonstrated the close connection between the intersubjective constitution of the self and the introduction of the minimal self. It is not his intention to marginalize the developmental or societal issues, rather he argues that the minimal self is the ground for the higher levels of experiential reality. In order to clarify misunderstandings, he offered, in one of his recent papers, a conceptual distinction between a "(more) minimal self of an infant", and the "(less) minimal self of an adult". The former tries to designate the undifferentiated presence of "phenomenology as such", a basic, undifferentiated dimension of phenomenal character: "A crying newborn is not a zombie bereft of experiences but a creature whose crying is expressing an experience of distress. The crying newborn is a subject of experience whatever else it might be" (Zahavi, 2017, p. 195). According to his thesis, the inner or private givenness of one's own experience has priority over the more concrete "first-personal experiential life". To be more specific, Zahavi, in reply to criticism, tries to conceptualize two almost indiscernible, interwoven levels of the minimal self: the first level is only about "having phenomenal states" per se, but the second layer of experiential life is far more wider than the first. The (less) minimal self is open to the processes of *organization* and *schematization* and it can be broadened to comprise the problems of personal identity and narrative self-understanding as well (Zahavi, 2017, p. 198). To be more precise, the (more) minimal self is the elementary fact of self-awareness and the (less) minimal self can be conceived as a layered concept that designates the unfolding, changing, and intersubjectively construed core of identity.

In order to avoid conceptual confusion, it is reasonable to suggest that the (less) minimal self can be reformulated as an *affective core self*. The self in question is *affective* in the sense that it comprises a basic form of self-awareness, and it is also a *core self*, that is, it is the quasi-stable experiential locus of identity that is intertwined with but not identical with narrative identity. The affective core self is not a reified or abstracted notion of the self, but rather the whirlwind of change that sometimes takes place gradually in normal circumstances or rapidly due to traumatic experiences. The affective core self is the flexible network of the implicit or affective unconscious – it can be placed between the (more) minimal self and the narrative self. On the one hand, the affective core self is the *subtle presence* or the holistic *affective tonality* of an individuated self-awareness, and, on the other hand, it is the condensation of dispositions, habits, and traits sedimented in the phenomenological unconscious. The affective core self is the obscure underground of self-constitution. This working definition tries to uncover the implicit layers of the self. However, at the same time, akin to the minimal self, it cannot be extracted from the embodied and situated life of the subject; it tries to capture the protean, changeable aspects of subjectivity without reification. The main inspiration for this kind of experiential self comes from phenomenology and psychoanalysis.

For addressing the problem of the self in general terms, let us examine the convergence between philosophy of mind and phenomenology. There is a marked tendency to explain phenomenal consciousness and the minimal self simultaneously.[34] For example, Owen Flanagan introduced the notion of low-level or *weak sense of self-consciousness* in his 1992 book *Consciousness Reconsidered.* In this early articulation of the minimal self, the connection between self-consciousness and phenomenal consciousness is already present: "…all subjective experience is self-conscious in the weak sense that there is something it is like for the subject to have that experience This involves a sense that the experience is the subject's experience, that it happens to her, occurs in her stream" (Flanagan, 1992, p. 194). In this passage the problem of *ownership* or *mineness* of experiences is also significant. An explicit reference to *mineness* occurs when Flanagan (1992) contends that: "The low-level self-consciousness involved in experiencing my experiences as mine…" (p. 194). By referring to William James, Flanagan argues that only the higher-order reflection suggests that there is always the act of "I think that" in each occurrent mental state. According to Flanagan, the basic form of consciousness is restricted to "having a subjective experience", that implies simply the appropriation of occurrent mental states. Consequently, the construction of a narrative self, in virtue of the self-representative capacity of language, happens at a further stage of development.[35] As was discussed in Sect. 2.1, Flanagan uses Luria's case study of a brain-damaged soldier and straightforwardly foreshadows Zahavi's conception of *mineness*: "We start at one end of the continuum with the dim and inchoate self/nonself distinction, the awareness of what is me and mine and what is not me and not mine. […] At the other end of the continuum we reach the weighty self that is the center of narrative gravity" (Flanagan, 1992, p. 199). In contrast to Zahavi, he does not accentuate the constitutive aspect of the low-level self-consciousness, but rather he puts emphasis on the role of autobiographical memory that is the prerequisite for the temporal expansion of the self (i.e., for appropriating the distant past and envisioning possible futures).

Zahavi unraveled a striking thematic and terminological convergence between phenomenology and philosophy of mind. Kriegel (2003), by referring to Alvin Goldman, introduced the notion of *intransitive self-consciousness* that also implies

[34] Zahavi summarizes the recent developments in philosophy of mind: "In recent years the issues of subjectivity, phenomenal consciousness, and selfhood have once again become central and respectable topics in analytical philosophy. After a long period of neobehaviorist functionalism it is nowadays almost commonplace to argue that the experiential or first-personal dimension of consciousness must be taken seriously, since an important and non-negligible feature of consciousness is the way in which it is experienced by the subject" (Zahavi, 2002, p. 7).

[35] "Maturation, living in the world, being socialized in a life form, and acquiring certain ways of thinking are required before one can start to build the complex set of representations that, taken together, come to constitute the narrative self" (Flanagan, 1992, p. 195).

the problem of phenomenal consciousness and the appropriation of experiences: "When we have a mental state self-consciously, there is a subtle awareness of self implicit in that state, whereby we are aware of ourselves as its *owners*. This is not just a matter of the traditional issue of ownership; it is a matter of ownership-awareness. It is not just that we *are* the owners; we are *aware* of being the owners, albeit implicitly and inattentively" (Kriegel, 2003, p. 104). Kriegel's intransitive self-consciousness is the *implicit awareness* of the self that occurs spontaneously in each act of perception. In other words, every perceptual experience implies the implicit awareness of owning, acknowledging, or appropriating that specific experience. Retrospectively, implicit awareness is often traced back to William James, who ruminated over the "warmth and intimacy" of our thoughts. In *The Principles of Psychology* (1890) he speaks of the warmth and intimacy of thoughts that belong to the Ego.[36] More specifically, James tries to give an answer to the psychological continuity of the self and he contends that the present self (including the spiritual and the material parts) unfolds either by a feeling which accompanies the current thought, or by the feeling of bodily existence. Then he concludes: "We cannot realize our present self without simultaneously feeling one or other of these two things" (James, 1983 [1890], p. 316). In this context the emphasis is not on the appropriation of experiences occurring in the experiential flow but rather on the continuity of the past and present selves. James introduces intriguing metaphors to demonstrate the bodily and feeling component of personal identity: he speaks of the "uniform feeling of 'warmth', of bodily existence", or of the "uniform feeling of pure psychic energy", then he comes up with the metaphor of "animal heat" of our dearest memories. Distant past experiences (such as memories of childhood) have lost their "animal heat"; that is, the warmth of connectedness to the present self is diminished or has been lost forever (James, 1983 [1890], p. 318). In my reading, James's reflections on personal identity are more closely tied to Schechtman's empathic access than to the "what-it-is-likeness" of appropriating experiences.

In contemporary studies the question arises: can we explain the problem of the self by utilizing the insights of the phenomenological tradition? For Zahavi, the notion of pre-reflective self-awareness seems to be the best candidate to solve the problems of ownership and diachronic identity. By contrast, Flanagan remarked that the experiential lived time in itself is not suitable to characterize the core of personal identity. In this respect, the experiential flow is interlaced with the oscillatory movement of narrative self-constitution that brings back the distant past into the present

[36] "… the thoughts which we actually know to exist do not fly about loose, but seem each to belong to some one thinker and not to another. Each thought, out of a multitude of other thoughts of which it may think, is able to distinguish those which belong to its own Ego from those which do not. The former have a *warmth and intimacy* about them of which the latter are completely devoid, being merely conceived, in a cold and foreign fashion, and not appearing as blood-relatives, bringing their greetings to us from out of the past. Now this consciousness of personal sameness may be treated either as a subjective phenomenon or as an objective deliverance, as a feeling, or as a truth. We may explain how one bit of thought can come to judge other bits to belong to the same Ego with itself; or we may criticise its judgment and decide how far it may tally with the nature of things" (James, 1983 [1890], p. 300, *emphasis added*).

and construes scenarios of the possible future.[37] The problem of pre-reflective self-awareness poses a challenge to the narrative self, but, simultaneously, it also constitutes an interdisciplinary dialogue between philosophy of mind and phenomenology. The main attempt of synthesis appeared in a co-authored paper. Zahavi and Kriegel (2016) argue that phenomenally conscious states directly imply pre-reflective self-awareness. For preliminary purposes, one can say that the phenomenal character of all conscious states involves the presence of the experiencing subject; for Zahavi and Kriegel (2016) this is a self-evident metaphysical and phenomenological fact as well. If we consider the experiences of perceiving an apple or remembering a banana, then it is true that they differ in their content and intentional type, however, both experiences occur *for us* and they have a latent sense of "what it is like to have them". The example is quite simple but enables the authors to introduce the composite jargon of "what-it-is-like-for-me-ness" (Zahavi & Kriegel, 2016, p. 36). However, the authors do not speculate on the nature and genesis of phenomenal consciousness in a naturalistic manner, but they only suggest that the first-personal perspective and the presence of the experiencing subject are the fundamental requirements for the phenomenal dimension of consciousness.

In line with this thought, Strawson accentuates that the phenomenological question of the self has priority over the ontological question (Strawson, 2000, p. 40), and he introduces the so-called "Experience/Experiencer Thesis" that means "Experience is impossible without an experiencer" (Strawson, 2011, p. 253). Consequently, if we are to define what the self is, then we have to look for "self-experience"; since self-experience is the source of the problem of the self. Zahavi (2016), in accordance with Strawson, argues that maintaining the thesis of the minimal (self-referential) subject is not only a conceptual or metaphysical claim but also a phenomenological. However, as Zahavi has shown, a major divergence ensues when Strawson claims that the self can be understood as an object, whereas phenomenology insists that the self is strictly the subject or pole of intentional acts. Despite the disagreement on the true nature of the self, due to traditional oppositions, Zahavi is quite optimistic and says that the conflict is mainly terminological and not substantial (Zahavi, 2016, p. 90). Strawson (2015), in a footnote to his paper *Self-intimation*, mentions a private conversation with Zahavi, where he also regards the above-mentioned disagreement as merely idiomatic (p. 27). Furthermore, Strawson shows the conceptual and metaphysical obstacles to revealing the structure and complexity of the minimal self:

> When we think hard and clearly about awareness we may come to see, with Aristotle and many others, that we say something true when we say that all awareness comports awareness of itself. This expression of this feature of reality, couched as it is in necessarily discursively articulated discursive thought or language, can seem paradoxical. But the reality it describes, and says something true about, is entirely unparadoxical, and indeed fundamentally simple in character. (Strawson, 2015, pp. 25–26)

[37] "Life and consciousness can be as streamlike as you want, but if memory is weak, if present thought is not powerful to appropriate what has gone before, no narrative can be constructed. There is simply the here and now" (Flanagan, 1992, p. 198).

To be aware of something and simultaneously to be self-aware is the main insight of the Husserlian paradigm of intentionality[38] and also the Cartesian *cogito*.[39] Despite the self-evident nature of self-awareness the problem of the minimal self is not without controversy. In the citation above, Strawson showed that the discursive clarification of this elementary phenomenological fact could lead to the borders of language. Strawson underscores that: (1) self-awareness in itself is intentional; and (2) awareness can objectify itself as one specific part of its own contents. Strawson also stresses the limits of language when he argues that the presence of self-awareness cannot be expressed without "relational idioms"; without the concepts of *structure* and *relation,* and ultimately one has to acknowledge the inadequacy of language (Strawson, 2015, p. 28). Strawson uses the term "Self" in accordance with William James, who claimed that the central problem of the Self is the question of how self-experience "feels" (Strawson, 1999, p. 103). Therefore, the self is basically self-experience. But what does it mean to talk about basic self-experience? When philosophers and psychiatrists speak of "sense of self" or the "experience of self or being", etc., they are often referring to ambiguous phenomena without tangible explanatory value. Nevertheless, one of the aims of the phenomenological undertaking is to run against the borders of inherited language and unravel the pre-intentional and pre-conscious levels of lived experience. As Husserl (1991), in his analysis of internal time-consciousness, and Ullmann (2010) have recently indicated, phenomenological inquiries are grounded in the *modality of intuition*; and the aim of the phenomenological investigation is to leave behind the habituated considerations through a perspectival shift.

The current work argues that the phenomenology of self-experience is a suitable candidate for a genetic-phenomenological approach. Traditionally, a genetic-phenomenological approach investigates the genesis of intentional objects in internal time-consciousness.[40] For Husserl, the genetic analysis reveals the essential laws

---

[38] "every experience is 'consciousness' and consciousness is consciousness of… But every experience is itself experienced [*erlebt*], and to that extent also 'conscious' [*bewußt*]" (Husserl, 1991, p. 291).

[39] In Descartes' philosophy there is a kind of ambivalence. Strawson shows that the term "conscious" in Descartes' usage is often refers to higher-order reflection, but in his reply to Hobbes he identifies consciousness with first-order awareness, therefore, Strawson came to the conclusion that Descartes was also the advocate of the same-order theorists (Strawson, 2015, p. 15). This means that there is no need for higher-order thought or perception for instantiating self-awareness.

[40] Husserl made a distinction between *static* and *genetic* phenomenology. The former aims to investigate the question of how the intentional acts constitute their objects, but in case of the latter, Husserl speaks of the process of *sedimentation*; that is, intentional structures have their own origin and history. The further purpose of genetic phenomenology is the exploration of the different forms and layers of intentionality. In this respect, the phenomenological reflection intends to disclose the constitutive layers of experience, and thus the pre-linguistic level will also be revealed (Zahavi, 2003, , p. 94). Association is one of the main interests of genetic phenomenology, since it plays a foundational role in the constitution of intentional and even pre-intentional objects. Matt Bower highlights the fact that genetic phenomenology can investigate the relation between *particular* intentional acts. For example, the pereption of a musical instrument inevitably occasions

of intentional objects. According to the genetic analysis, consciousness is a complex and dynamic process: "the stream of consciousness is a stream of constant genesis" (Husserl, 1998, p. 137). In parallel, the aim of phenomenology is to examine the genesis and development of the sense of self as well. However, the genetic-phenomenological investigation of self-experience is a paradoxical undertaking, since it is inclined to collide with the static analyses of philosophy of mind. Of course, considerable work has been directed at the synchronic and diachronic unity of consciousness in philosophy of mind as well.[41] The aim of a genetic-phenomenological approach is to arrive at the periphery of our ordinary perception and to investigate the very process of the constitution of sense.

Ullmann has drawn attention to the fact that the praxis of philosophy is almost impossible in the realm of chaos from which meaning and sense are constantly emerging. The genetic-phenomenological undertakings always bring up the danger of arbitrary constructions. Ullmann illuminates the dangers of constructive phenomenology: "Thus, on the one hand, one projects back an articulated intuition onto the unarticulated source of meaning, or, on the other hand, one projects something into this unattainable beginning" (Ullmann, 2010, p. 226). The affective realm of self-experience poses a challenge even for phenomenologically oriented thinkers. Let us recall Jopling's considerations regarding the coexistent nature of affectivity and narration. Affectivity, or the *desiderative* ground of consciousness, as Maiese stated, is the engine as well as the subject of schematization processes. As a result, this constitutive dimension of consciousness is inevitably colored and distorted by the above-mentioned constructive elements. On Husserlian grounds, we can apply a certain kind of phenomenological reduction in which the upper layers of the psychologically understood *person* are peeled off in order to reveal the point of origin or the locus of self-experience, i.e., subjectivity as such. In this work I intend to show that personal identity presupposes the affective identity. Or, to put it differently, the (less) minimal self has its own affective unconscious. Paradoxically, the aim of the current investigation is to unravel the inner complexity of the minimal self.

Zahavi and Kriegel (2016) brought together the problem of phenomenal consciousness and self-awareness in a constructive way and crossed the border between philosophy of mind and phenomenology. From phenomenological standpoint, Zahavi came up with the neo-Sartreian claim that: "the first personal givenness of the experience can be said to entail a built-in self-reference, a primitive

---

the thought of the instrument's owner. That is, genetic phenomenology draws attention to the pre-egoic associative processes. On the other hand, genetic phenomenology discloses the regularities between different intentional types. The aim of the investigation in this context is that how could a certain intentional act bring about another kind of intentional act. Furthermore, Bower draws attention to the interesting fact that Husserl made references to developmental psychology in the Cartesian Meditations (Bower, 2015, pp. 84–85). It seems to be the case that the role of *affect* in goal-directed behavior is a common theme in Husserlian phenomenology and also in developmental psychologyl.

[41] cf. Dainton (2004, 2008), Bayne (2010), Bayne and Chalmers (2003).

experimental self-referentiality" (Zahavi, 2002, p. 16). Zahavi enthusiastically defended this *experiential minimalism* thesis from the late 1990s onwards and claimed that parallel views can be found in the analytic literature as well (Zahavi, 2020, 2017). At first glance, the idiom "first personal givenness" seems to be a very ambiguous expression; however, it simply refers to the previously mentioned formal notion of the minimal self, i.e., to the capacity that enables us to differentiate between our own and others' intentional states. Drawing on this implicit knowledge the corollary ensues that the subject does not conflate reality with imagination (or hallucination). Intentional acts (e.g. perceiving, imagining, remembering, etc.) have self-illuminating capacities, meaning that the subject is pre-reflectively aware of the type and nature of the ongoing intentional act without conscious, reflective deliberations. In daily life we live in the holistic network of intertwined and intermingled intentional acts and this particular stream of experience is organized by implicit rules that bolster the coherence of perception and action. When the implicit organization of the stream is damaged by internal or external factors, a confusion between intentional states can occur especially in pre-psychotic and full-blown psychotic states. Strawson and Zahavi share the view that the experiential flow has a complex microstructure that can be characterized by the ongoing dynamics of intentionality. Zahavi proposes that the Husserlian idea of intertwined intentional types can serve as an explanatory framework:

> First, we have an intentional act of a specific type (a perception) which is directed at a specific object (a frog). Then we retain the intentional act-type (the perception), but replace the frog with another object (a banana). In a final step, we replace the perception with another act-type (a recollection) while retaining the second object. By going through these variations, we succeed in establishing that an investigation of our experiential life shouldn't merely focus on the various intentional objects we can be directed at, but that it also has to consider the different intentional types or attitudes we can adopt. This is all trivial (Zahavi, 2012, p. 149).

One of the basic tenets of Husserlian phenomenology is that consciousness is not only the pure act-pole of intentional directedness, but it can also be depicted as the – often chaotic – mixture of intentional states. Under ordinary circumstances, one can never experience his or her consciousness as an empty pole. As we shall see in the following sections, the empty or hypervigilant states of consciousness are often characterized as altered states of consciousness rather than normal mental states.[42] In our day-to-day living, ruminations, rememberings, reminiscences, and reveries occur to us at the periphery of consciousness. These occurrent mental states are explored by Husserl through the analysis of "intertwinings" (*Verflechtungen*). The

---

[42] It goes without saying that investigating the nature of consciousness, or defining the notion of consciousness in general, is a culturally relative undertaking. As discussed previously, Wilkes (1981, 1988, 1995) argued that our notion of consciousness has its own genealogy, it is the result of contingent cultural and linguistic changes. Or, to come up with another frequently occurring issue: in Eastern Buddhist and Indian philosophy there is room for the experience of *empty consciousness*, that is, a special type of consciousness without mental content or intentional directedness. From the viewpoint of Western phenomenology, which is based on intentionality, this special kind of altered state of consciousness seems inconceivable and unfathomable.

crux of his observation is that, on the one hand, experience is basically composed of holistic Gestalts (i.e., it is meaningless to presuppose independent rays of sense data), and, on the other hand, there is no room for isolated lived experiences. The following quote from Husserl's *Phenomenological Psychology* illustrates the holistic character of our experiential realm:

> For the single datum is a mere abstraction in the psychic. A feeling, a mood, an emerging thought, a hope which makes itself felt, etc. – nothing of the sort is ever an isolated lived experience; it is what it is in the psychic milieu, in its intertwinings, its motivations, indications, etc.; and these are moments of the nexus, of the psychic function, which are lived together inseparably. Thus the great task is to analyze systematically and describe in their typology the-many-sided intertwinings which unite in the respective unity of a structural nexus. "Structure" designates the complex intertwining which belongs to every concrete phase of the streaming psychic life. Even the series of phases which successively give unity to the life-stream has its typology. The psychic nexus is a nexus of efficacy, a nexus of development, and is governed throughout by an immanent teleology which can be exhibited analytically (Husserl, 1977 [1925] pp. 5–6).

Dermot Moran argues that the problem of "intertwining" has different layers. In the first place, "intertwining" refers to the above-mentioned entanglement of intentional acts. A perceptual cue can give rise to a distant memory, or an occurring fantasy bursts into the actual perceptual field in daydreaming and so on. The two other interpretations of the term allude to the lived body and time-consciousness. According to Moran: "Intertwining is also used by Husserl to express the relations between protentions and retentions *within* specific lived experience" (Moran, 2013, p. 297). Finally, Moran highlights the fact that the problem of intertwining is particularly relevant from the perspective of pre-reflective experience. Merleau-Ponty was strongly influenced by the intertwined nature of the lived body and consciousness. Moran shows that even the characteristically Merleau-Pontian terms of "chiasm" and "wild being" can be traced back to Husserlian inspirations (Moran, 2013, p. 290). There are several pathological and non-pathological situations in which one can vividly experience the subjectively lived body as an objectified body. But in normal circumstances we have a pre-reflective sense that our lived body (*Leib*) is intact, and we dwell in it. My body is pre-reflectively given as *mine*.[43] It would take me too far afield here to delve into the problems of embodiment. The crucial issue here is only that the body (i.e. the *Leib*) plays an important role in the holistic network of intentional acts.

---

[43] The phenomenology of embodiment brings us back to the *Leib-Körper* relation. Moran describes particularly extreme examples from the literature of psychopathology to describe the anomalous, reifying experiences in which the *Leib* loses its organic, "ensouled" nature: "There are also parts of my body that are almost completely 'external' (e.g. hair, nails) that can be removed from my body; albeit they can be experienced as deeply personal, e.g. sudden alopecia can be traumatic. If I lose feeling in a limb it may take on this character of pure physical entity and indeed there are pathological situations of body dysmorphic disorder (BDD) where people reject parts of their body as ugly, or alien limb syndrome, where people become convinced that a part of their body is not their own but an alien limb controlled by others. Sufferers of the latter condition may even seek amputations to remove these 'alien' appendages. On a more 'normal' level, I can seek to remove a mole from my skin, and so on, because it is alien to me. External and internal are inextricably and fluidly *intertwined* and I dwell into a kind of interleaving between these dimensions" (Moran, 2013, p. 295).

Zahavi maintains the idea that the common denominator in the flow of intentional states is the first-personal character of experiences. Experiences are unveiling themselves from our first-person perspective and they intimately belong to us. Zahavi, in this respect, often refers to "ownership", "mineness", "for-me-ness" and "first-personal givenness". Guillot (2017), in her impressive and meticulous paper, sheds light on the terminological obscurity and confusion surrounding the problem of self-awareness that boils down to the question of "what remains constant across experience" (Guillot, 2017, p. 25). Guillot has reservations against the thesis that first-personal experiences directly entail self-awareness. But Zahavi counters the objection by placing emphasis on the Janus-faced aspect of intentional experiences: experiences are *for* the subject and they are something different *from* the subject. Zahavi's thesis is often misunderstood, therefore he painstakingly argues that the acceptance of the presence of pre-reflective self-awareness leaves no room for solipsism, rather it indicates the "otherness" of the experience. Pointedly, the heated debates on terminological differences can be avoided if we accept the brute fact that there is "a very tight constitutive link between self and experience"; that is, the self is part and parcel of the experiential flow in a non-objectifying manner (Zahavi, 2017, p. 9).[44] In sum, one cannot deny the *for-me-ness* or *mineness* of experiences without denying the simple fact of the first-person perspective. The self as the subject of experience is not a separate entity, higher-order state, or transcendental ego but rather the immanent constitutive process of the experiential flow (Zahavi, 2012, p. 150). There is no self over and above the experiential flow, the first-person perspective is seamlessly embedded into the flow. This minimalist notion of the self substantiates the diachronic continuity of consciousness. Zahavi (2005) introduces his view on Husserlian grounds: "Thus, the phenomenological ego is not something that floats above manifold experiences; rather, it is simply identical with their unified whole. […] To put it differently, the stream of experiences is self-unifying. To understand its unity, we do not need to look at anything above, beyond, or external to the stream itself" (p. 35). Despite the intuitive appeal of the self-unifying function of intentional consciousness, the "engine" that establishes the diachronic unity (or length-intentionality) of the self remains hidden. Where is the locus of the process of synthesis that gives rise to the implicit sense of unified consciousness? Zahavi goes on to argue that "unification is not due to the synthesizing contribution of the ego; on the contrary, such a contribution is *superfluous* since the unification has already taken place in accordance with intraexperiential laws. To put it differently, the stream of experiences is self-unifying" (p. 35).

How can we describe the phenomenal "feel" of the minimal self, if it is not identical with a specific qualitative state? How can we characterize the what-it-is-like-*for-me*-ness ingredient of experience? How can we characterize the rudimentary

[44] There is no room in this work to provide a detailed account of the heated debates between philosophy of mind and phenomenology concerning the nature of the minimal self. Zahavi (2017, 2020) has provided a detailed account of the contemporary critiques of the minimal self. The aim of this book is only to point out the practical, phenomenological, and psychoanalytic significance of the minimal self.

sense of self-awareness, and does it necessarily entails an implicit qualitative experience? Is it possible that the qualitative character of "owning" experiences is a tangible phenomenon or is it only a theoretical fiction? If we reconsider the Husserlian critique against naturalization of consciousness and the dangers of genetic-phenomenological approach, it can be objected that reification is latent in these conceptualizations. It is not surprising that even philosophers of mind cast doubt on the elementary aspect of *mineness* of experiences. For example, Slors and Jongepier (2014) contend that: "We claim that the mineness of experience is not given with the internal structure that is common to all individual experiences. Rather, we claim, it is the product of what we shall call the *external* structure of experience, i.e. the way in which each experience is connected with and embedded in a context of other experiences" (p. 194). The claim is reminiscent to the phenomenological problem of "intertwining". As stated previously, it is hard to find an act of pure perception, pure recollection, or pure emotion in the intentional life of consciousness. One can discern intentional experiences among others with varying degrees of intensity or vividness, but the process of intensification presupposes the holistic network of intentional types. The ownership or mineness of experiences cannot be extracted from the stream of experiences.

In order to untangle the dilemmas of the qualitative aspects of experience it is worth considering two approaches to the unity of consciousness. Kozyreva (2017) highlights the fact that in contemporary philosophy of mind there is a marked transition in the usage of *qualia*. At first, qualia denoted the qualitative properties of subjective experiences, but in the recent vocabulary there is an identification between qualia and consciousness. That is, awareness implicitly entails phenomenal consciousness – however, this inference is anything but not self-evident (p. 60). Of course, there are several definitions of qualia. Chalmers (1996) defined them as qualitative characters of mental states or qualitative feelings. Nagel (1974) spoke of the subjective character of experiences that associated with the first-person point of view. Kozyreva concludes that a close link was established between subjectivity and (phenomenal) consciousness and qualia is associated with phenomenal properties (Kozyreva, 2017, p. 61). However, the reification of such qualitative states leaves open the question of the distinction between the conscious and non-conscious qualitative states, and there is a certain *elementarism* in explaining the unity of consciousness in terms of qualia. For example, Bayne and Chalmers (2003) presupposes a single encompassing consciousness that subsumes different experiences (perceptual, bodily, etc.) including their qualitative character. Kozyreva argues that while this kind of unifying consciousness may give rise to the synchronic unity of consciousness, in which mental states as well as their qualia are superimposed on each other, the temporal unification of consciousness remains absent (Kozyreva, 2017, p. 64). She argues that: "…subjectivity cannot be reduced to a certain quality of experience; subjectivity, rather describes the totality of experience" (p. 65).

Kozyreva contends that the tradition of transcendental philosophy and especially the notion of Husserlian time-consciousness offers a valuable alternative to explain the unity and totality of experiential life. For Husserl, and phenomenologically oriented thinkers in general, the totality of experiential life is not a snapshot of the actual representations but rather an open totality constituted by internal time-consciousness. Her critical view is that the connection of experiences, via rules of associations, is the essential ingredient of conscious life. The Husserlian paradigm of time-consciousness has the unique ability to take into account the *formal* as well as the *affective synthesis* of experiential life. The synthesis of sensory data in the specious present and the affective-associative connection between past and present are implicit characteristics of time-consciousness (Kozyreva, 2017, pp. 66–67). By introducing the notion of *affective identity* Kozyreva reconfigures the problem of pre-reflective self-awareness. The focus shifts from searching for the qualia of what-it-is-like-*for-me*-ness to the underlying passivity in the process self-constitution. The notion of affective identity gives a new method for the analysis of the minimal self. First, not only synchronic but also diachronic unity is explained by this descriptive undertaking; second, the analysis of the self turns from the formal notion of the minimal self to the more *content-based approach* in which the problem of passivity and the unconscious is highly relevant. In this respect, the characterization of pre-reflective self-awareness as the what-it-is-like-*for-me*-ness of experiences can be divided into the different branches of *formal* and *affective* identity. This line of thinking will be developed further with the help of psychoanalytic insights in Chap. 4.

In conclusion, Zahavi's notion of the minimal self implies a descriptive approach that lies at the intersection of phenomenology, cognitive science, and philosophy of mind. The connection between the minimal self and passivity has been highlighted by Kozyreva, but Zahavi himself considered other alternatives for describing the minimal self. The sophisticated distinction between the (more) minimal self and (less) minimal self is of particular importance for the current analysis. The problem of *self- or auto-affection* is attached to the problem of the minimal self. These concepts are often associated with the desiderative and sensual aspects of the lived body. From genetic-phenomenological perspective, self-affection is the hidden engine or core of personhood and, consequently, it bolsters the idea of the minimal self. We can also formulate a rudimentary answer to the genesis of phenomenal consciousness based on self-affection, that is, personhood necessarily involves the *feeling of being a (minimal) self.* In the contemporary approaches this is an acceptable theoretical way to reconcile the problem of self-affection (i.e., having phenomenal experiences *per se*) with the problem of phenomenal consciousness. Whereas self-affection and phenomenal consciousness stems from different philosophical traditions the mystery of life (i.e., sentience) and subjectivity lies at their heart. This synthetic idea is clearly expressed by Thomas Fuchs who accentuates the continuity between living (*Leben*) and experiencing (*Erleben*). As I have indicated above, there is an intertwining between the different modalities of intentionality, and it is reasonable to speak of a more basic level of intertwining (*Verflechtung*) between life and consciousness:

> An experience only becomes conscious as such when it acquires a certain degree of intensity. Nonetheless, it was my experience from the outset. In other words, there is a continuum from (a) unconscious life as in dreamless sleep to (b) awakening consciousness, implying a pre-reflective self-awareness, further to (c) peripheral consciousness (being only

> marginally aware of something, e.g. being hungry), up to (d) full-blown waking alertness and attention, including finally the possibility of reflective self-consciousness (explicitly directing one's attention to oneself as an object) (Fuchs, 2017, p. 301).

Fuchs suggest that the phenomenal feeling component of self-awareness is gradated by the intensity of self-affection. The engine of life and the intensity of affections that sustains and maintains even the dynamics of intentional acts and the minimal self as well. Self-affection will be the obscure but also tangible ground of the minimal self that lies behind the formal self-referential loop of pre-reflective self-awareness. And what happens when the feeling tone of self-affection is altered or diminished? There is a widespread assumption in psychopathological studies that the minimal self (here understood as an experiential self) has the propensity for disturbances and full-blown disintegration, which implies that the coherence of the self is highly sensitive to psychosocial conflicts and endogenous alterations occasioned by brain lesions or by different kinds of physical and psychical trauma. But before we delve into the phenomenological psychopathology of self-disorder it is necessary to clarify the different aspects of the minimal self in greater detail.

Ullmann (2015) argued that the minimal self, as the basis of self-experience and self-knowledge, is an abstract and empty conception. However, as I have shown previously, Zahavi demonstrated that the self in question is not an abstract identity-pole that gives unity and coherence to the flow of experiences in a Kantian manner (Zahavi, 2005, p. 104). It is not enough to argue that the self is a necessary precondition for coherent self-experience since the notion that every experience is self-centered seems to be an indisputable fact in normal circumstances. As the previous sections of the paper demonstrated, a more dynamic reading of the minimal self is conceivable based on the role of self-affection and on the thesis that the self is the integral part of internal time-consciousness; that is, it has an immediate experiential reality (Zahavi, 2005, p. 106). Interestingly, both Zahavi and Ullmann argue for the fundamental role of a "phenomenological self" in an indirect way from the standpoint of the narrative self. Both authors refer to MacIntyre who argued that "Stories are lived before they are told – except in the case of fiction" (MacIntyre, 2007, p. 212).

For Zahavi, the key question is the very nature of subjectivity and thus he introduces the notion of Husserlian intentionality to philosophy of mind. The minimal self is by no means restricted to sheer self-referentiality. Considering the problem of the self, Zahavi developed several layers of defining strategies: (1) primarily, the self as pre-reflective self-awareness refers to the first-personal givenness of experiences. Zahavi argues that experiences "bring me into the presence of different intentional objects" (Zahavi, 2005, p. 122). (2) Thesis (1) also implies that the accomplishments of the subject and the impression of the object are inseparably intertwined in the intentional relation – which, of course, echoes the Husserlian paradigm of intentionality (Zahavi, 2005, p. 123). (3) Furthermore, he argues that the pre-reflective sense of *mineness* is an integral part of the stream of consciousness and there is no room for a worldless ego-pole here; the main question is not that who is the *subject* of experiences, but rather, the very nature of "subjectivity of experience" (Zahavi, 2005, pp. 125–126). Naturally, Ullmann and other interpreters rightly pointed out that the sheer distinction between the experiences of mine and others' is too empty to characterize a self; but Zahavi argues that there is no self without content and every intentional type also entails an experiential difference in their phenomenal characters. Both intentional objects and intentional acts (i.e. the constitution of intentional objects) have their peculiar phenomenal character, and these kinds of experiential qualities are incomprehensible without the capacity of self-affection. Based on Michel Henry's philosophy, Zahavi argues that the diachronic continuity of the self is anchored in an "interior self-affection" of the living present (Zahavi, 2005, p. 116).

Zahavi counters the critique that the minimal self is only an abstraction: "…we will never encounter the minimal self in its purity. It will always already be embedded in an environmental and temporal horizon. It will be intertwined with, shaped and contextualized by memories, expressive behaviour and social interaction, by passively acquired habits, inclinations, associations, etc." (Zahavi, 2010, p. 6) There are several strategies in the literature to give a phenomenal, temporal and embodied thickness to the minimal self. As was discussed in Sect. 2.3, one of the strategies to flesh out the minimal self is the introduction of the *core self*. The notion of the experiential core self echoes Damasio's *core self*, the latter is a special kind of underlying activity that manifests itself in every conscious content in a "pulsative fashion". Another important feature of the core self is its "remarkable degree of structural invariance" (Zahavi, 2005, p. 139). By referring to Damasio's core self Zahavi is trying to bolster the idea of the stable minimal self with neurobiological findings. However, from the first-personal phenomenological inquiry one is unable to conceive the roots of self-affection or *phenomenalization* per se. The most intriguing phenomenological challenge is not to reduce the self to naturalistic or psychopathological results, but rather to sensitively describe the complexity of conscious life in mundane and altered circumstances. I maintain that Ullmann's *affective unconscious* can be seen as a significant addendum to the discussion of the minimal self. Zahavi endorses the multi-dimensional account of the self, therefore, the role of the affective unconscious might be a significant addendum beside the developmental investigations of self-other emotions.[45] In order to get a detailed picture of the affective unconscious, chapters I will focus on the intricate problems of the unconscious in the following and examine the psychopathological utilization of the minimal self in the ipseity-disturbance model (IDM).

[45] Zahavi argues that the affective attunement and the adaptation to other's attitude are significant themes around the self; these investigations can be placed between the minimal and narrative self. Therefore, it is useful to supplement the minimal and the narrative self with the investigation of interpersonal relations, i.e., with the *interpersonal self* (Zahavi, 2014). However, Zahavi often emphasizes that the formally understood minimal self is immune to the environmental and societal affects.

References

Bayne, T. (2010). *The unity of consciousness*. OUP.

Bayne, T., & Chalmers, D. (2003). What is the unity of consciousness? In A. Cleeremans (Ed.), *the unity of consciousness: Binding, integration, and dissociation* (pp. 23–58). OUP.

Bower, M. (2015). Genetic phenomenology, cognitive development, and the embodied/extended mind. *Journal of Consciousness Studies, 22*(9–10), 83–108.

Chalmers, D. (1996). *The conscious mind: In search of a fundamental theory*. OUP.

Conway, M. A. (1992). A structural model of autobiographical memory. In M. A. Conway, D. C. Rubin, H. Spinnler, & W. A. Wagenaar (Eds.), *Theoretical perspectives on autobiographical memory* (pp. 167–194). Kluwer Academic Publishers.

Conway, M., & Pleydell-Pearce, C. W. (2000). The construction of autobiographical memories in the self-memory system. *Psychological Review, 107*(2), 261–288.

Conway, M. A., & Rubin, D. C. (1993). The structure of autobiographical memory. In A. F. Collins, S. E. Gathercole, M. A. Conway, & P. E. Morris (Eds.), *Theories of memory* (pp. 103–137). Lawrence Erlbaum Associates.

Dainton, B. (2004). The self and the phenomenal. *Ratio, 17*(4), 365–389.

Dainton, B. (2008). *The phenomenal self*. OUP.

Damasio, A. (1999). *The feeling of what happens: Body, emotion and the making of consciousness*. Vintage Book.

Dennett, D. C. (1991). *Consciousness explained*. Penguin Books.

Dennett, D. (1992). The self as a center of narrative gravity. In F. S. Kessel, P. M. Cole, & D. L. Johnson (Eds.), *Self and consciousness: Multiple perspectives* (pp. 103–115). Lawrence Erlbaum Associates.

Fivush, R., & McDermott Sales, J. (2006). Coping, attachment, and mother-child narratives of stressful events. *Merrill-Palmer Quarterly, 52*(1), 125–150.

Flanagan, O. (1992). *Consciousness reconsidered*. The MIT Press.

Fuchs, T. (2017). Self across time: The diachronic unity of bodily existence. *Phenomenology and the Cognitive Sciences, 16*(2), 291–315.

Gallagher, S. (2000). Philosophical conceptions of the self: Implications for cognitive science. *Trends in Cognitive Sciences, 4*(1), 14–21.

Goldie, P. (2003). One's remembered past: Narrative thinking, emotion, and the external perspective. *Philosophical Papers, 32*(3), 301–319.

Goldie, P. (2012). *The mess inside: Narrative, emotion, and the mind*. OUP.

Guillot, M. (2017). I me mine: On a confusion concerning the subjective character of experience. *Review of Philosophy and Psychology, 8*(1), 23–53.

Husserl, E. (1977 [1925]). *Phenomenological psychology. Lectures, summer semester 1925* (J. Scanlon, Trans.). Martinus Nijhoff.

Husserl, E. (1991). *On the phenomenology of the consciousness of internal time* (1893–1917) (J. B. Brough, Trans.). Kluwer Academic Publishers.

Husserl, E. (1998). Static and genetic phenomenological method (A. J. Steinbock, Trans.). *Continental Philosophy Review, 31*(2), 135–142.

Hutto, D. D. (2017). Memory and narrativity. In S. Bernecker & K. Michaelian (Eds.), *The Routledge handbook of philosophy of memory* (pp. 192–204). Routledge.

James, W. (1983 [1890]). *The principles of psychology*. HUP.

Jopling, D. A. (2000). *Self-knowledge and the self*. Routledge.

Kozyreva, A. (2017). *Phenomenology of affective subjectivity: Analyses on the pre-reflective unity of subjective experience*. Universitäts Bibliothek.

Kriegel, U. (2003). Consciousness as intransitive self-consciousness: Two views and an argument. *Canadian Journal of Philosophy, 33*(1), 103–132.

Krueger, J. W. (2013). The who and the how of experience. In M. Siderits, E. Thompson, & D. Zahavi (Eds.), *Self, no self?: Perspectives from analytical, phenomenological, and Indian traditions* (pp. 27–55). OUP.

MacIntyre, A. (2007). *After virtue: A study in moral theory* (3rd ed.). University of Notre Dame Press.
McCarroll, C. J. (2018). Looking at the self: Perspectival memory and personal identity. *Philosophical Explorations, 22*(3), 259–279.
Menary, R. (2008). Embodied narratives. *Journal of Consciousness Studies, 15*(6), 63–84.
Mink, L. O. (1970). History and fiction as modes of comprehension. *New Literary History, 1*(3), 541–558.
Moran, D. (2013). The phenomenology of embodiment: Intertwining and reflexivity. In R. T. Jensen & D. Moran (Eds.), *The phenomenology of embodied subjectivity* (pp. 285–303). Springer.
Nagel, T. (1974). What is it like to be a bat? *The Philosophical Review, 83*(4), 435–450.
Nelson, K. (2003). Narrative and the emergence of a consciousness of self. In G. D. Fireman, T. E. McVay, & O. J. Flanagan (Eds.), *Narrative and consciousness* (pp. 17–36). OUP.
Ricoeur, P. (1992). *Oneself and another* (K. Blamey, Trans.). The University of Chicago Press.
Schechtman, M. (1996). *The constitution of selves*. Cornell University Press.
Schechtman, M. (2001). Empathic access: The missing ingredient in personal identity. *Philosophical Explorations, 4*(2), 95–111.
Schechtman, M. (2007). Stories, lives, and basic survival: A refinement and defense of the narrative view. *Royal Institute of Philosophy Supplement, 60*, 155–178.
Schechtman, M. (2009). Personal identity. In J. Symons & P. Calvo (Eds.), *The Routledge companion to philosophy of psychology* (pp. 634–646). Routledge.
Schechtman, M. (2011). The narrative self. In S. Gallagher (Ed.), *The Oxford handbook of the self* (pp. 394–418). OUP.
Schechtman, M. (2014). *Staying alive: Personal identity, practical concerns, and the unity of a life*. OUP.
Schechtman, M. (2016). A mess inside: Emphatic access, narrative, and identity. In J. Dodd (Ed.), *Art, mind, and narrative: Themes from the work of Peter Goldie* (pp. 17–34). OUP.
Slors, M. V. P., & Jongepier, F. (2014). Mineness without minimal selves. *Journal of Consciousness Studies, 21*(7–8), 193–219.
Stocker, K. (2020). Mental perspectives during temporal experience in posttraumatic stress disorder. *Phenomenology and the Cognitive Sciences, 19*(2), 321–334.
Strawson, G. (1999). The self and the SESMET. *Journal of Consciousness Studies, 6*(4), 99–135.
Strawson, G. (2000). The phenomenology and ontology of the self. In D. Zahavi (Ed.), *Exploring the self: Philosophical and psychopathological perspectives on self-experience* (pp. 39–54). John Benjamins.
Strawson, G. (2011). The minimal subject. In S. Gallagher (Ed.), *The Oxford handbook of the self* (pp. 253–278). OUP.
Strawson, G. (2015). Self-intimation. *Phenomenology and the Cognitive Sciences, 14*(1), 1–31.
Tengelyi, L. (2007b). *Erfahrung und Ausdruck: Phänomenologie im Umbruch bei Husserl und seinen Nachfolgern*. Springer.
Ullmann, T. (2010). *A láthatatlan forma*. L'Harmattan.
Ullmann, T. (2015). A narratív, a traumatikus és az affektív szubjektivitás. In I. Bujalos, M. Tóth, & T. Valastyán (Eds.), *Az identitás alakzatai* (pp. 21–37). Kalligram.
Ullmann, T. (2017). Phenomenology of experience and the problem of the unconscious. In M. Gabriel, C. Olay, & S. Ostritsch (Eds.), *Welt und Unendlichkeit: Ein deutsch-ungarischer Dialog in memoriam László Tengelyi* (pp. 141–161). Verlag Karl Alber.
Velleman, J. D. (2006). *Self to self: Selected essays*. Cambridge University Press.
Wilkes, K. V. (1981). Multiple personality and personal identity. *The British Journal for the Philosophy of Science, 32*(4), 331–348.
Wilkes, K. V. (1988). *Real people: Personal identity without thought experiments*. OUP.
Wilkes, K. V. (1995). Losing consciousness. In T. Metzinger (Ed.), *Conscious experience*. Imprint Academic.
Williams, H. L., Conway, M. A., & Cohen, G. (2008). Autobiographical memory. In G. Cohen & M. Conway (Eds.), *Memory in the real world* (3rd ed., pp. 21–90). Psychology Press.

Wollheim, R. (1984). *The thread of life*. Yale University Press.
Zahavi, D. (1998). Brentano and Husserl on self-awareness. *Études phénoménologiques, 14*(27–28), 127–168.
Zahavi, D. (2002). First-person thoughts and embodied self-awareness: Some reflections on the relation between recent analytical philosophy and phenomenology. *Phenomenology and the Cognitive Sciences, 1*, 7–26.
Zahavi, D. (2003). *Husserl's phenomenology*. Stanford University Press.
Zahavi, D. (2005). *Subjectivity and selfhood: Investigating the first-person perspective*. MIT Press.
Zahavi, D. (2007). Self and other: The limits of narrative understanding. *Royal Institute of Philosophy Supplement, 60*, 179–202.
Zahavi, D. (2010). Minimal self and narrative self: A distinction in need of refinement. In T. Fuchs, H. C. Sattel, & P. Henningsen (Eds.), *The embodied self: Dimensions, coherence and disorders* (pp. 3–12). Schattauer.
Zahavi, D. (2012). The time of the self. *Grazer Philosophische Studien, 84*, 143–159.
Zahavi, D. (2014). *Self & other: Exploring subjectivity, empath, and shame*. OUP.
Zahavi, D. (2016). Analytic and continental philosophy: From duality through plurality to (some kind of) unity. In S. Rinofner-Kreidl & H. A. Wiltsche (Eds.), *Analytic and continental philosophy: Methods and perspectives. Proceedings of the 37th International Wittgenstein Symposium* (pp. 79–94). Walter De Gruyter.
Zahavi, D. (2017). Thin, thinner, thinnest: Defining the minimal self. In C. Durt, T. Fuchs, & C. Tewes (Eds.), *Embodiment, enaction, and culture: Investigation the constitution of the shared world* (pp. 193–201). MIT Press.
Zahavi, D. (2020). Consciousness and (minimal) selfhood: Getting clearer on for-me-ness. In U. Kriegel (Ed.), *The Oxford handbook of the philosophy of consciousness* (pp. 1–19). OUP. (in press).
Zahavi, D., & Kriegel, U. (2016). For-me-ness: What it is and what it is not. In D. O. Dahlstrom, A. Elpidorou, & W. Hopp (Eds.), *Philosophy of mind and phenomenology: Conceptual and empirical approaches* (pp. 36–56). Routledge.

Part II
Varieties of the Phenomenological Unconscious

Chapter 3
The Unconscious in Psychoanalysis and Phenomenology

Abstract This chapter is devoted to the comparative analysis of the phenomenological and the psychoanalytic unconscious. The aim of the chapter is to examine the convergences and divergences between these two forms of the unconscious and to extend the Husserlian concept of affective relief. The chapter compares the role of repression in the Freudian unconscious and in Husserl's analysis of the living present. The chapter will also examine the harmonizing attempts that try to connect the Freudian and the Husserlian notions of the unconscious. Finally, a partial synthesis between the Freudian and the Husserlian unconscious is proposed.

Keywords Unconscious · Phenomenological unconscious · Repression · Affective relief · Passive synthesis · Affective core self

There are several seemingly irreconcilable positions between psychoanalysis and phenomenology. According to Moran (2017), Husserl's main motivation was the exploration of the intentional structure of consciousness (*Bewusstseinsstorm*) and the experiential flow of consciousness (*Erleibnisstrom*). Husserl's methodological considerations (i.e., reduction and *epoché*) and the transcendental framework stands in stark contrast with the Freudian mechanistic naturalism that comprises the hidden forces of the "id" (Moran, 2017, pp. 5–6). Nevertheless, a significant number of strategies have recently emerged in order to harmonize the two paradigms. The most common interests are the problems of association and passive synthesis, the role of memory and fantasy-activity in self-constitution, the desiderative foundations of the embodied and embedded subject, and the descriptions of dreams and anomalous self-experiences. Despite these thematic convergences, there is no direct connection between Freud and Husserl, although Husserl left some comments on psychoanalysis in *Ideas II* (1989).

In the last few decades, the relationship between phenomenology and psychoanalysis has become more complicated than ever. While Husserl himself had a keen interest in unraveling the layers of the empirical person, the original field of inquiry of phenomenology has been the world-constituting *ego* and its intentional accomplishments. Moran has shown that Husserl did not reduce the problem of the self to

L. Horváth, *The Affective Core Self*, Contributions to Phenomenology 130,
https://doi.org/10.1007/978-3-031-56920-3_3

the Cartesian *cogito* or to the transcendental ego, but also meticulously analyzed the social embeddedness and the naturalistic roots of personhood. According to Moran's thesis, the analysis of the bodily and habitualized accomplishments of the person was an integral part of the genetic phenomenological turn. For the mature Husserl, the person is a *free, rational agent* with egoic intentional acts on the one hand, and an *embodied subject* with a specific psychophysical structure on the other hand. Moreover, the person has different layers behind or above the egoic pole of intentionality: Husserl speaks of the modification of the self (e.g., in a dreams) and of fantasized selves. The consciousness of the living present is always surrounded by other horizons of consciousness; that is, the distant past and the imagined or feared future surrounds the living present (Moran, 2017, p. 16). In sum, Moran claims that, Husserl's phenomenological analysis of the (layered) person is roughly commensurable with the psychoanalytic dissection of the psyche (Moran, 2011, 2017).

And there is another aspect of the person that is central to the minimal self. Husserl tried to unravel the most basic layer of self-constitution when he addressed the problem of self-temporalization (*Zeitigung*). The ego not only constitutes intentional objects in time-consciousness, but it also, counterintuitively, constitutes itself through self-affection and association. The analysis of internal time-consciousness reveals the ongoing activity of primal association; that is, a "primal sensibility" unfolds in self-affective way. Moran complements the Husserlian constitutive processes with *passive reproductions*, that is, we have the propensity to relive the shocking moments of our lives without conscious deliberation. One memory can trigger another in virtue of a spontaneous association, or distant emotions can be reactivated with an increasing (affective) force (Moran, 2017, p. 22).[1] The primal sensibility of the experiential flow is always traversed by the countercurrent of associative awakenings of the past or with the imaginings of the possible future. The deepest layer of the ego is a stratified experiential flow, surrounded by the horizon of associations.

Naturally, there is a remarkable methodological difference between phenomenology and psychoanalysis. On the one hand, phenomenology is, basically, an epistemological undertaking based on the paradigm of intentionality, on the other hand, psychoanalysis was founded on the results of clinical practice and psychopathological observation. Thus, we have found two incommensurable starting points of inquiry in these paradigms. Freud examines the human soul in a clinical context and places great importance on the autonomy of unconscious drives, attitudes, forces, instincts, etc. For Freud, posthypnotic suggestion, hysterical symptoms, dream, fantasy, obsessions and delusions are all point to the underlying activity of the unconscious. Against philosophers, Freud argues that the psyche cannot be equated with the "conscious", rather he allows the existence of the unconscious, the hypothetical reservoir of latent contents. In *Beyond the Pleasure Principle* (Freud, 1990a [1920]) he asserts that "Psycho-analytic speculation takes as its point of departure the

[1] Moran claims that Husserl has never considered the problems of trauma and psychopathology, but the passivity of reproductive associations could straightforwardly lead to the problem of traumatic experiences.

impression, derived from examining unconscious processes, that consciousness may be, not the most universal attribute of mental processes, but only a particular function of them" (p. 18). Freud (1990a [1920]) claims that the conscious part of the mental apparatus is bombarded by "perceptual excitations" stemming from the external world and by "feelings of pleasure and unpleasure" brought about by the mental apparatus (p. 18). If we determine consciousness, primarily, as a perceptual-consciousness, then we are not too far from the standpoint of descriptive phenomenology. But the gap between the two paradigms is apparent when we pay attention to the problem of memory traces. Freud contends that psychoanalytic practice throws light on the intriguing nature of memory: the most powerful and enduring memory traces, or, more precisely, their excitatory processes, cannot to enter into consciousness. And the processes of "something becoming conscious" and "leaving behind a memory-trace" are also incompatible, therefore, excitations must be transmitted into another system in order to constitute a memory trace (p. 19). In *A Note on the Unconscious in Psycho-Analysis* Freud (1958 [1912]) denounces the idea of the philosophical unconscious. That is, the dynamic unconscious stands in stark contrast to the general notion of latent ideas: "The term unconscious, which was used in the purely descriptive sense before, now comes to imply something more. It designates not only latent ideas in general, but especially ideas with a certain dynamic character, ideas keeping apart from consciousness in spite of their intensity and activity" (p. 262). The dynamic unconscious was inspired by Bernheim's experiments with posthypnotic suggestion. As a result, Freud argues that: "We are led from the purely descriptive to a dynamic view of the phenomenon. The idea of the action ordered in hypnosis not only became an object of consciousness at a certain moment, but the more striking aspect of the fact is that this idea grew active: it was translated into action as soon as consciousness became aware of its presence" (p. 24). According to Giampieri-Deutch, Freud's discovery of the (dynamic) unconscious allowed for the detailed explanation of the seemingly well-known phenomena of "latent ideas", which had been addressed in philosophical literature by Leibniz, Kant, and Helmholtz (Giampieri-Deutsch, 2012, p. 247).

The difficulties early memories may also indirectly point to the dynamic unconscious: "Thus an unconscious conception is one of which we are not aware, but the existence of which we are nevertheless ready to admit on account of other proofs or signs" (Freud, 1958 [1912], p. 260). Furthermore, the *naturalist-positivist world-view* of Freud and the Husserlian *transcendental background* gives rise to seemingly insurmountable difficulties to the phenomenological interpretation of the unconscious. Nonetheless, in recent literature, it is a widespread method to demonstrate the thematic convergences between Husserl and Freud: both Husserl and Freud were highly interested in the so-called *affective-associative synthesis* that underlies the conscious activity of the ego.[2] According to Brudzińska, a phenome-

[2] The term affective-associative synthesis refers to the associative processes operating in the periphery of consciousness. The working of these presupposed processes was defined by Husserl as passivity or passive genesis, and these subliminal processes are of pivotal importance in several phenomenologically inspired models of the unconscious.

nological reading of psychoanalysis offers insight into the constitution of the self and the shared world: "A phenomenological reading of psychoanalysis also promises crucial insights into the processes and dynamics of passive genesis and into the processes of subjectivity forming at the most elementary levels of passive synthesis, levels of affective-associative generation that will likely prove very helpful for understanding the constitution of our shared world" (Brudzińska, 2012, p. 24). In a similar way, Askay and Farquhar (2006) argue that both Freud and Husserl engaged in an *archaeological undertaking* in order to uncover the hidden compartments of the person and to gain knowledge about the affective-associative history of the ego (p. 165).[3] Interestingly, there is a mutual silence between the two towering figures: Husserl only referred to depth psychology in his later work, and Freud had an aversion to Husserl's abstract epistemology and theory of perception. An intriguing historical fact that Alfred Adler has attended Husserl's lectures on passive synthesis in which the problem of repression and the unconscious was addressed. Freud's close friend, Ludwig Binswanger was also heavily influenced by Husserl, and, according to the authors, it is certain that Binswanger referred to Husserl during his conversations with Freud (p. 167). However, the authors also point to the irreconcilable standpoints of the two paradigms. From a Husserlian perspective Freudian psychoanalysis is filled with dualistic and physicalistic presuppositions and failed to discover the domain of "pure meanings" or the "life-world" through a radical, unprejudiced, phenomenological reflection. From Freudian perspective, the phenomenological investigation of consciousness seems to be a form of idealism and leads to theoretical alienation from everyday life (pp. 167–168). Despite the contradictory frameworks, there are remarkable convergences: both Husserl and Freud tried to shed light on the unity of body and psyche, and they also placed great significance on the role of the unconscious influences (p. 171). Husserl argued that intentionality is deeply grounded in the body; that is, the responsive body gives rise to the "hyletic substrate" of consciousness. In the *Crisis* (1970) he argues that the lived body is co-constituted in each instance of the perceptual object-consciousness. In other words, we are not disembodied spirits, but rather our perceptual experiences are centered around our lived body (Husserl, 1970, p. 331). Askay and Farquhar (2006) argues that the bodily grounding of meaning is equally important

[3] Despite the significant differences, Askay and Farquhar listed ten common features between psychoanalysis and phenomenology: "In general, both Husserl and Freud (1) undertook a regressive, archaeological inquiry into the affective histories of individuals and into the hidden recesses of the psyche; (2) acknowledged the existence and importance of 'unconscious' intentionalities; (3) affirmed that entire associative chains of ideas can run their course without ever emerging into consciousness, and that they can help us to recover unconscious memories; (4) claimed that consciousness and the unconscious mutually modified one another; and (5) asserted that the ego is at the mercy (at least at times) of the underlying processes of the unconscious. In addition, both held that the unconscious (6) is rooted in instinctual life; (7) involves dynamic forces that continually conflict, and affect consciousness; (8) retains ideas interminably; (9) involves processes/ideas that are atemporal; and (10) is also a dimension of the ego. Furthermore, Freud and Husserl held that ideas require a certain force to reach the threshold of consciousness and when they do, they may only be fragmentary or confused in nature" (Askay & Farquhar, 2006, 166).

for Freud and Husserl. On the one hand, Husserl admits the important role of the lived body in meaning-bestowing processes, and, on the other hand, Freud shows that the *primary processes* are prone to generate appropriate fantasy-images for satisfying the bodily instinctual needs, if there is no other possible way for gratification. Therefore, the authors argue, both Husserl and Freud consider instincts and drives to be fundamental underlying mechanisms of the psyche and intentionality (p. 173). In *Ideas II* (1989), Husserl speaks of blind rules and operations of associations, feelings, drives, and "tendencies which emerge from obscurity". These passive tendencies exert an influence on consciousness. In the end, the authors claim that the ruminations about the passive-associative processes correspond to Freud's archeological efforts and Freud himself would have written such passages (Askay & Farquhar, 2006, p. 173).

However, a few cautionary remarks are necessary. Husserl and Freud agreed that life is governed by drive-related and instinctual strivings and the aim of human life is to achieve the balance of conflicting forces and to satisfy different levels of desires. Moran (2017) mentions that Freud's drive-concept is not unambiguous, as Lacan has already noted that the German notion of "Trieb" was translated into English as "instinct" by James Strachey. In addition, Husserl frequently identifies drives (*Triebe*) and instincts (*Instinkte*) and applies the general term of tendencies (*Tendenzen*) (p. 13). It is of crucial importance to see that Husserl introduces the notion of "instincts" into the paradigm of intentionality. Contrary to the Darwinian or Freudian usage of the term, he defines instincts as pre-intentional activities that may arise without conscious effort or conscious deliberations. Furthermore, Moran highlights that Husserl is uncomfortable with the concept of *drive* and uses it in a very broad sense.[4] Husserl refers to "drive habitualities" that cover the network of instincts and intentions that arise without consciousness and form a "network of habits" (p. 13). Therefore, it would be premature to conclude that the blind operations and drives, which constantly influence consciousness, are perceived in the same way by Husserl and Freud.[5] However, in the following sections we will explore the convincing theories of Smith, Fuchs, and Kozyreva that phenomenologically reformulate the depth psychological unconscious. Despite the harmonizing attempts between the two paradigms, it must be acknowledged that there are significant differences between the intentional paradigm of Husserl (including the contrast between consciousness and dreamless sleep) and the compartmentalized notion of the Freudian psyche.

---

[4] Moran defends the incommensurability between the Freudian and Husserlian terminology: "It would be a fruitless exercise to try to find in Husserl analogues of all the key Freudian notions (both discuss 'instincts' and 'drives' and do not sharply distinguish between them – although instincts generally are seen to belong more to biological animal nature), and the matter is further complicated by Freud's evolving conception of drives" (Moran, 2017, p. 15).

[5] It is worth mentioning that Husserl declares that he cannot work with "unclarifiable instincts", that means that he demarcates his own view from Theodore Lipps' conception of instinct. Moran shows that Husserl uses the term "instinct" in a broad sense, that sometimes refers to the drive to existence and, in other cases, to tendencies of interest (Moran, 2017, pp. 13–14). As we shall see in the analysis of affective relief, the sensual field is permeated by tendencies and instincts (Husserl, 2014, p. 26).

Considerable work has been directed to investigate the status of the unconscious in phenomenology and psychoanalysis. According to Holenstein's interpretation, what counts as unconscious to Husserl is the *pre-conscious* to Freud (Holenstein, 1972, p. 322, cf. Moran, 2017, pp. 5–6).[6] In *A Note on the Unconscious in Psycho-Analysis* Freud typifies the broad category of "latent ideas". He claims that there are, on the one hand, the "foreconscious ideas", which can appear and reappear in consciousness, and, on the other hand, there are "unconscious ideas (proper)". We come to know unconscious ideas indirectly: through neurotic symptoms, errors in memory and speech, forgetting names, and lapsus linguae, etc. These phenomena are indirect signs of the so-called "active unconscious ideas" (Freud, 1958 [1912], p. 263).

Another frequently cited passage can be found in Husserl's *Analyses Concerning Passive and Active Synthesis.* Here, Husserl (2001) speaks of "background lived-experiences to which the ego is not present and 'in' which it does not reside"; these background experiences comprise affective tendencies as well as the "stirrings of the will. Owning to these tendencies, insights, memories, and decisions may come to the sight of the ego (p. 19). Moreover, Husserl also mentions depth psychology in the context of the so-called "unconscious intentionality". In this respect, he gives room for repressed emotions of love, humiliations, resentments, and behaviors motivated by them (Husserl, 1989, p. 308). Askay and Farquhar argued that Husserl tried to keep a distance from the Freudian economic and dynamic model of the unconscious and did not consider the possibility of the censor. However, he agreed with Freud on the fundamental role of affectivity in conscious life. As the authors put it: "…for Freud, the will/id is the fundamental condition of all mental life, and the intellect and consciousness are nothing without that underlying affective force." And Husserl recognized that "…the ideas of consciousness engage and are affected by underlying instincts and drives" (Askay & Farquhar, 2006, pp. 176–177).

It is a surprising development in contemporary phenomenology that certain alternatives attempt to integrate the *active* Freudian unconscious into a phenomenological framework. As we shall see, especially in Chap. 4, the analysis of time and trauma will give us the impression that repression has a role in phenomenologically inspired psychodynamic as well. However, traditionally, repression and resistance, due to their clinical consequences, heavily resist to the phenomenological interpretation that focuses on the accomplishments of consciousness. Perhaps it is not an exaggeration to say that in some cases unconscious impulses, which can be transformed via repression and resistance, behave like *intrapsychic prime impressions*. These untamed impressions constitute an alien alterity-experience in the immanent experiential flow of time. Freud contended that there is a close analogy between the external perceptual object and the ego-alien, unconscious content, however, at the

[6]One can find the elaboration of this observation in Ricoeur's book on Freud (Ricoeur, 1970, p. 392).

same time, he insisted that there is no sharp line between the compartments of the psyche; for example, the contours of conscious and unconscious are blurred in cases of mental illnesses and mystical practices (Freud, 1965 [1933], p. 99). Freud draws attention to impulses in the ego that are *alienated* from consciousness.

> Apart from the circumstance that what belongs to consciousness is not always in consciousness but can also be temporarily latent, observation has shown that much which shares the attributes of the system Pcs [preconscious] does not become conscious; and, further, we shall find that the entry into consciousness is circumscribed by certain dispositions of attention. Hence consciousness stands in no simple relation either to the different systems or to the process of repression. The truth is that it is not only what is repressed that remains alien to consciousness, but also some of the impulses which dominate our ego and which therefore form the strongest functional antithesis to what is repressed (Freud, 1957b [1915], pp. 192–193).

In the light of the above citation, there is a striking similarity between Husserl and Freud in their descriptions of the affective rivalry within the ego. However, on closer inspection, the tripartite structure of the mental apparatus and the process of censorship are absent in the Husserlian notion of the phenomenological unconscious. The affective rivalry and the latent instinctual character of the ego can certainly be seen as a common ground between the two paradigms; but in itself it is merely a *superficial analogy* that conceals the conspicuous differences between the two paradigms.[7] In my view, the most important and intriguing aspect of the unconscious lies in the dilemma of internal alterity that was also addressed by Freud and Husserl, furthermore, the ego-alien tendencies are associated with the problems of affective (and traumatic) subjectivity. In one of his papers, in *The Unconscious*, Freud defines the unconscious contents as primitive formations:

> The content of the Ucs [unconscious] may be compared with a primitive population in the mental kingdom. If inherited mental formations exist in the human being – something analogous to instinct in animals – these constitute the nucleus of the Ucs. Later there is added all that is discarded as useless during childhood development, and need not differ in its nature from what is inherited. A sharp and final division between the content of the two systems, as a rule takes place only at puberty (Freud, 1957b [1915], p. 127).

As the quotation above suggests, the dichotomy of consciousness and unconsciousness is due to biological and anthropological consequences, i.e., developmental considerations are of pivotal importance. In this same paper, Freud remarks that the topographical separation between the conscious (Cs) and unconscious (Ucs) systems in the mental apparatus leads to strange consequences. Freud considers such cases in which he talks to a patient about an uncovered repressed idea. In these cases, the patients frequently dismiss the idea offered by the therapist, and the integration of the unconscious memory of the repressed experience takes place after lifting the repression. Freud (1957b [1915])

[7] Part IV will elaborate further in greater detail the dynamics of affective rivalry.

considers two explanatory scenarios: (1) the *functional description* suggest that the transformation of the idea is only a change in the state of the same idea (i.e. the unconscious idea transformed into conscious idea through reflection); (2) the *topographical description* is more complicated, when the mental act has been transposed from the Ucs to the Cs (or to the preconscious), a fresh registration of the idea occurs – thus, the second inscription involves a "fresh locality" (pp. 106–107). Freud concludes that the conscious and unconscious ideas are not separate records of the same content: "To have listened to something and to have experiences something are psychologically two different things, even though the content of each be the same." (p. 109).

The notion of the "id" was of central importance in the development of psychoanalysis. The indirect knowledge about the "id" paves the way to an inconceivable inner alterity: "Symptoms are derived from the repressed, they are, as it were, its representatives before the ego; but the repressed is foreign territory to the ego – internal foreign territory – just as reality (if you will forgive the unusual expression) is external foreign territory" (Freud, 1965 [1933], p. 71). In order to emphasize the latent *otherness* in the unconscious, Freud borrows Nietzsche's concept of the "id" that was also suggested by Groddeck.[8]

3.1 The Passive and Active Unconscious

It is well known that Eugen Fink, in the appendix to Husserl's *Crisis* (1970), addressed the problem of the unconscious and made cautionary remarks about depth psychology. Fink argued that the unconscious of the naturalistic psychology is governed by philosophically naïve theory of consciousness. According to him, depth psychology, characterized by therapeutic interventions and biological naturalism, is a threat to phenomenology. By contrast, phenomenology, viewed from psychoanalytical perspective, shows the signs of "consciousness idealism". For Fink, the notion of the *libido* and the dynamics of instincts and drives are considered to be mythical theories based on the ambiguous notion of the "true essence of life" (Husserl, 1970, p. 386). He further argues that the psychoanalytical and other naturalistic inquiries omitted the analysis of the intentional structure of consciousness, therefore, they are unable to precisely define the unconscious in comparison to

[8] "We perceive that we have no right to name the mental region that is foreign to the ego 'the system Ucs', since the characteristic of being unconscious is not restricted to it. Very well; we will no longer use the term 'unconscious' in the systematic sense and we will give what we have hitherto so described a better name and one no longer open to misunderstanding. Following a verbal usage of Nietzsche's and taking up a suggestion by Georg Groddeck, we will in future call it the 'id'. This impersonal pronoun seems particularly well suited for expressing the main characteristic of this province of the mind – the fact of its being alien to the ego" (Freud, 1965 [1933], p. 90).

consciousness.[9] However, as was pointed out above, Husserl has already addressed the problem of the unconscious in his *Analyses Concerning Passive and Active Synthesis*. While the main subject of his investigation was conceptual thinking (inside the framework of transcendental logic), he also turned to the problem of the genesis of the living-streaming life of the subject. This was the aim of his genetic phenomenological undertaking (Askay & Farquhar, 2006, p. 179). In the *Analyses*, Husserl explicitly addresses the problem of the influence of the no-longer conscious ideas. This topic will be of particular importance in Part IV, where the analysis of the interrelated aspects of time-consciousness and the unconscious ensues.

In recent literature, one can find several – often contradictory – positions regarding the role of the phenomenological unconscious. In comparison to Freudian psychoanalysis, Husserl's analysis of the pre-egoic passive synthesis from the standpoint of intentional consciousness seems to be a far more speculative or constructive undertaking. To put it more powerfully, it is highly questionable that the ego has sufficient reflective capacity to unravel all the automatic, pre-intentional and pre-egoic activities. The ego cannot take a glance at its own constituting processes because it is founded on those constitutions. However, from psychoanalytic perspective, it is highly plausible to suggest that patients, i.e., persons in distress or in any kind of existential crisis, are able to grasp and verbally represent their internal conflicts. Owing to the therapeutic relation a psychointegrative process can take place, and patients can reflect on their own cognitive and affective reactions, such us the repetitive behavior of obsessive patients. It is customary to describe the Husserlian analyses concerning passive synthesis as a theoretical undertaking, however Rogozinski (2010) proposed a more visceral *epoché*, a "savage" reduction, in which one tries to sink into the realm of pure immanence, into the undifferentiated chaos from which meaningful Gestalts emerge (p. 139).

What is the purpose of savage reduction? Rogozinski tries to bring about a highly concentrated mindset in which the usual perceptual objects disintegrate and snippets of perception are beginning to move on chaotically. Rogozinski tells us: "What is presented as single object is endlessly divided and dispersed into many fugitive silhouettes, tactile impressions, and primary sensations of color and sound. The landscape is dissolved into countless colored impressions. The form of the object is dislocated, and splinters into a multiplicity of profiles that overlap and intersect without completely obscuring one another" (Rogozinski, 2010, p. 139). Think about the case when one is concentrating on a specific object, and, due to the strenuous

[9] For Fink, not only consciousness is pregiven in the practical immediacy of everyday life, but a more detailed analysis is needed for altered states of consciousness as well: "For the unconscious, too, as well as for consciousness, there exists the illusion of everyday, given immediacy: we are all familiar, after all, with the phenomena of sleep, of fainting, of being overtaken by obscure driving forces, creative states, and the like. The naivete of the current theory of the "unconscious" consists in the fact that it engrosses itself in these interesting phenomena which are pregiven in everyday life, undertakes an inductive empirical inquiry, proposes constructive "explanations," and is tacitly guided all along by a naive and dogmatic implicit theory about consciousness of which it always makes use in spite of its demarcation [of these phenomena] from phenomena of consciousness, which are also taken in their everyday familiarity" (Husserl, 1970, p. 386).

concentration, the contours of the object become murky, or levitating color-impressions occur in the field of our vision. Or think about the frequent experience of lucid dreamers who concentrate upon a "dream-object" and they surprisingly begin to experience the meltdown of that object. Rogozinski speaks of "splintering or diffraction to infinity", in this state perceptual objects disintegrate into colored impressions and multiplicity of profiles (p. 139).

It is not difficult to find a Husserlian inspiration in these lines. Not only the hyletic substrate of experience can reveal itself in the process of the "savage reduction", but also an "indeterminate halo" takes shape around our perceptual objects. Here, Rogozinski arrives at the problem of intertwinement above analyzed. According to his analysis, the immanent sphere of the ego is a "field always in movement", an "incessant flux" in which perceptual objects and memories are inextricable intermingled. In other words, Rogozinski recognizes the phenomenon of intertwinement in the medium of time:

> There is an indeterminate halo that surrounds my perception; there is the other side, the reverse that I do not currently see but that is presented indirectly to me by the part of the table that I do see; and there is also the table that I imagine, the table that I remember, and a whole world of memories awakened within my perceptions: Piazza San Marco and all of Venice are resuscitated by the irregular cobblestones in the courtyard of the Guermantes. Each time it is always the same table, and yet it is never the same table: the echoes and resonances are innumerable, and this irreducible diversity of the modes of givenness make the immediate unity of the thing fragment (Rogozinski, 2010, p. 139).

It is important to clarify that Rogozinski is not referring to the disturbing cases of ego-dissolution, or to any other kind of psychotic states. The crucial point of his analysis is that the sense of stable identity emerges from a pre-reflective chaos. As he writes:

> This [the transcendence of the everyday ego] supposes that the field of immanence is not pure chaos; that regularities and constant concordances come to order the flux of appearances; that different perspectives, series of divergent impressions, may also be able to converge on, intersect with, and cross one another; and that each point of intersection is constituted as a pole of identification, allowing for always larger and more stable unities to be formed. In this way, order will emerge from chaos (Rogozinski, 2010, p. 140).

Rogozinski (2010) does not arrive at the "id", but rather speaks of splintered primary sensations and diffracted egos that occur in the savage reduction. There cannot be even a fragment of experience without egoic character: "Just as with my affects, so too are impressions the same as my consciousness: *I am* my despair; I am an ego-in-despair just as I am this icy sensation" (p. 144). Following Husserl, Rogozinski does not rule out the layered concept of the ego and claims that field of immanence is a stratified terrain that contains the accumulated and sedimented layers of experiences that are superimposed on one another (cf. p. 145). He has also discovered the phenomenology of self-affection in Freud's psychoanalysis. The kinaesthetic experiences (sensations of movement, effort, resistance, or tension) and the resurgence of desires are considered to be the clear signs of self-affection. Freud stated that instinctual drives are indifferent to their objects. Rogozinski suggests that both Husserl and Freud have recognized the self-givenness of impressions prior to the otherness or transcendence of perception: "Of all my sensings, the affective, instinctual, or motor impressions are the closest to self-givenness" (p. 144). The phenomenon of self-affection and wish-fulfillment deserves a more detailed comparative study. For Freud, it is a crucial to characterize the sphere of instincts that resists to the representational capacity of consciousness. How can phenomenology deal with the non-representational layer of consciousness? In the following chapters on the phenomenological unconscious we will encounter with this problem.

Beyond the harmonizing efforts, the main dichotomy between phenomenology and psychoanalysis remains intact because of the methodological discrepancies: Husserl, following in the footsteps of Descartes, developed the method of bracketing and reduction in order to analyze the living present, whereas Freud unwittingly used hermeneutical methods in therapeutic situations. Despite the contextual differences, Zahavi (1999) argues on the one hand, that we can speak of a kind of "depth-phenomenology" owing to Husserl's reflection on passivity, and, on the other hand, he regards nonsensical the notion of "unconscious experience"; that is, the reminiscence of the self-contradictory notion of unconscious consciousness. Zahavi also argues, against Searle, that an unconscious mental state cannot be exactly like a conscious state, which implies intentionality and egocentricity but lacks conscious quality. Consciousness is not an inessential feature of an intention or emotion (p. 206). Zahavi places the Husserlian unconscious into the core of self-awareness. He argues that the "impenetrable elements" of the unconscious are inside, and not behind of our conscious experience (p. 210). Zahavi does not adhere to a reified notion of the unconscious. Instead, the unconscious seems to be a not yet thematized, pervasive affective infrastructure that influences consciousness. It makes no sense to talk about unfelt or unconscious pain. We cannot imagine "unconscious experiences" without their first-personal givenness including their phenomenal character (the what-it-is-likeness of experience). Therefore, the notion of unconscious experience is as nonsensical as the unconscious consciousness (p. 214). The merit of genetic phenomenology is that it bring to the foreground of the "persisting influence" of the past and "the very formation of concepts and habitualities"; that is, the exploration of the "obscure underground" of intentional life. In this regard Zahavi distinguishes between surface phenomenology and *depth phenomenology*. As he puts it: "The moment phenomenology moves beyond an investigation of object-manifestation and act-intentionality, it enters a realm that has traditionally been called the unconscious" (Zahavi, 1999, p. 207).

Let us turn back for a moment to the dilemma of "unconscious affect or emotion". Zahavi argued that consciousness is an inalienable element of emotions (e.g. pain experience). His Sartrean argument claims that the crucial difference is not based on the conscious-unconscious dichotomy but rather the opposition between the *thematic* and *unthematic* that matters. That is, when I have, for example, a toothache or any other kind of discomfort, I can divert my attention to something else. But in this situation the pain experience is not completely eliminated, it is continuously present in a non-thematic or pre-reflective way. As Zahavi puts it: "Just as I do not stop being prereflectively aware of my body when I converse with a friend, the pain does not necessarily cease to exist as felt just because the conversation distracts my attention from it; it simply becomes a part of the lived intentionality" (Zahavi, 1999, p. 213). The pain experience remained a first-personal givenness for the subject – with a specific phenomenal character[10] – and not transformed into an unconscious activity that lies out of the range of first-personal access. However, Freud in *The Unconscious* (1957b [1915]) does not deny that consciousness can be attributed to emotions or affects. In fact, he argues that the process of affect-transformation is a crucial mechanism of the psychic apparatus and it cannot be reduced to the thematic-unthematic distinction, because it stems from the quantitative nature of instinctual life.

As a result, the problem of affectivity and instinctual life deepens the problem of the unconscious and poses a challenge for phenomenologically inspired models. One can get acquainted with instinctual impulses via their "ideational presentations", that is, through *affective states* or *ideational representation.* Nonetheless, Freud does not deny the natural conviction that emotions are basically conscious: "It is surely of the essence of an emotion that we should feel it, *i.e.,* that it should enter consciousness. So for emotions, feelings and affects to be unconscious would be quiet out of the question" (Freud, 1957b [1915], pp. 109–110). However, the psychoanalytic practice discloses ambiguous phenomena in the psyche. For example, Freud (1957b [1915]) speaks of unconscious love, hate, anger, and even more perplexing phenomena such as "unconscious consciousness of guilt", or "unconscious anxiety" (p. 110). Affectivity is an elusive phenomenon with a protean character. For Freud, dreamwork and neurosis clearly demonstrate the transformation of affect; that is, an affect can remain in the same state, suppressed, or it can be transformed into anxiety. Moreover, the fate of affectivity is ruled by instinctual impulses. The role of repression is to govern the development of affect or to suppress it. In this regard, the meaning of *unconscious affect* can be clarified: "In every instance where repression has succeeded in inhibiting the development of an affect we apply the term 'unconscious' to those affects that are restored when we undo the work of repression" (p. 110). However, the *unconscious affect* can be contrasted with the *unconscious idea.* While the former exists only as a potential disposition in the same system without further development or transformation, the latter "continues after repression" as an "actual formation" in the unconscious" (p. 111).[11]

To recapitulate, from the viewpoint of the dynamic unconscious, Freud does not deny that an affect or emotion can be perceived by the subject. Freud does not speak of self-contradictory "unconscious experiences", but rather he addresses the

[10] Contrary to Zahavi, Sebastian Gardner supposes that we can speak of "unconscious emotions" by subtracting the fact of manifestation but retaining the efficacy of their phenomenal character (the way how they feel). Thus, an unconscious emotion can be efficacious by virtue of *how it feels.* He argues that phenomenal properties (unconscious pain, pleasure, anxiety) have a "heightened causal value" in the unconscious because they are instinctual, infantile, and fantastic (Gardner, 1991, p. 155). The argument relies on some kind of bottom-up (causal) determination that captures more faithfully the puzzle of the unconscious than the thematic-unthematic distinction.

[11] For preliminary purposes, let me point to the fact that the dispositional nature of the unconscious is of particular importance for various ideas of the phenomenological unconscious.

problem of affect-transformation that cannot be diluted by the opposition of thematic-unthematic experiences. The main problem is not the possibility of unconscious experiences in the unconscious, but rather that neurosis and dreamwork distort and misrepresent affective states due to repression:

> To begin with, it may happen that an affect or an emotion is perceived, but misconstrued. By the repression of its proper presentation it is forced to become connected with another idea, and is now interpreted by consciousness as the expression of this other idea. If we restore the true connection, we call the original affect "unconscious", although the affect was never unconscious but its ideational presentation had undergone repression. In any event, the use of such terms as "unconscious affect and emotion" has reference to the fate undergone, in consequence of repression, by the quantitative factor in the instinctual impulse (Freud, 1957b [1915], p. 110).

In my view, one of the main challenges for the phenomenologically oriented models of the unconscious is to address the problem of affect-transformation that was discussed by Freud with energetic and mechanistic metaphors. Let me remind the reader that Fink has warned against the speculative-mythical theories of the unconscious and the libido, and thus his critique thwarted the phenomenological interpretation of affect-transformation.

Another criticism of the naturalistic interpretation of the psyche comes from the classical phenomenological-hermeneutical reading of psychoanalysis. Brudzińska highlighted the fact that Ricoeur (1970) and Hebarmas (1971 [1968]) examined Freud through the lens of interpersonal communication and they regarded the therapeutic situation as a symbolic meaning-bestowing process. The main idea of psychoanalysis, according to phenomenological-hermeneutical approaches, is to discover "sense-connections". The hermeneutical approaches dismissed the idea of drive-type energies: "How meaning comes from energy is a question that falls by the wayside, though" (Brudzińska, 2012, pp. 29–30). But Brudzińska claims that the symbolic interpretation of meaning in hermeneutical approaches does not rule out entirely the possibility of different phenomenological interpretations of the psychoanalytic assumptions. I am in full agreement with Brudzińska's proposal, and I will try to investigate the sub-layers of meaning-bestowing processes through body memory and with the help of the phenomenological unconscious (These interconnected issues are to be addressed in Chap. 4). According to Brudzińska (2012), psychoanalysis examines a special kind of "experiential consciousness" with a unique intentional structure. Psychoanalysis may offer different ways for phenomenology to unravel the constitution of the self and the shared lifeworld; thus, it can also be understood as a "lifeworld phenomenology" (p. 30, 46).

There is a consonance between Brudzińska and Zahavi in rejecting the reified view of the unconscious. The unconscious cannot be understood as a kind of residuum of repressed affects, but rather it is a "productive complex" that comprises "tendencies, strivings and forces which come to light psychically and become accessible to study in the form of an affective wish representations" (Brudzińska, 2012, pp. 45–46). The main challenge for phenomenology is to reinterpret the problem of "strangulated" (*eingeklammert*) affect. Recent approaches tend to employ a genetical-phenomenological framework to integrate the dilemmas of drives and affectivity. Brudzińska (2012) defines strangulated affects as such contents that have "affective power" without explicit representation; non-representational affective states are separated from consciousness, i.e., conscious recollection, and accompanied by hysterical symptoms. The aim of the therapeutic intervention is to find a representation for the repressed affect, and in the cases of successful procedures the extension of consciousness ensues (p. 31). Freud rejected the method of hypnotic abreaction and developed the techniques of free association and talking cure, with the help of Breuer's and Anna O's contribution. According to Brudzińska (2012), the therapeutic considerations of Freud, that is, the meticulous analysis of the internal reality of patients through dream, memory, and imagination, pulls as toward the phenomenological problem of *passive genesis*. In the recent analyses of the phenomenological unconscious, the common ground between psychoanalysis and phenomenology is *passive genesis*, i.e., the problem of affective-associative synthesis (pp. 31–32; cf. Kozyreva, 2017, pp. 70–113). The phenomenon of passive genesis does not refer to the disembodied level of the pre-intentional layer of consciousness. Brudzińska argues that phenomenology, guided by the results of psychoanalysis, could gain insights into the psychosomatic nature of intentional accomplishments.[12] In this respect, both Freud and Husserl, can be seen as progenitors of the embodied mind paradigm, since, as we shall see, embodiment and affectivity (especially the desiderative and self-affective aspects of conscious life) play an important role in characterizing the laws and patterns of intentional consciousness.

The task of Part IV will be to explore the dispositional and traumatic layers of the unconscious and link them to the phenomenological problems of self-affection and time-consciousness. In addition, affect-transformation, that closely connected to time-consciousness, has a pivotal role in self-constitution. The above analysis has shown that the dynamic unconscious poses a challenge to phenomenology, but, as we shall see, several alternatives of hybridization can be envisioned. Before delving deeper into the problems of the phenomenological unconscious and self-constitution, it is worth bear in mind that the dynamic unconscious cannot be identified with the passive genesis of meaning and cannot be eliminated by the Husserlian paradigm of association. Nevertheless, the notion of the *affective relief*, which also had its origin in Husserlian phenomenology, can constitute a viable solution to the phenomenological interpretation of the dynamic and repressed unconscious.

[12] Somatic states also play a fundamental role in the intentional life of consciousness: "The conception of and praxis in the psychoanalytic setting in my opinion enables the recuperation of insights concerning the psychosomatic basis of mental achievements, in the sense of the inner psychical, comprehending experience, not in the sense of an external causal genetic analysis of the bodily influences on psychic experience. Moreover, we gain intuitions for handling the question as to how the somatic state is an effective moment in intentional accomplishment" (Brudzińska, 2012, p. 33).

3.2 Affective Relief and Repression

The problem of affectivity plays a crucial role in Husserl's *Analyses Concerning Passive and Active Synthesis.* The aim of the genetic phenomenological investigations is to unravel the affective-associative syntheses that constitute the sublayer of conscious intentional relation to the world. According to Kozyreva's (2017) interpretation, Husserl saw an inseparable connection between associative and affective syntheses. An "affective associative connection" can restore the affective vivacity of a past event. Kozyreva contends that this special relation to the past might function in an a-temporal manner. The phrase a-temporal here means that the affective structuration of experience can be contrasted with the formal temporal syntheses of inner time-consciousness (p. 134). The problem of affective syntheses is closely related to the dilemmas of forgetting and remembering. For Husserl, passive synthesis can be understood as a *pre-egoic activity* that constantly schematizes and enlivens the conscious phenomenal field. Husserl has discovered the constant flow of unnoticed associations at the periphery of conscious attention. While he was inspired by Gestalt psychology, his phenomenological investigation is not based on empirical studies; rather he applies layers of reductions and the "abstractive" method in order to explore the problem of association and to distance himself from everyday intuitions or from mechanistic psychology (cf. Mishara, 1990, p. 34).[13] Husserl made a distinction between association and originary association (*Urassoziation*), the former is the integral part of time-consciousness as reproductive and anticipatory association, the latter denotes a deeper unconscious mechanism embedded in the flowing land living present. The originary association is constituted by the synthesis of similarity, continuity and contrast; these are the fundamental rules of the pre-temporal passive synthesis (Mishara, 1990, p. 38). It is important to note, again, that association and affectivity go hand in hand when Husserl tries to describe the structure and the dynamic of the living present. In terms of synthesis, he distinguishes between *affective synthesis* (i.e., conscious or becoming conscious) and *pre-affective synthesis*, which is entirely unconscious, although the subliminal pre-affective fusion (*Verschmelzung*) of hyletic data occurs as an excitation for the ego.

Pre-affective unity formation is an integral part of association.[14] Husserl differentiated between various levels of association and different scholars have offered

[13] Mishara highlights the fact that the aim of Husserl's so-called *abstractive method* is to unravel the deepest layers of experience; thus, it can labelled as an "archaeological effort". This tendency is a common ground between Freud, Husserl, and Kafka. They all rely on the German Romantic literary tradition where the metaphor of the unconscious frequently occurs (e.g. as "night" and "mine") (Mishara, 1990, p. 36).

[14] Let me emphasize, again, that pre-affective or pre-egoic constitution is governed by the laws of unity and contrast. Pre-affective formations are totally egoless tendencies, and they can be approached only by the "abstractive reduction", that is, they cannot appear in clear intuition, they are only the result of inferences based on well-developed, well-known perceptual Gestalts. Kozyreva argues that, in this case the limit of the describable is reached (cf. Kozyreva, 2017, p. 123).

different categorizations. Here, I adhere to Kozyreva's categorization who differentiated between three types of associations: (1) reproductive; (2) anticipatory; (3) primordial (or *Urassoziation*). In the subconscious level, affective syntheses and pre-affective unity formation are the two main functions of the primordial association (Kozyreva, 2017, pp. 100–101). Our perceptual field is permeated with the rivalry and competition of affective tendencies. On the one hand, our perceptual field is constituted by the ongoing competition between perceivable stimuli and conscious responses. On the other hand, another competition can be observed among subconscious affective tendencies. Kozyreva argues that a more elementary affective tendency always present against the background of "concurrent affective tendencies" (Kozyreva, 2017, p. 118; cf. Zahavi, 1999, p. 119).

It is important to note that Husserl's metaphors are distinctly different from the psychophysical descriptions. He does not emphasize the role of a specific stimulus, but rather speaks of relative height or contrast in the so-called *relief structure* of the perceptual field.[15] Mishara describes the relief structure of perception in the following way:

> Every sensory stimulus as it enters the field of the impressional present exerts an attractive force or pull on the ego in a "tendency" to be noticed. Husserl calls this "affection." This affective "pull (*Zug*)" can only be subsequently neutralized by the ego's "turning towards" (*Zuwendung*), adverting its glance, to the affecting stimulus. The ego thereby exercises, as it were, a "counter-pull (*Gegenzug*)" which balances the original force. When the stimulus passes out of the living present, it no longer exerts the same affective force, but diminishes in its potentiality to affect the ego (Mishara, 1990, p. 38).

The ongoing syntheses of affectively charged associations constitute the topological surface of the living present. Mishara summarizes Husserl's passive synthesis with a geological metaphor: the accomplished affective syntheses constitute the highest peaks of the relief structure. These are the Gestalts that can burst in the sphere of consciousness. However, the result of pre-affective synthesis are not available to conscious awareness; metaphorically speaking, they constitute the valleys of the relief structure (Mishara, 1990, p. 39).

The crucial question to ask is whether there is a connection between the Husserlian phenomenological unconscious (i.e., passive syntheses) and the Freudian Unconscious? Mishara does not directly compare the Freudian unconscious and the Husserlian marginal consciousness (i.e., the phenomenological unconscious). However, he offers a striking comparison with regard to traumatic experiences:

> The ego is awoken into its noticing activity of turning towards by the affective force of the greatest prevailing contrast value relative to the others in the sense-fields. When the ego is first awoken by a contrasting affective force, it is at the mercy of the hyletic forces. Just as

[15] Furthermore, Husserl does not employ Pavlovian or other reflex-model to explain the orientation of the ego; rather the so-called "dialogical terms" are prominent in his papers. Another departure from empirical findings is that Husserl, in contrast to psychophysics, does not consider the notion of *absolute threshold*. For Husserl, as Mishara has shown, the threshold between conscious and unconscious processing is determined by the actual pattern of affective forces at any given moment (Mishara, 1990, pp. 39–40).

> the Freudian ego is not master of its own house, so "the ego is not master of its own field. It can be completely overpowered by it," as in traumatic experiences of sudden physical or emotional pain. It is attracted or repulsed by the "instinctual preferences," or sudden contrasts, in the very process of becoming conscious of them. The first relationship of the ego to its field is neither neutral nor voluntary, but one of exposure and vulnerability (Mishara, 1990, p. 39).

Mishara argues that in cases of intense physical and emotional trauma the continuous flow of experiences can be halted and the ego is trapped in the immutable now. In connection with this phenomenon, the role of retroactive awakening will be utilized in order to explore phenomenologically the preconscious tendencies of trauma processing (see Part IV).

For present purposes, let us examine the problem of the affective relief. On the one hand, the affective relief is constituted by the syntheses of instinctual drives, sedimented dispositions, and desiderative preferences; on the other hand, it is capable of modifying and schematizing perceptual experiences. To complicate matters further, Husserl is often ambiguous about the nature of affection. First, he speaks of external affective forces exerting an influence on the ego. Simply put, the ego has the tendency to turn toward excitations that come the external world. In this regard, Fuchs, based on Merlau-Ponty's phenomenological unconscious, has reformulated the psychic mechanism of *projection*. Instinctual drives may schematize the surroundings according to the possible ways of satisfaction (Fuchs 2012, p. 77). The Husserlian notion of affectivity is also compatible with this idea, but it cannot be reduced only to the projection of the drive-motivated fantasies. Behnke (2008) emphasizes that Husserl's *affectivity* is a double-sided concept: in the perceptual field we encounter varying degrees of affective power, but affectivity does not consist solely of alluring saliences, but also presupposes the sensitivity and vulnerability of the perceiver (p. 48). Behnke speaks of "sedimented general readiness" that comprises the various responses to the affective impact of the environment.[16]

As we have seen, there is a constant interplay and competition between the affective forces in the process of the constitution of the perceptual field. Therefore, affectivity, including the pre-affective synthesis of hyletic data, is embedded in our perceptual field, and these processes cannot be identified with Freud's primary processes or with the fundamental conflict between the reality and pleasure principle. Nevertheless, as Nicholas Smith (2010) has shown, Husserl in his late manuscripts seriously considered the role of the Freudian unconscious in the constitution of living present (further elaboration ensues). In sum, the phenomenological unconscious, i.e., the affective relief, can be regarded as the temporally extended and affectively charged medium or horizon of perceptual experience. From this

---

[16] "But above and beyond one's response on any particular occasion, there can be a sedimented general readiness not only to receive or to reject – to be open for the affective invitation or closed off from it – but also to be moved and go along with the movement, or to refuse to do so. Affectivity in the sense of the experiencer's ability to be affected in this double way is thus a practical condition of possibility for the affective power of the non-I to come to fruition in an actual affective event" (Behnke, 2008, p. 48).

Husserlian perspective, the phenomenological unconscious is the sedimented system of meaning structures, and its lowest layer is the null zone of affectivity (Husserl, 2001, p. 216; cf. Kozyreva, 2017, p. 159).

Husserl's *Analyses* focuses on the temporal and spatial synthesis of sensual (hyletic) matter; that is, on the formation of perceptual Gestalts and the horizontal structure of. The result of his *Analyses,* however, was not only a special theory of perception, but also an attempt to clarify the relationship between instinctual drives and perceptual interests. The ego is stimulated by affective forces that are responsible for the contrast between shape and background, or they push or pull toward different intentional objects. This subconscious mechanism implies the thematic and pre-thematic distinction: "In every living present that is looked upon universally, there is naturally a certain relief of salience, a relief of noticeability, and a relief that can get may attention. In this case, we accordingly distinguish between background and foreground. The foreground is what is thematic in the broadest sense." (Husserl, 2001, p. 215). Husserl tries to dive into the underground of intentional objects; he wants to describe the "constitutive process of becoming". In this respect, he begins to talk about pre-intentional hyletic objects, such as "object phases" and "sensible points", which are elementary building blocks in the process of object-constitution (Husserl, 2001, p. 213). The rivalry of affective forces is governed by the process of *suppression*: in the case of intense concentration or in hearing a loud noise of an explosion, certain affections sink into the background and others come to the foreground. The act of suppression often works without conscious awareness due to pre-affective syntheses. At the same time, he remarks that the repressed affective force is not completely terminated, but continues to live on in a state of latency (cf. Mishara, 1990, p. 43).

Now, it is clear that Husserl defines affection in a distinctly different way from Freud. for Husserl, affection does not refer primarily to the instinctual needs of amorous behavior that takes place during the child's primary object-choice period. Moreover, Husserl's starting point in his *Analyses* is the living present which is "structured as ready-made" and has its own affective formation (Husserl, 2001, p. 214). Furthermore, the living present is constituted by the incessant flow of primal impressions, which are also the source of affectivity: "The primordial source of all affection lies and can only lie in the primordial impression and its own greater or lesser affectivity. The lines of affective awakening, or again, the lines of the maintenance or propagation of affectivity proceed from there" (Husserl, 2001, p. 217). Another synonym for affection in Husserl's *Analyses* is "affective height", or the "fresh" impressional moment of the now. Husserl invites us to examine a concrete sensible datum, for example a protracted sound. The extended tone gives us a perfect example to illustrate the working of the Husserlian internal time-consciousness. Husserl argues, albeit not without hesitation, that in conscious intuition we can be aware of the modification of primal impressions into "fresh retentions". The latter term refers to the fact that a small part of a tone remains closely connected to the present moment, in other words, the present has a temporal extension despite the incessant retentional flow in which primal impressions are transformed into retentions. However, despite the conservative aspect of fresh retention the temporal (and affective) modification of the sensible datum cannot be halted forever. The actual impression reaches its affective height and begins to diminish into the oblivion of the affective emptiness (to the phase of *nil*, as Husserl has stated). Then, a new sensible impression gains vivacity and sticks to the currently diminishing sound. However, Husserl (2001) makes a striking observation: when the intuitability of the new sensible datum becomes nil (i.e. the fresh impression completely fades away), the affective force remains intact. The primal impression loses its contours, but despite the clouding over of the perceptual object in the retentional flow the affective force remains intact at the periphery of consciousness (p. 218).

In this context, the unconscious is a *marginal consciousness* characterized by the hidden laws of synthesis that govern the constitution of intentional objects. Affectivity is distributed horizontally in the living present and diachronically in time-consciousness.[17] In order to get a clear picture of affectivity, I adhere to Kozyreva's succinct definition: "By affectivity, therefore, I understand the varying vivacity of subjective experience in what concerns not only is impressional organization but equally its connectedness with the past and its openness towards the future" (Kozyreva, 2017, p. 119). That is, from phenomenological viewpoint affectivity may or may not include the problem of "strangulated affect". An opposition begins to take shape between the vertical and horizontal conception of the unconscious. As we shall see, Thomas Fuchs demonstrated the usefulness of this elementary distinction in explaining the workings of implicit body memory.

The main feature of the affective relief is its horizontal structure that schematizes perception by means of contrast and difference. As Mishara puts it: "The unconscious may be initially thought in Husserlian terms as the bottommost, or deepest, parts of the affective relief structure, which enable differences between coexistent and successive *Gestalten*" (Mishara, 1990, p. 52). Brudzińska argues that we can observe this relief structure from the outside as well as from the inside. From the outside the relief is the precondition of affectivity; a terrain on which affective elements can be noticed by the I. From the inside the relief implies "the *subjective preparedness* or *tendency* to permit one's being affected" (Brudzińska, 2012, p. 44). Brudzińska (2012) points to the schematized nature of responsiveness: the I that reacts to the impulses of hyletic data is not an empty transcendental pole but rather an embodied subject. That is, the reactive schemas to the prominences of the relief structure are determined by bodily drives, biographical or situational context, or the other person. Brudzińska also highlights the biological, anthropological, and cultural determinants of the reacting I (p. 44). In the *Analyses* Husserl tried to explicate the pre-affective tendencies of the unconscious. He distinguished between actual affection and deeper affective tendencies – the latter implies the sphere of potentiality on which the competition of affective tendencies takes place. Kozyreva observes that Husserl also speaks of "affectively charged predispositions" including feelings, drives, and instincts (Kozyreva, 2017, p. 117). With Husserl's own words: "On the

[17] A more detailed analysis of the relation of affectivity and time-consciousness will be elaborated in Part IV.

one hand the emergent affection is functionally co-dependent upon the relative size of the contrast, on the other hand, also upon privileged sensible feelings like a passionate desire founded by a prominence in its unity" (Husserl, 2001, p. 198).

The general idea behind of the above-mentioned claims is that the synchronic unity of our perceptual field is co-dependent on the affective and pre-affective syntheses that constantly take place in internal time-consciousness. The main conclusion here is that the affective relief indirectly points to an underlying schematization process: the affective predispositions, the rules of responsiveness, the reactive schemas can be understood as the *implicit rules* of the affective relief. If we think of the affective relief structure in terms of the correlation of consciousness and perceptual object, then, on the object side, we can discern the multitude of perceptual objects with their own affective allure, and, on the subjective side, we can imagine an affective unconscious that governs or crystallizes not only *what* can be seen or felt, but also *how* one can see or feel in certain situations. The affective allure (i.e., when the ego turns toward an intentional object) of the environment and the underlying rivalry of affective tendencies (which constitute prominences via the laws of similarity and contrast) are basic organizing principles of the phenomenal field. Without these kinds of passive syntheses, the phenomenal field and the whole conscious life would be a chaotic flux of disconnected and disintegrated experiences. In my view, the affective relief is of tremendous significance in the explanation of self-constitution. The affective relief can be considered to be the pre-reflective subsoil of self-awareness and an alternative route for explaining the genesis of phenomenal consciousness. Therefore, I side with Kozyreva in pursuing the "affective constitution of the pre-reflective selfhood".[18] Furthermore, as we shall see in the discussion of the ipseity-disturbance model in Chap. 4, the Husserlian concept of passive syntheses could play a significant role in explaining the processes of disintegration and reintegration of the self.

The affective relief is the pre-thematic dimension of the perceptual field. How can we reconcile the tension between the dynamic unconscious and this kind of phenomenological unconscious (i.e., the marginal consciousness)? Two lines of argument will be considered. One argument concerns the process of suppression in affective and pre-affective syntheses, and another argument discusses the late Husserl's sporadic considerations on psychoanalysis and Freud.

The affective relief is created by contrasting affective tendencies and constitutes the field of prominences in our phenomenal field. Husserl's recurring example is the blast of an explosion, which rearranges the acoustic field and represses other sensual fields at once. Suppression can occur in more subtle ways. For instance, Husserl mentions the situation in which we hear the sounds of a passing car but also the tones of a song and in the end the song wins out (Husserl, 2001, p. 197). The relief

[18] "The role of affectivity acquires special original meaning when applied to such issues as the affective constitution of the pre-reflective selfhood; the formation of affective unities; and the clarification of the affective dimension of memory-related phenomena (namely, retentional modification, recollection and constitution of the past). Moreover, in the larger perspective, it amounts to a new approach to consciousness, the unconscious, and subjectivity itself" (Kozyreva, 2017, p. 205).

structure is the affective undulation of the living present; therefore, repression does not push out hyletic elements into the "cohesive sphere of the distant past" but only forces them into the background. Husserl suggests that the suppressed (i.e., covered up) affections lie in the background of consciousness, waiting to be awakened. However, this kind of repression in the affective relief structure could also lead to a paradoxical result; it does not annihilate the rival hyletic data in question, but rather it increases the vivacity of the experience:

> In this case, a special repression takes place, a repression of elements, which were previously in conflict, into the 'unconscious,' but not into the integrally cohesive sphere of the distant past; by contrast, in the living conflict, repression takes place as a suppression, as a suppression into non-intuitiveness, but not into non-vivacity – on the contrary, the vivacity gets augmented in the conflict, as analogous to other contrasts (Husserl, 2001, 514–515).

In Husserl's examples a louder tone makes clearer the other softer tones, or we can more easily notice a string of light when one light-source is suddenly becoming brighter than the others. There are two main dimensions of the *propagation of affection*. On the one hand, Husserl speaks of the horizontal effect of high contrast that illuminates other suppressed elements of a Gestalt, on the other hand, the temporal propagation of affection is the crucial aspect of the universal phenomenon of association. The above-analyzed processes highlight the important role of affectivity in organizing the impressional sphere into meaningful intentional objects. Therefore, it is important to note that, repression in the affective relief is drastically distinct from the Freudian repression. Husserl claims that "repression takes place as a suppression", therefore it is necessary to draw a sharp line between Freud's theory of *repression*, which presupposes a compartmentalized psychic apparatus, and Husserl's theory of *suppression*, which exerts its influence in the phenomenal field. In analyzing the problem of repression in more depth, we will also consider the late Husserl's comments on psychoanalysis.

3.3 The Implicit Unconscious

While Husserl was inspired by the transcendental philosophy of Kant, the genetic phenomenological turn leads to the problem of the layered concept of self and self-constitution. Husserl has also arrived at the problem of the unconscious and the instinctual dimension of the empirical person. With regard to the affective relief, however, we have seen that Husserl's phenomenological unconscious is basically the marginal consciousness. The concept of the affective relief is closer to Leibniz's *petites perceptions* (subconscious, unapprehended perceptions) than to Freud's unconscious. However, as was discussed above, the concept of the affective relief depicted an instinctually schematized perceptual field, in which prominences (i.e., fields of interest) are schematized by sedimented habits, dispositions, desires, and moods. At first glance, these considerations can be compared to Freudian libidinal interests and cathexes. However, sexual energies do not play a key role in Husserl's affective relief concept. The idea of affective relief is more in line with the discoveries of Gestalt psychology than with psychoanalysis. Another discrepancy arises when we realize that there is only one consciousness for Husserl; there is no sign of the Freudian duality between perceptual-consciousness (Pcs) and the unconscious (Ucs). The Freudian tripartite structure is also absent, of course. In the end, the two models of mind are practically incommensurable. Despite the irreconcilable differences, Nicholas Smith, in his ground-breaking work, has provided a compelling argument for reconsidering the Freudian unconscious (including repression) in phenomenological terms. Smith draws attention to Husserl's unpublished manuscripts in which he specifically addresses the problem of the Freudian repression.

Before delving onto the Husserlian analysis of repression, it is worth considering Freudian repression, which is inseparable from and sustained by *resistance* that can be indirectly observed in the analytic praxis. Smith (2010) claims that repression is not an unambiguous term in the Freudian oeuvre since several shifts of meaning can be observed: (1) first, repression means a kind of defense mechanism, a "double forgetfulness" that arises in hysterical amnesia; that is, one has to forget the act of forgetting of a distressing idea or affection; (2) later, repression was used in a wider sense referring to several defensive processes of consciousness; (3) another usage subsumes repression under the wider context of defense-mechanisms that protect the ego from the demands of drives; and, finally, (4) Smith mentions the general definition of repression that was elaborated in Freud's metapsychological papers. In the last sense, repression "denotes the general effort of pushing something away from consciousness, and its purpose is simply to avoid unpleasure by keeping these now unconscious representations away from consciousness. In these texts, repression is analyzed as being the perhaps most important vicissitude of the drive (*Triebschicksal*)" (p. 44).

Furthermore, Smith accentuates that Freud in the *Ego and the Id* (1990b [1923]) regards the repressed as "the prototype of the unconscious" and in *Repression* (1957a [1915]) he presupposes that originary (or primal) repression (*Urverdrängung*) determines the first representative of the drive and constitute the "first pole of attraction" for secondary repressions. Thus, the originary repression works as an "invisible magnetism" for subsequent repressions. To sum it up, the process of original (or primal) repression is beyond observation and it is the prototype of the unconscious for Freud. However, the originarily repressed but unfathomable drive-representation constitutes a fixation point in the psyche, which also serves as a point of attraction for repeated representations in dreams and hysteria (Smith, 2010, pp. 45–46). Smith (2010) observes that in other places Freud talks about many originary repressions which constitute several fixations in the psyche (p. 45). Smith demonstrates that if we examine the problem of repression proper (i.e. secondary repression), then we have to reconsider the ideal case of the virtually inaccessible unconscious. But, from therapeutic insights, the situation ensues that the repressed is moderately accessible; that is, "phenomenological retrieval" is possible since the repressed entails a "constant striving to break through to become conscious" (p. 58). In *Repression* Freud speaks of the dynamics of constant pressure and counter-pressure between the repressed and consciousness.[19] Smith argues that the significant facet of repression is *dual motility*: there is a repulsion from the direction of consciousness in order to compartmentalize the repressed, but, at the same time, the originarily repressed functions as attraction point for associations and dream content (p. 46). Smith argues that Freud caried out a "phenomenology of extended consciousness" because of the impossibility of ideal repression (Smith, 2010, p. 49). That is, consciousness has a clear relationship to the repressed, since the act of pushing away the repudiated content presupposes a connection between the conscious and the unconscious. Smith (2010) concludes that repression is best understood phenomenologically as an "intra-psychic resistance against presentifying consciousness" (p. 49). This conclusion suggests that the repressed content strives for manifestation and this mechanism can be demonstrated by the peculiar psychosomatic symptoms of hysteria or dream content. Nonetheless, the modified return of the repressed poses a challenge for phenomenology. Smith accepts the possibility of dissociation between the repressed affect and the conscious idea; that is, what (later) expresses itself as repressed content is not the same as the originarily repressed. The intriguing question is whether a phenomenological framework could account for the modification of the repressed?

An intriguing turn can be observed in Husserl's *Ideas II* (1989) and in his late manuscripts, when he takes into account the intentionality of drives and the habit structure of the personal ego. Talia Welsh (2002) agrees with Zahavi that it is appropriate to distinguish between surface and depth phenomenology. However, she has reservations against the harmonizing attempts; that is the alterity of the repressed cannot coherently be integrated into the phenomenology of passivity. Welsh states that the Freudian unconscious behaves quite independently from consciousness and often works against the ego's aims and ambitions (Welsh, 2002, p. 166). According to Welsh's interpretation, the sedimented remnants of the past are dead; intentionality and affective force is absent in the unconscious domain. Furthermore, the condition of possibility for the awakening of the past always springs out from the living present; there is a stimulus or clue that initiates the associative connection with the distant past. Welsh's description of access to the past is consistent with contemporary memory research that demonstrates the importance of retrieval cues in remembering (cf. Schacter, 1996, pp. 58–62).

In contrast to Smith, Welsh considers the Freudian unconscious as a reservoir "barred from entering into conscious life" that is at the same time active and affective. Therefore, the dynamic unconscious cannot be identical with the unthematic or pre-thematic sphere of consciousness, that is, with the flip side of consciousness

[19] "We may imagine that what is repressed exercises a continuous straining in the direction of consciousness, so that the balance has to be kept by means of a steady counter-pressure. A constant expenditure of energy, therefore, is entailed in maintaining a repression, and economically its abrogation denotes a saving. The mobility of the repression, incidentally, finds expression also in the mental characteristics of the condition of sleep which alone renders dream-formation possible. With a return to waking life the repressive cathexes which have been called in are once more put forth" (Freud, 1957a, [1915], pp. 89–90).

(Welsh, 2002, p. 174). For Welsh, the Husserlian unconscious encompasses only the Freudian pre-conscious, but cannot be identified with the level of the inaccessible unconscious. Welsh further argues that a more glaring discrepancy is the fact that Freud characterizes the genesis of the unconscious (i.e., the id) through the child's developing sense of self. When the object of pleasure is out of reach, a fissure takes place in consciousness; the ideational representative of the instinctual drive cannot reach the doorway of consciousness and primal repression occurs. While the free-floating energy of the repressed drive for pleasure remains active, the ideational representation is hidden from the sight of consciousness. In other words, consciousness is *ignorant* of the unconscious and vice versa, as a result, the subject may experience psychosomatic symptoms. However, the activations of libidinous energy and the emergence of hysterical symptoms are not necessarily the signs of a communication between the conscious and the unconscious. The Freudian ego strives for coherence and tries to adapt to the symptoms, but there is no way for a "harmonious unity" between the demands of the id and the reality principle of the ego (Welsh, 2002, pp. 175–176).

The unconscious is not a reservoir of nonconscious mental states; rather it is irrational, alien to the ego, and totally devoid of self-awareness. The main contradiction between Smith and Welsh lies in the characterization of the nature of the unconscious. Smith tries to find a common ground between Husserl and Freud, when he argues that what is repressed is not "dead" but remains affectively efficacious. However, Welsh's "radical alterity" thesis, as Smith refers to it, states that the Freudian unconscious is not open to investigation (Smith, 2010, pp. 54–55). The crux of their arguments is that passivity cannot be equated with the deepest layer of the Freudian unconscious. Moreover, for Welsh (2002) the Husserlian one-system is incommensurable with the Freudian dual system, but Smith seeks the opportunity to extend the Freudian rudimentary notion of the "conscious" by means of the paradigm of intentionality. Smith's radical claim is that the entire practical-clinical endeavor of psychoanalysis presupposes the Husserlian conclusion that "what is repressed remains in consciousness" (p. 54). Is it possible to incorporate the Freudian unconscious into the sphere of Husserl's marginal consciousness? Whether the affective relief structure is suitable to represent the unfathomable depths of the unconscious (and especially the id)? Smith and Kozyreva, and even Welsh offered intriguing proposals to address these problems. Their arguments are based on the fundamental role of association in Husserlian phenomenology.

As we have seen, it is a fundamental Husserlian thesis that unconscious associations are the genetic foundations of the higher-level intentionality and repression is a necessary mechanism of the living present. However, as discussed above, for Husserl repression is only an alternative formulation of suppression that regulates the perceptual sphere.[20] The starting point of Smith is that the deeper genetical-phenomenological investigation of consciousness may give rise to the phenomeno-

[20] Smith enumerates the frequently used terms of Husserl in this context: *Verdeckung* (covering over), *Deckung*, *Unterdrückung*, *Hinunterdrückung* (suppression), and *Hemmung* (inhibition) (Smith, 2010, pp. 56–57).

logical clarification of the Freudian repression (Smith, 2010, p. 56). According to both Smith and Kozyreva, the passive (i.e., unnoticed) associations are not affectively dead or completely empty intentions (these are misleading Husserlian metaphors). As I understand it, both Smith and Kozyreva have developed a special kind of *implicit unconscious* that tries to demonstrate the presence of vivacity and affectivity below the level of conscious awareness. They attempted to conceptualize the non-representational aspects of consciousness. As we have previously seen, in the affective relief structure of the phenomenological unconscious, the sensual elements show a fluctuation due to the suppression or enhancement of certain hyletic elements. Therefore, the sedimented "dead shapes" are prone to become efficacious again; that is, for Husserl, dead shape only means that certain experiential content is out of the sphere of streaming life or rendered non-intuitive (Smith, 2010, p. 55). Husserl argued in his D14 manuscript that the sedimented sphere of passivity is imbued with implicit motivations for reproductive realization or modification. Based on a thorough textual analysis, Smith concluded that Husserl came close to the Freudian repression by introducing a special form of unconscious activity:

> The sedimented "unconscious" can thus also function as the initiator of new processes of awakening. With the givenness of this reversed, genetic mode of ordinary intentional modification, we accordingly have movements of intentional modification being initiated from both directions, i.e. going out from the present towards that which is past, but also from the past to the present (Smith, 2010, p. 240).

The feature of unconscious activity becomes a common denominator between Freud and Husserl. The Freudian return of the repressed is one manifestation of this "passive activity", and the oscillatory movement of intentional modifications, initiated from the past or from the present, is another mode of activity that presupposes the passive infrastructure of the unconscious (Smith, 2010, p. 240). In general terms, the major difference between the Freudian and Husserlian unconscious is blurred, since both conceptions presuppose an unconscious with an implicit tendency to become conscious. Kozyreva (2017) argues along similar lines when she claims that every remembering (i.e. reproductive consciousness) has an implicit-affective motivational basis. These alternatives of the phenomenological unconscious suggest the idea that sedimented mental contents remain vivacious and affectively charged even in a non-representational form. Smith speaks of "sedimented concrete complexes" with a Freudian allure, and highlights that Husserl, in his D14 manuscript further, develops the problem of unconscious motivations first elaborated in *Ideas II*:

> Husserl analyzes the possibility of sedimented concrete complexes (lived experiences, feelings, thoughts and kinaesthesia etc.), whose affective force does not diminish with time. Contrary to normal retentional procedure these complexes live on, actively engaging other sedimented events and attracting attention also from presentifying consciousness; this strange living on Husserl calls "perseverance" (*Perseveranz*) (Smith, 2010, p. 228).

As we have seen, the repression of affection is central to Husserl's concept of affective relief. The relief, drastically speaking, is the field of consciousness including its marginal dimension. For a mental state to be conscious is to be part of the relief. Smith and Kozyreva, in different ways, imagine the unconscious as an implicit system in which affective rivalry is prominent. An unbearable pain suppresses or covers over all other aspects of the relief, but the suppressed affects are still potentially present. Smith and Kozyreva accept the possibility of repressed affective forces in the relief structure of consciousness. As Smith (2010) puts it: "Repressed feelings and drives can according to this analysis 'persevere' in the unconscious and still exercise an affective calling upon the I, which means that a dynamic relationship between consciousness and the unconscious has been indicated also from the point of view of feelings and drives (and not only perception)" (p. 235). This "active" passivity clearly is also present in the D14 manuscript: "The 'unconscious' stream of memories, of the sedimented in continuous overlapping, is not eternal frozen immobility: 'it does not remain at rest'" (p. 235). But how can we take into account phenomenologically the deepest layer of the unconscious which is beyond any recollection or representation? One way is to dismiss the idea of the Freudian *censor* and to deny the possibility of inaccessible content; this line of thinking frequently appears in the harmonizing attempts.

As previously discussed, Welsh tried to take seriously into account the problem of complete alterity or otherness of the unconscious. Her alternative explanation dismisses the spatial metaphors of the unconscious and relies on time-consciousness. Welsh suggested that in order to tackle the problem of the *barred* unconscious, we have to introduce a new layer of retentions: the so-called sedimented but "active repressed retentions" (Welsh, 2002, p. 181). Welsh (2002) provided an alternative picture of the psyche based on the analysis of internal time-consciousness. There are three different layers of retention: first, there is the level of "near retentions" embedded in the typical potential-retentional structure of consciousness; second, the sphere of "inactive retentions" can also be envisioned at the deeper level. The latter is identical with the Freudian preconscious and the Husserlian unconscious. And third, there is the layer of "active repressed retentions", which poses a challenge to phenomenology since it can only be approached indirectly by means of clinical, psychotherapeutic observations. Welsh underscores the clinical and practical aspects of the Freudian psychoanalysis and states that the Husserlian unconscious is unable to incorporate the behavioral (and thus therapeutic) aspects of the unconscious (p. 181). Unlike Welsh, Smith highlighted the oscillatory relation between conscious and unconscious processes and found in the Husserlian manuscripts the thesis of the perseverance of affects, drives, and motivations, which are constantly striving for conscious recognition. In contrast to Smith, Welsh draws attention to the irrationality of the id and to the bewildering symptoms of hysteria, phenomena that are beyond the reach of the ego, which constantly strives for coherence and adheres to the reality principle. Smith gives an alternative answer to the conflict between the practical and theoretical approaches to the unconscious. Interestingly, there is an explicit reference to psychoanalysis in Husserl's *Ideas II*, where he ruminates over the passive motivations and the obscure background of consciousness.[21] Smith comes up with the idea that a division of labor can be imagined between the phenomenologist and the psychoanalyst. The psychoanalyst can reveal the "buried motivations" and the phenomenologist can give a general account of the role of association. As a result, the Freudian method of free association may be a specific form of transcendental aesthetics:

> Phenomenology could here show that "free association" is a part of a scientifically understood transcendental aesthetics, by inscribing it in a larger and more coherent framework. Such a framework would consist of originary association at the lowermost level of constitution which accounts for the ever new structuration of the living present. This would also interact with what becomes sedimented, and given that the psychoanalytical "repressed" could be interpreted as something that lives on in sedimentation, it would also be a part of unconscious motivations that associatively communicate with the living present (Smith, 2010, p. 224).

Smith has even found the problem of the strangulated affect in one of the manuscripts of Husserl. In a short section of the manuscript E III 10, Husserl ruminates over the transformation and inhibition of the sexual drive and explicitly refers to Freud. Smith (2010) accentuates that: "...Husserl notes, the repeated experience that fulfilment of a drive is inhibited (for whatever reason) may eventually become a habituality, a second nature, so that the whole consisting of drive, kinaesthesia and feeling gets 'jammed up' (*eingeklemmt*) and its protentional strive forwards is hindered to varying degrees" (p. 164). Smith (2010) observers that Husserl is in the vicinity of psychoanalysis in this short text; he contemplates on the perseverance and plasticity of drives, and, more importantly, he states that a drive-complex may not be obliterated ("crossed out"), but rather leads to illness as a kind of habitual dissatisfaction (p. 168).

Smith (2010) tries to integrate the seemingly barred unconscious into the passive-associative network of the living present. The dormant sphere or obscure background of the unconscious might contain the repressed in the latent forms of habits, dispositions, preferences, motivations, anxieties, and moods. Smith himself accentuates that "all previous lived experiences have an afterlife in that they partake in the continuous 'new formation or reformation of dispositions' in the form of habits, memory, and the orientation of convictions, feelings and will; all at ready disposal of association" (p. 225). Smith's division of labor between phenomenology and psychoanalysis integrates certain aspects of the unconscious and certain methods of psychoanalysis; however, radically anomalistic (including psychopathological) self-experiences are also worthy of phenomenological consideration. In my view, the notion of affective relief is of tremendous importance in unraveling the inner complexity of the self. As I have indicated in the previous chapter, the affective

[21] "What is specific therein is motivated in the obscure background and has its 'psychic grounds', about which it can be asked: how did I get there, what brought me to it? That questions like these can be raised characterizes all motivation in general. The 'motives' are often deeply buried but can be brought to light by 'psychoanalysis'. A thought 'reminds' me of other thoughts and calls back into memory a past lived experience, etc. In some cases it can be perceived. In most cases, however, the motivation is indeed actually present in consciousness, but it does not stand out; it is unnoticed or unnoticeable ('unconscious')" (Husserl, 1989, p. 234).

relief not only schematizes *what* can be seen or felt but also influences the *how* of experience. If we turn our attention to the habitual and dispositional structure of the unconscious, we can offer a tangible notion of the relief structure. And it is here where the problem of repression and its relation to memory recurs. In the following chapter I will discuss in greater detail the dispositional and habitual aspects of the unconscious that lead to the constitutive role of the lived body. According to my thesis, the notion of implicit body memory complements the phenomenological understanding of the unconscious. Moreover, it can give us the "phenomenological thickness" of the minimal self and also offers opportunities for enriching the notion of traumatic subjectivity.

References

Askay, R., & Farquhar, J. (2006). An intelligible yet enigmatic mutual silence: Freud and Husserl. In R. Askay & J. Farquhar (Eds.), *Apprehending the inaccessible: Freudian psychoanalysis and existential phenomenology* (pp. 165–189). Northwestern University Press.

Behnke, E. A. (2008). Husserl's protean concept of affectivity: From the texts to the phenomena themselves. *Philosophy Today, 52*, 46–53.

Brudzińska, J. (2012). Depth phenomenology of the emotive dynamics and the psychoanalytic experience. In D. Lohmar & J. Brudzińska (Eds.), *Founding psychoanalysis phenomenologically* (pp. 23–52). Springer.

Freud, S. (1957a [1915]). Repression. In J. Riviere (Trans.), *Collected papers* (Vol. 14, pp. 84–96). Basic Books.

Freud, S. (1957b [1915]). The unconscious. In J. Riviere (Trans.), *Collected papers* (Vol. 14, pp. 98–136). Basic Books.

Freud, S. (1958 [1912]). A note on the unconscious in psycho-analysis. In J. Riviere (Trans.), *Collected papers* (Vol. 12, pp. 260–267). Basic Books.

Freud, S. (1965 [1933]). The dissection of the psychical personality. In J. Strachey (Ed. & Trans.), *New introductory lectures on psycho-analysis* (pp. 71–100). W. W. Norton & Company.

Freud, S. (1990a [1920]). *Beyond the pleasure principle* (J. Strachey, Trans.). W. W. Norton & Company.

Freud, S. (1990b [1923]). *The ego and the Id* (J. Riviere, Trans.). W. W. Norton & Company.

Fuchs, T. (2012). Body memory and the unconscious. In D. Lohmar & J. Brudzińska (Eds.), *Founding psychoanalysis phenomenologically* (pp. 69–82). Springer.

Gardner, S. (1991). The unconscious. In J. Neu (Ed.), *The Cambridge companion to freud* (pp. 136–160). CUP.

Giampieri-Deutsch, P. (2012). Perception, conscious and unconscious processes. In F. Barth, P. Giampieri-Deutsch, & H.-D. Klein (Eds.), *Sensory perception. Mind and matter* (pp. 245–264). Springer Verlag.

Habermas, J. (1971 [1968]). *Knowledge and human interest*. Boston Press.

Holenstein, E. (1972). *Husserls Phänomenologie der Assoziation. Zu Struktur und Funktion eines Grundprinzips der passive Genesis bei Edmund Husserl*. Nijhoff.

Husserl, E. (1970). *The crisis of European sciences and transcendental phenomenology* (D. Carr, Trans.). Northwestern University Press.

Husserl, E. (1989). *Ideas pertaining to a pure phenomenology and to a phenomenological philosophy, Second Book: Studies in the phenomenology of constitution* (R. Rojcewicz, & A. Schuwer, Trans.). Kluwer Academic Publishers.

Husserl, E. (2001). *Analyses concerning passive and active synthesis: Lectures on transcendental logic* (A. J. Steinbock, Trans.). Kluwer Academic Publishers.

Husserl, E. (2014). *Grenzprobleme der Phänomenologie. Analysen des Unbewusstseins und der Instinkte. Metaphysik. Späte Ethik* (Texte aus dem Nachlass 1908–1937). Springer.
Kozyreva, A. (2017). *Phenomenology of affective subjectivity: Analyses on the pre-reflective unity of subjective experience*. Universitäts Bibliothek.
Mishara, A. (1990). Husserl and Freud: Time, memory and the unconscious. *Husserl Studies, 7*, 29–58.
Moran, D. (2011). Edmund Husserl's phenomenology of habituality and habitus. *Journal of British Society for Phenomenology, 42*(1), 53–77.
Moran, D. (2017). Husserl's layered concept of the human person: Conscious and unconscious. In D. Legrand & D. Trigg (Eds.), *Unconsciousness between phenomenology and psychoanalysis* (pp. 3–23). Springer.
Ricoeur, P. (1970). *Freud and philosophy: An essay on interpretation*. YUP.
Rogozinski, J. (2010). *The ego and the flesh* (R. Vallier, Trans.). Standford University Press.
Schacter, D. L. (1996). *Searching for memory: The brain, the mind, and the past*. Basic Books.
Smith, N. (2010). *Towards a phenomenology of repression – A Husserlian reply to the Freudian challenge*. Stockholm University Press.
Welsh, T. (2002). The retentional and the repressed: Does Freud's concept of the unconscious threaten Husserlian phenomenology? *Human Studies, 25*, 163–183.
Zahavi, D. (1999). *Self-awareness and alterity: A phenomenological investigation*. Northwestern University Press.

Chapter 4
Body Memory and the Unconscious

Abstract This chapter is the direct continuation of the previous chapter in which the problem of the affective identity and affective relief was discussed. The current chapter argues that body memory can flesh out and expand the concept of the affective relief structure. According to Fuchs, body memory constitutes the horizontal unconscious of a person in the lifeworld. In line with this thinking, the chapter argues that there is a close connection between the relief structure of the living present and body memory. In this chapter the affective impact of traumatic experiences and the notion of the affective unconscious will resurface again.

Keywords Body memory · Unconscious · Horizontal unconscious · Affective unconscious · Affective relief

As we have seen in the previous chapter, Smith argued that Husserl discovered the mutual relationship between the conscious and the unconscious. The affective relief structure of the living present is modulated by suppression that operates at the level of the hyletic data as well as at the level of practical beliefs or convictions. Smith has shown that Husserl refers to Freudian insights in the manuscript D14. Husserl argued that the realm of sedimentation is not the level of "eternal frozen immobility", but rather the overlapping stream of unconscious memories (Smith, 2010, p. 235). In the unconscious "faraway-associations" constantly emerge and, due to the process of affective awakening, distant memories may intrude into the living present. The associative connections with the long-forgotten past endow us with the sense of familiarity and allows us to process and integrate new experiences (Smith, 2010, p. 236).

Smith and Kozyreva emphasize the crucial role of feeling and affection by which the sedimented content gains energy to spring forth to the level of consciousness. Following in the footsteps of Husserl, Smith observes that there is another level of constitution under the fundamental structure of time-consciousness (i.e., below the stream of the primal impression-retention-protention); that is, sedimented habits, acquired evaluations, emotions, and interests constitute the so-called "originary affection", which cannot be identified with the network of higher-level emotive acts.

L. Horváth, *The Affective Core Self*, Contributions to Phenomenology 130,
https://doi.org/10.1007/978-3-031-56920-3_4

The originary affection, the structure of habits and interests "thickens out" the living present (Smith, 2010, p. 238).

In a similar vein, Kozyreva speaks of an *affective identity* that underlies the formal identity of time-consciousness: one can assume an identity that is not based on temporality or reflective consciousness, but rather on the affective connections between past and present (Kozyreva, 2017, p. 137). The relationship between time-consciousness and affectivity will be analyzed in greater detail in Part IV. For preliminary purposes, let us introduce the thesis that affective identity (including the affective unconscious) generates the *core*, or, so to speak, the *phenomenal thickness* of the minimal self, which may undergo radical alterations due to subversive experiences or remains relatively stable during our life history. According to this proposal, the habitual and dispositional structure of the phenomenological unconscious constitutes a unique type of phenomenal consciousness. This individuated phenomenal consciousness can be regarded as a (less) minimal self. That is, there can be the experience of what is it like to possess an individual (i.e., personal and unique) affective identity. As we have previously seen, Zahavi argued that the subtle experience of what-it-is-like-for-me-ness connects the formal notion of the minimal self with phenomenal consciousness (i.e., the phenomenal character of our self-awareness is based on the for-me-ness or mineness of experience). What I intend to propose is that affective identity reveals not the *formal* but the *personal* (i.e., individuated) ways of experiencing ourselves and the world. The concept of affective identity constitutes a bridge between the environmentally and socially schematized (less) minimal self and phenomenal consciousness.

A simple example can illustrate the difference between the (more) minimal self and the (less) minimal self. What is it like to watch a sunset? The formal notion of the (more) minimal self suggests that there is a subject who lives through the experience of the sunset and there is a general and subtle phenomenal character that accompanies the mode of the givenness of the sunset. The for-me-ness of the intentional object is the universal characteristic of intentional life. However, the (less) minimal self with its affective infrastructure makes the experience thicker and personal. I see the sunset and it seem to be a phenomenological fact that the experience belongs to me as the experiencing subject. On the other hand, the scene of the sunset may (or, of course, may not) generate a passive associative process in which positive or negative memories may arise with their accompanying affective colorization. I focus my attention on the scene of the sunset, and I suddenly find myself in the distant past. The actual scenery may facilitate the occurrence of recent or distant memories. The simple intentional object of the sunset is enriched and colorized by the occurring mental states, including recollections, reflections, and affections. Suddenly the penumbra of childhood memories may cover over the actual experiences as in the case of Proust's madeleine experience. A similar retroactive effect can be imagined in the case of activities such as riding a bicycle or driving a car and so on. For example, I sit in my car and drive effortlessly but on certain occasions I suddenly feel the frustration of the novice driver and reproach myself for my mistakes that I made as a beginner. What I want to emphasize is that the phenomenal character of our experience is schematized by the affective unconscious.

This thesis is in line with Schechtman's concept of the narrative self. Schechtman argues that the self is constituted in the "soup" of a narrative where each mental state is connected to and modified by others. The retrospective and prospective processes of our narrative also give personal flavor (i.e., phenomenal character) to our self:

> The narrative is like the soup into which experiences are thrown, seasoning and altering one another – the past is reinterpreted and experienced in a new light in virtue of the present; the expectation of the future gives a different taste to current experience; and future experiences will have their character within the context of the whole. The experience of the person is thus had by an extended narrative subject, and not by a time-slice (Schechtman, 1996, p. 144).

I agree with Schechtman contention that "the phenomenological life of an individual who narrates" differs significantly from those who do not narrate. However, by emphasizing affective identity and affective unconscious, I would like to complement Schechtman's top-down approach with a bottom-up approach. As I understand it, Schechtman's concept of the implicit narrative, including the Freudian insights to which she refers, is also a bottom-up strategy for underpinning the narrative self. What I would like to suggest is that the processes of body memory and the Freudian Nachträglichkeit can also contribute to the constellation of affective identity that underlies narrative identity. In this chapter we focus on the horizontal unconscious and body memory, which can bolster and complete the idea of the affective relief structure.

The origin of affective identity can be traced back to Husserl's theory of the affective relief, which also presupposes the network or habits, interests, cares, and concerns. That is, the notions of originary affection and affective relief point to the same latency of an affective identity that is one of the main constituents of personhood. The following quotation shows that Smith has found the roots of affective identity in the realm of sedimentation which is the ground of every associative awakening:

> According to Husserl's dynamic conception of the psyche which holds that it is in a constant state of change – *das Seelenleben ist nach Wesensnotwendigkeit ein Fluss* – all previous lived experiences have an afterlife in that they partake in the continuous "new formation or reformation of dispositions" in the form of habits, memory, and the orientation of convictions, feelings and will; all at the ready disposal of association. (Smith, 2010, p. 224–25)

The concept of body memory gives way to an indirect approach to examine the process of sedimentation and the layers of sedimentations as well. As we shall see, body memory gives us the possibility to compare the reproductive and affective processes of remembering. In this section, I will briefly analyze the alternative view of the phenomenological unconscious that emerges from the notions of body memory and habituality. The claim that body memory can complement the notion of affective relief is supported by the fact that Husserl also integrated the tendencies of the lived body into the affective relief (cf. Brudzińska, 2012, p. 43).

4.1 Affective Relief and Body Memory

Fuchs distinguished between the vertical and horizontal unconscious; the former referring mainly to the depth psychological unconscious and the latter to the phenomenological unconscious. Of course, there are several other subcategories of these two general types. According to this proposal, the notion of the horizontal unconscious represents a special kind of affective identity that closely connected to the lived body, habits, and the lifeworld in general. A crucial question to ask is whether the notion of the horizontal unconscious is capable of integrating Freudian repression? The preliminary answer is that there is a possibility of integration through the phenomena of implicit knowledge and implicit memory. The idea of body memory, which is inspired by studies of implicit memory, reformulates the mechanism of repression. Both Smith and Fuchs draw on Husserl's phenomenology of passivity in the characterization of the unconscious. Fuchs provides the synthesis of Husserl's analysis of passivity and implicit memory. Body memory and the horizontal unconscious incorporated and phenomenologically reinterpreted the process of the "return of the repressed".

In short, body memory constitutes such an unconscious that operates in the interpersonal realm and has a dispositional structure. The origin of body memory dates back to Husserl and Merleau-Ponty and contemporary scholars regard it as a complex and multifaceted phenomenon. Casey (2000) distinguished three main aspects of body memory, namely the habitual, the traumatic, and the erotic body memory. Fuchs further developed the concept and introduced other types of body memory. The six forms of body memory are the procedural, situational, intercorporeal, incorporative, pain, and traumatic memory (Fuchs, 2012a, p. 12). Casey mentions that Merleau-Ponty, inspired by Bergson, introduced the notion of habitual body, which roughly implies the bodily forms of recollection, but he missed the opportunity to explicitly define body memory: "If Merleau-Ponty fills the void left gaping in Heidegger's *Being and Time* – where the role of the body, through implicit throughout, is never thematized – his own text exhibits a no less glaring lacuna in its bypassing of body memory." (Casey, 2000, p. 147). For Fuchs, body memory provides the opportunity to reconsider certain issues of implicit memory, and, in turn, to interpret certain bodily-related phenomena as the manifestations of implicit body memory. According to Fuchs, the incessant flow of experiences and the subjective feeling of familiarity (i.e., the sense of self) are supplemented by the hidden workings of (implicit) body memory. Body memory is a special kind of implicit memory that manifests itself not only in *procedural* skills (e.g. driving a car or cycling), but also in *intercorporeal* and *interpersonal* situations.[1] Fuchs regards the so-called *situational memory* as a special form of implicit memory:

> Implicit memory is not confined to the body itself. It extends to the spaces and situations in which we find ourselves. It helps us to get our bearings in the space of our dwelling, in the neighborhood, in our home town. Bodily experience is particularly linked to interiors, which, over time are imbued with latent references to the past and with an atmosphere of familiarity. Dwelling and habit (in German *Wohnen* and *Gewohnheit*) are both based on the memory of the body (Fuchs, 2012a, p. 13).

Body memory expresses the general vibe or familiarity of experiential life; that is, certain experiences can be inscribed into the intermodal body memory and play an identity-constituting role. In this respect, Fuchs has broadened the category of implicit memory. Our general moods and reactive patterns can be regarded as dispositional structures acquired through the mechanisms of implicit and procedural learning. Interpersonal relations and situations are schematized by the subconscious activity of implicit body memory. As Fuchs (2012a) argues: "What we have forgotten has become what we are." (p. 13) Therefore, body memory can be considered to be an intercorporeal memory since it comprises "implicit relational styles" or "schemes of being-with" (Stern, 1985), which stem from the early caregiver-infant relationship and shape the whole embodied personality. Fuchs highlights the importance of developmental studies:

> With the progress of developmental research, we can now better comprehend the history of intercorporeal memory. This research has shown that motor, emotional, and social development in early childhood does not proceed on separate tracks, but is integrated through the formation of affective-interactive schemata. [...] This early intercorporeality has far reaching effects: early interactions turn into implicit relational styles that form one's personality. [...] These implicit relational styles are also expressed in the habitual posture of the body. Thus, the submissive attitude toward an authority figure implies components of posture and motion (bowed upper body, raised shoulders, inhibited motion), components of interaction (respectful distance, low voice, inclination to consent), and of emotion (respect, embarrassment, humility) (Fuchs, 2012a, pp. 14–15).

In short, body memory refers to the implicit structure of perception and behavior and presupposes that day-to-day living is permeated with latent habits and dispositions owing to repeated and interconnected experiences (Fuchs, 2012b, p. 69). However, body memory also allows us to notice the impact of the dynamic unconscious in the lifeworld. As previously discussed, the living present or phenomenal field is not an empty, or neutral horizon, but rather it can be characterized by the rivalry and conflict of affective forces. The phenomenal field is inseparable

[1] For the crucial role of implicit memory in consciousness studies see Sect. 1.3. Let me remind the reader that Schacter defines implicit memory in the following way: "... implicit memory: when people are influenced by a past experience without any awareness that they are remembering." (Schacter, 1996, p. 161) Schacter claims that research on implicit memory has revolutionized memory research and inspired researchers to reconsider the nature of memory, and to develop new experiments to show the effect of previous experiences without the awareness of remembering. In agreement with Fuchs, Schacter interested in cases where implicit memory influences what we think and what we do in day-to-day living (Schacter, 1996, p. 162).

from the schematization process of the affective relief. Smith argued that an intentional state is conscious when it is brought to the fore within the affective relief structure where affections are incessantly covering over and suppressing one another (cf. Smith, 2010, p. 235). That is, the formation of the phenomenal field is governed by an affective and cognitive schematization process. Fuchs draws an analogy between the phenomenal field and body memory. The Phenomenal field "denotes the spatially and temporally extended subjective experience at a given moment, referring both to the center and the margins of awareness, to what is explicitly or only implicitly given" (Fuchs, 2019, p. 2). So far, Fuchs' notion of the phenomenal field bears a resemblance to the affective relief, analyzed above. In the Husserlian affective relief structure affectivity played a central role in shaping the terrain of the perceptual field. By contrast, Fuchs (2019) puts more emphasis on the role of *bodily affectivity*. Background feelings permeate and color the environment owing to their atmospheric quality, and the body is connected to the environment through fundamental sensorimotor relations and affordances.[2] The field is also schematized by social interactions involving bodily resonances, intercorporeality, and interaffectivity (p. 2).

How, then, can we characterize the relation between the affective relief and body memory? I argue that Fuchs' body memory can substantiate the affective relief structure, since dispositions, habits, drives, and traits are not instantiated in the interiority of the subject but anchored in the body; that is, the relief structure is grounded in the projective capacities of the lived body (*Leib*). In this sense, Fuchs' considerations show a stronger commitment to Merleau-Ponty than Husserl. Body memory does not refer to the representational capacity of the mind, but rather to the implicit horizonal structure of the self. The concept of body memory is an important contribution to the phenomenological unconscious because it refers to the preservation of dispositions and habits. Fuchs argues that the unconscious is not a hidden reservoir behind the back of consciousness, but rather an implicit meaning-bestowing process that takes place through bodily behavior. What I would like to add to Fuchs' considerations is that implicit body memory is embedded into the affective relief structure of perception. In the following chapter the ideas of the horizontal unconscious, repression, and body memory will be examined in greater detail. This analysis may lead us to a zone of convergence between phenomenology and psychoanalysis.

[2] The term "affordances" refers to the process in which the human observer "learns to detect" the values and meanings of the perceived objects. The detection of affordances is a process of categorization that is crucial from both phylogenetic and ontogenetic perspectives. Gibson defines the affordances, offered by objects in our environment, in the following way: „I have coined this word as a substitute for *values*, a term which carries an old burden of philosophical meaning. I mean simply what things furnish, for good or ill. What they *afford* the observer, after all, depends on their properties. The simplest affordances, as food, for example, or as a predatory enemy, may well be detected without learning by the young of some animals, but in general learning is all-important for this kind of perception." (Gibson, 1983, p. 285)

4.2 The Horizontal Unconscious of Body Memory

Fuchs throws light on the inherent dualism in Freudian theory. On the one hand, according to Freud the ego originates from the body, and the development of the partial drives is closely related to the body – and also to bodily needs. However, Fuchs accentuates that the drives are not the features of the subjectively lived body, but rather objective-somatic quantities and drive-energies that are distributed across the various levels of the psychic apparatus. Therefore, Fuchs concludes, the body is mainly the "primary projection field for the psyche" in which imagined or symbolic meaning emerge (Fuchs, 2012b, p. 71). Due to the economic model of repression a radical separation ensues between the pre-conscious and the unconscious, and the latter is characterized as an internal foreign country.

Fuchs (2012b) sides with Merleau-Ponty in order to "extend subjectivity in the *horizontal* dimension"; that is, he introduces a special kind of phenomenological unconscious that stands in stark contrast with the inherent dualism attributed to Freudian metapsychology. In the horizontal dimension of the lifeworld the unconscious does not reside in the inner sanctum of the psyche but rather manifests itself in the behavior and the structures of a person's *lived space* (p. 74). Merleau-Ponty developed the concept of the "habitual body in his seminal work, *Phenomenology of perception*". In Fuchs's view the habitual body is the precursor of body memory which can be defined as the *invisible network of predispositions*. In Fuchs' words: "The body is thus the ensemble of organically developed predispositions and capacities to perceive and to act, but also to desire and to communicate. Its experience, anchored in the body memory, spread out and connect with the environment like an invisible network, which relates us to things and to people." (Fuchs, 2012b p. 73) The implicit system of body memory is characterized as the condensation or synthesis of repeated situations:

> If, following Merleau-Ponty, we view the body not as the visible, touchable and sentient physical body but first and foremost as our capacity to see, touch, sense, then body memory designates the totality of these bodily predispositions as they have developed in the course of our development – in other words, in their historical dimension. In body memory, the situations and actions experienced in the past are, as it were, all fused together without any of them standing out individually (Fuchs, 2012b, p. 73).

By quoting Merleau-Ponty, Fuchs argues that unconscious fixations are based on traumatic experiences and the lived space is schematized by the "intercorporeal presence" of the forgotten or repressed experiences, which unnoticedly constrain the possibilities of action in the daily life of the person (Fuchs, 2012b, p. 78). Body memory, located behind the back of consciousness, is an intricate system and comprises several layers of sedimented skills, dispositions, capacities, and desires. Fuchs states that body memory "constitutes a sensomotoric, libidinous and interactive field in which we, as embodied beings, constantly move and conduct ourselves."

(Fuchs, 2012b, p. 74) In this sense, moods and existential feelings[3] are anchored in the body; that is, the repulsive or attractive forces of affectivity in the day-to-day living are strengthened by the projective capacity of the lived body. Without intersubjective regulation, the lived body's auto-affective (i.e., projective) activity could disorganize or even disintegrate the self, leading to exaggerated fantasy activity or even to symptoms of psychosis such as delusions and hallucinations. We are not encapsulated selves, but we dwell in the world of intercorporeality, and Fuchs claims that the "implicit relational styles", which are one of the most important components of body memory, are acquired through interactions with the environment and others.

Fuchs' body memory is not only based on studies of implicit memory but also on Kurt Lewin's field psychology (Fuchs, 2012b, p. 74). Lewin introduced the concept of *life space*, which is centered around the bodies of persons. The life space is not a neutral spatiotemporal dimension, but rather a qualitatively rich phenomenal field. The "field forces" of aversion and attraction are deeply embedded in the life space and demarcate the zones of freedom and prohibition. Memorable experiences are engraved into body memory and later pre-reflectively schematize and colorize the atmospheric qualities of the life space. Developmental considerations play a crucial role here: the child's spontaneous impulses interfere with parental imperatives. Another telling example is the child's exploratory behavior: the child gravitates toward the mother or moves away bravely in order to find something interesting (Fuchs, 2012b, p. 77).

In Fuchs' framework, the unconscious surrounds the body, permeates it, or can be found before our eyes in the modulations of the lived space. He comes to the conclusion that the unconscious is not an "independent intra-psychic process" or a barred compartment of the psychic apparatus, but rather the horizontal dimension of the lived body and lived space, filled with libidinous tendencies and value judgments. The affectively loaded life space bears a certain resemblance to Husserl's affective relief structure. As it was previously argued, while Husserl mainly focused on the Gestalt formations in the perceptual field and tried to disclose the layers of object-constitution, he also accentuated the role of the body and affectivity in the constitution of the lifeworld. In my reading, the broad category of body memory is compatible not only with the notion of the life space, but also with the affective relief structure. However, the main question remained unanswered: is there any way to integrate the role of *repression* into the system of implicit body memory? As has been shown, the concept of the affective relief reduces repression to suppression or rivalry between affective components of the phenomenal field. Nevertheless, the affective relief is the extension of consciousness with inexplicably interwoven layers of passivity, leaving no room for a separate unconscious reservoir. Fuchs, in a similar vein, locates the mechanism of repression in the phenomenal field and speaks of "blind spot in self-awareness" and the dark side of our personality (Fuchs,

[3] I will discuss the problem of Ratcliffe's existential feelings in more detail in Chap. 5. For preliminary purposes, Ratcliffe defines existential feeling as a background feeling of experiences, which are inseparable from bodily feelings (Ratcliffe, 2008, p. 7).

2012b, p. 77). For him, the unconscious is the non-thematic, implicit layer of experience situated not behind the back of consciousness, but rather in front of our awareness.

In the case of the Husserlian affective relief structure the role of repression was marginal and limited to the rivalry of affective tendencies located in our perceptual field. In contrast to Husserl, Fuchs characterizes the phenomenal field from a psychopathological point of view and develops an alternative theory of repression: the mechanism of repression is not located in the organic body or in the psychic apparatus; rather, it manifests itself in the sensations of the lived body and in the perception of the surrounding world. For example, Fuchs argues that even a seemingly simple repressed wish has a *field structure*. Fuchs illustrates this phenomenon by recounting Heinrich von Kleist's short story about an abstinent soldier who suddenly begins to hear the names of various brandies in the ringing of a cathedral bell and quickly relapses. On the one hand, Fuchs admits that this example can be interpreted as the sudden emergence of a repressed wish, but on the other hand, it can also serve as an illustration of a case in which hidden bodily drives find their way back to the ego in an indirect way (Fuchs, 2012b, p. 75).

The implicit body memory may have several opportunities to organize or hinder day-to-day living. Prominent examples are the child's *avoidance behavior* and the *zones of prohibitions*. In a wider anthropological context, Fuchs highlights the importance of *taboo zones*: the taboo is a negative curvature around the shared life space that implicitly influences behavior and structured by societal demands and prohibitions. Violating the laws of taboo can lead to fear, guilt, and shame without awareness of mechanism of conditioning. The impact of affectivity on cognition is obvious in these examples. Fuchs claims that "barriers" of the lived space are affectively loaded, and they give rise to unconscious but bodily felt forces of repulsion. Last but not least, projection can also serve as an example of "blind spots" located at the very center of conscious awareness. For Fuchs, the prerequisite of projection is a defense mechanism and an ambiguous perception that triggers the characteristic behavior of the person:

> Here the beam in one's own eye becomes the splinter in another's eye, in other words, one perceives in others the impulses and motives against which one has built defences in oneself. Naturally, this perception is also ambiguous, since the excessive zeal with which the impulses in others are disapproved derives its energy precisely from the efforts one has to make to neutralize one's own impulses. The blind spot in self-awareness – and here Freud is doubtlessly right – does not result from a mere "overlooking", but from active and emotionally charged repression. Nevertheless, this repression remains the work and the effort of the subject herself, not of a mechanism outside her (Fuchs, 2012b, p. 77).

However, as the quote above suggested, the repression in the case of projection is not due to the unconscious, but rather to the subject's *unacknowledged knowledge*. Fuchs concludes that consciousness in not transparent; rather the field of awareness is schematized by unintentional and unperceived forces of the implicit unconscious. We can find affectively schematized zones and places in the living present that immediately repel or attract conscious awareness.

It seems plausible to suggest that there are strong correlations between the Husserlian affective relief structure and implicit body memory. Both ideas presuppose a phenomenal field which is schematized by instinctual, drive-related need, and modulated by habits and aversions. Both of them presuppose and emphasize the underlying factors of the unconscious; that is, the layers of our basic habits, skills, and preferences (including aversions and blind spots), which has their own developmental history and trajectory in the life of the person. However, a glaring distinction ensues when psychopathology comes to the foreground. Husserl contended that the anomalous self-experiences (e.g. madness) are the limit problems of phenomenology, and the investigation of the affective relief structure focuses mainly on the rivalry of affections and the genesis of intellectual convictions. Contrary to the affective relief, which is conceived as a version of the phenomenological unconscious, the concept of body memory directly addresses the issues of psychopathology. According to the framework of body memory, phobias and other irrational fears and delusions can manifest themselves in the person's bodily behavior and lived space. Furthermore, body memory has much to offer concerning the problem of trauma, which will be addressed in Part IV along with further considerations of affective identity.

A succinct definition of body memory can be found in Summa's paper: "Body memory coincides with implicit memory insofar as the latter is lived through by a bodily subject. Body memory, thus, embraces the totality of our subjective perceptual and behavioral dispositions, as they are mediated by the body." (Summa et al., 2012, 418). In my view, body memory and the affective relief are two interconnected forms of the phenomenological unconscious. Both ideas open the way to the non-representational realm of subjectivity. Body memory can be subsumed under the label of implicit or non-representational remembering (cf. Kozyreva, 2017, p. 143). However, we will see that emotional body memory and traumatic memory are situated on the grey area between active[4] and implicit remembering; that is, we can observe an implicit meaning-bestowing phenomenon in the realm of passivity or marginality. As Summa and her colleagues (2012) describe: "rather than being a re-presenting or presentifying act of recollection, body memory designates the prethematic impact of preceding bodily experiences on the meaningful, and yet, implicit, configuration of our actual experience." (p. 418) Moreover, while implicit memory

---

[4] Kozyreva draws a close analogy between the phenomenological term of active remembering and explicit memory: "In both disciplines, the definition of implicit memory is dependent on the definition of explicit memory. Allegedly, it is generally agreed that explicit memory corresponds to the recollection or active remembering of a past event. It is assumed that a subject is aware of such recollection. In cognitive psychology, explicit remembering is further clarified as a form of autobiographical, declarative memory, or episodic memory. In phenomenology, especially in Husserl, explicit remembering belongs to the class of the so-called reproductive presentifications, that is to intuitions in which absent (i.e. past) objects are presentified, as opposed to intuitive presentations (such as perceptions) which designate intentions of present objects." (Kozyreva, 2017, p. 178)

refers to a quasi-independent brain system,[5] body memory is a more general and holistic term: "it is a form of lived experience, which is constantly re-actualized and implicitly lived through by a bodily subject" (Summa et al., 2012, p. 425). How does body memory manifests itself from first-person perspective? In order to answer this question the mutual relationship between the organic body (*Körper*) and the lived body (*Leib*) will also be addressed.

4.3 Body Memory and the Affective Unconscious

In this chapter, I argue that the emergences of body memory are the indirect signs of an affective unconscious that may constitute the sedimented layer of our affective identity. Affective identity refers to the affective core self that is closely related to Zahavi's (less) minimal self. As it was outlined in Sect. 2.4, the (less) minimal self is open to developmental changes and can also be expanded or modified by narrative self-understanding. Therefore, it has a special kind of phenomenological thickness and complexity in comparison to the formal (more) minimal self. Our belief system can be altered in a subtle or in a radical way during lifetime. In a similar vein, our affective identity, including the system of implicit affective and behavioral reactions, can also undergo subtle or radical changes over the course of our life history. For example, the irritability and impulsivity of a young adult will recede into the background, giving way to patience and equanimity. Or take for example the case of traumatized subjects. Traumatized individuals may feel the loss of trust in others and in the world around them. As they begin to see the world through the lens of the traumatic experience their affective identity also changes. The change in the affective identity of the mature and the traumatized subject involves the subtle or radical change in the implicit network of preferences, habits, dispositions, and desires.

How can we approach to the implicit layers of the affective unconscious that constitute the affective core self? Certain features of implicit memory[6] has been reinterpreted in phenomenological terms by Kozyreva (2017) and she also introduced the notion of *affective memory*. The main function of affective memory is to

[5] It is worth considering that not everyone in current memory research agrees with the multiple-systems approach. Roediger argues that the dissociation between explicit and implicit memory may be the result of the methodological differences in experiments (Roediger, 1990). According to Summa et al. (2012), current experiments show the signs of empirical underdetermination since the results can be equally considered from the viewpoint of system and the processing approach (p. 425).

[6] As it was discussed in Sect. 1.3, implicit memory encompasses the strange accomplishments of amnesic patients, who are able to acquire certain skills without prior explicit recollection of the process, or exhibit aversive behavior as if they could remember painful events (e.g., electric shock or a pinprick) (Schacter et al. 1993). In general terms implicit memory shows some kind of preconscious information processing. We can speak of implicit processing when people are influenced by past experience without explicit recollection, or "without awareness that they are remembering" (Schacter, 1996, p. 161).

invisibly schematize the self.[7] Affective memory refers to the affective presence of the past in the present. That is, the past has "to be affectively present despite its temporal distance and to have a strong impact on the ongoing experience." (Kozyreva, 2017, p. 137). Kozyreva's (2017) another important contribution is the introduction of the notion of *affective identity*. Affective identity complements the discussion of personal identity, and it refers to a special kind of "affective connectivity" between the present and past experiences (p. 137.).

Following these considerations, the affective unconscious can be regarded as the non-representational presence of the past in the present. The affective unconscious comprises affective reactions and dispositions, and functions as an implicit (pre-intentional) organization that colorizes and transforms the atmosphere of the lived space through the subjectively felt lived body. Both Fuchs and Kozyreva argue that the dynamic unconscious can be reformulated with the phenomenology of the lived space. The atmospheric qualities of the lived space, including the zones of avoidance and interest, delineate a kind of horizontal, spatially extended and qualitatively felt, unconscious. Their notion of the horizontal unconscious is inspired by Merleau-Ponty's phenomenological unconscious: the repressed is not identical with a trace or an image of a traumatic event, but rather manifests itself in the present as the "style of being" (Kozyreva, 2017, p. 156, cf. Merleau-Ponty, 2012, p. 85).

In addition to the spatial and atmospheric qualities, another important manifestation of the affective unconscious lies in the hidden, implicit personality traits and dispositions to behave. These traits and dispositions can be indirectly observed or inferred from bodily behavior. Fuchs argues that the problem of personhood is not limited to the questions of persistence or sameness of personality, but also involves the history of embodied experience:

> …the embodied self or self-as-subject does not only imply an abstract selfhood, persistence or sameness, but also displays a qualitative form of identity across the life span. It emerges not from the store of explicit knowledge about oneself or one's biography, but from a history of embodied experience which has accumulated and sedimented in body memory and as such is implicitly effective in every present moment." (Fuchs, 2017, p. 308)

The embodied, qualitative form of identity can diverge from explicit self-knowledge. The embodied "history" of the person can contain the remnants of traumatic experiences, which may reappear in enactments in everyday or therapeutic circumstances (Fuchs, 2012a, p. 19). We can assume the elements of otherness in the non-representational sublayer of personality.

Ullmann (2015) argues that the person is always a differentiated or individuated form of being, but despite the apparent coherence of the person, there are good reasons to assume that we are often defined by our *blind spots* in our consciousness

[7] Kozyreva argues for the role of implicit memory to understand pre-reflective self-experience: "Similarly, research on implicit memory in cognitive psychology has shown that its influence extends beyond mere bodily or perceptual experience and includes feelings, behavior, conceptual thinking, and the interaction with other people. It is precisely this perspective that justifies the position of implicit memory as a constitutive dimension of the pre-reflective self-experience, and thereby puts it in the center of the current research." (Kozyreva, 2017, p. 175)

(p. 33). There could be several occasions in our day-to-day living when we try to reflect on why we have acted in certain ways and what has happened to our general habitual tendencies. Illuminating examples can be those occasions of irritability when we show a burst of anger or hatred toward someone who has no idea what he or she has done. Of course, affectivity cannot be reduced to suddenly emerging feelings, emotions, and sentiments. Nor can affectivity cannot be identified with the desiderative projections of the psyche in the Freudian sense.

Bortolan (2020) highlights that, in the phenomenological tradition, affective states not only designate occurrent responses, but also refer to "long-lasting and dispositional states" with evaluative purposes (p. 75). In a similar vein, Ullmann argues that affectivity is not identical with the self-affection of the lived body, but rather comprises an "implicit structure of sense". At first glance, the term "structure of sense" may imply a representational structure, but the affective unconscious is rather the "invisible nervation" of the visible behavior in the Merleau-Pontian sense (Ullmann, 2015, pp. 35–36). In a similar way to Fuchs, Ullmann defines affectivity as a network of emotions and moods that are dimly present in our being-in-the-world. Affectivity is endowed with a horizonal structure, a space of possibilities that constitutes the secondary or unconscious affectivity of our mental life. The typical manifestations of the affective unconscious are our usual reactive patterns in certain situations; these patterns are passive and automatic and they appear without conscious deliberation. Ullmann sees the phenomenological unconscious as an "invisible affective pattern", and he does not want to reduce it to one specific type of psychological unconscious (e.g., to the dynamic or cognitive unconscious). He also argues that the affective unconscious is basically a relational unconscious, since its origin and existence is based on interpersonal and especially traumatic relationships (Ullmann, 2017, p. 159). According to Ullmann, the broader sense of traumatic subjectivity means that we are all traumatized subjects in one way or another. The narrower sense of traumatic subjectivity presupposes a specific traumatic event in life history. The concept of the traumatic subject has its own aporias, but the general idea behind the concept is that traumatic experiences play a crucial role in the constitution of the affective unconscious (Ullmann, 2017, p. 153).

Similarly, for Fuchs, trauma means a not yet integrated or interpreted event that manifests itself as a foreign body in the lived body:

> The traumatic event is an experience that may not be appropriated and integrated into a meaningful context. As in pain memory, mechanisms of avoidance or denial are installed in order to isolate, forget, or repress the painful content of memory. The trauma withdraws from conscious recollection, but remains all the more virulent in the memory of the lived body, as if it were a foreign body. At every turn, the traumatized person may come across something that evokes the trauma (Fuchs, 2012a, 17).

It is reasonable to supplement the concept of traumatic subjectivity with the role of the lived body. Following in the footsteps of Ullmann, we could say that the affective unconscious is not only a volatile sphere of moods and emotions around us, but also the malleable medium of traumatic events. And the trauma is often traceable back to (1) outer events or (2) inner non-representable wishes or moods (Ullmann, 2017, pp. 158–159). However, Ullmann's notion of affectivity also mirrors certain psychoanalytical insights. For example, with respect to affectivity, he argues that the feeling of our own being (i.e., self-affection) is always strongly connected to the basic affections of pleasure and pain. But, for him, affectivity is not the sheer propagation or transmutation of libidinal energies; rather, he emphasizes the *meaning structures* of affectivity that modulate and govern the self and its interpersonal relations. On the one hand, Ullmann resists to the temptation to reduce the affective unconscious to the activity of the lived body, but, on the other hand, his concept is also closely related to Fuchs' implicit body memory. I would like to suggest that Ullmann's notion of the affective unconscious, i.e., the "implicit meaning structure", might be anchored in body memory; that is, the manifestations of body memory might give an indirect insight into the schematization process of the affective unconscious. There is a tangible convergence between Fuchs and Ullmann regarding the importance of hidden, implicit meaning structures in the constitution of the self.

My suggestion is that Fuchs' *horizontal unconscious* can be seen as an alternative of the affective unconscious. In this respect, the affective unconscious is at once a corporeal and intercorporeal unconscious.[8] Ullmann does not specify the meaning structures of the affective unconscious, he only accentuates their presence in our daily life and their passive-automatic effect on the person. Contrary to Ullmann, Fuchs as a psychiatrist, argues that vaguely felt emotions or impulses are often the "reverberations of forgotten or repressed contents as well as forebodings and anticipations of a possible future" (Fuchs, 2012a, p. 20). As was mentioned above, Fuchs claims that body memory is not only the *disposition* of perception and behavior, but also the *carrier of life-history* (Fuchs, 2012a, p. 20.).[9] He argues that psychotherapeutic

[8] By contrast, Ullmann's argument often bears the marks of Heideggerian philosophy of moods, since he does not try to reduce the affective unconscious to the tangible dimension of life, or to the lived body.

[9] My approach is basically a theoretical-phenomenological investigation, but the problem of affect-transformation is inseparable from its therapeutic origins. Fuchs has mentioned Gendlin's (1981) focusing therapy in relation to body memory, from which the concept of "felt sense" stems from. One can also find intriguing description of affect-transformation in Levine's (2015) work. Levine tries to characterize the materialization of episodic memories triggered by faint implicit memories. Through his rich phenomenological analysis of his encounter with a childhood friend, he eloquently describes a case in which an uncertain feeling of familiarity slowly crystallizes into an autobiographical memory of early school experiences. He also maps the various, interrelated memory systems. His analysis shows that there is an ongoing interaction between faint (emotionally laden and procedurally expressed) implicit memories and episodic memories. He defines implicit memories as "hot" and "compelling" memories in contrast to the "warm" episodic and "cold" declarative memories (p. 21). An important characteristic of implicit memories is that they are not individuated representations: "These memories cannot be called up deliberately or accessed as 'dreamy' reminiscences. Instead, they arise a collage of sensations, emotions, and behaviors. Implicit memories appear and disappear surreptitiously, usually far outside the bounds of our conscious awareness." (Levine, 2015, p. 21) Concerning Proust's famous madeleine-reminiscence, Levine (2015) speaks of an "implicit trigger". According to his interpretation, the retrieval of childhood memories was eventuated by the sensory experience of the tea and the madeleine, which triggered the network of subconscious procedural, episodic, and emotional processes (p. 29). In a similar vein, Fuchs interprets Proust's experience as the process of excavating an autobiographical

techniques (e.g., Gendlin's focusing therapy) can crystallize a *meaning core* from the bodily "felt sense"; and these emerging meaning cores are often connected to distant autobiographical episodes. The concept of "felt sense" originates from Gendlin (1981) who describes it as a special kind of bodily awareness:

> The process I am going to teach you in this book, the inner act, is a perfectly natural one. But as our language contains no words to describe it, I have had to invent the needed words. I call the process focusing. It is a process in which you make contact with a special kind of internal bodily awareness. I call this awareness a *felt sense*. A felt sense is usually not just there, it must form. You have to know how to let it form by attending inside your body. When it comes, it is at first *unclear*, fuzzy. By certain steps it can come into focus and also change. A felt sense is the body's sense of a particular problem or situation. A felt sense is not an emotion. We recognize emotions. We know when we are angry, or sad, or glad. A felt sense is something you do not at first recognize— it is vague and murky. It feels meaningful, but not known. It is a bodysense of meaning. When you learn how to focus, you will discover that the body finding its own way provides its own answers to many of your problems. The process brings change. (Gendlin, 1981, p. 10)

From this therapeutic perspective, the "felt sense" is not identical with any kind of occurrent emotion, but rather it is a kind of "bodysense of meaning".[10] In other words, the felt sense can be seen as the lived body's implicit or intuitive response to certain events, a special kind of implicit and also often retrospective appraisal of intersubjective or even traumatic situations.

In general, body memory serves as an umbrella concept in the synthesis of affective life. It seems to be the case that the intangible affective allure of situations, the random eruption of impulses, the unbearable (social) pressure of a situation, a hypersensitivity to certain situations, and intuitive impressions about other people, etc., became tangible, characterizable phenomena by means of body memory. At first glance, body memory may seem to be a reified notion that combines several implicit processes, but the psychotherapeutic underpinnings of the notion highlight the relationship between different memory systems in generating meaningful content. That is, the lived body is not the reservoir of images, or it is not a hidden "image" within the lived body. Instead, body memory refers to the habitual ways of

---

content that lies at the core of an implicit bodily experience: "Proust's madeleine-memory thus hides within a complex of bodily sensations and implicit, only intuited recollections and meanings. I would like to call such a complex a meaning core. It is a nodal point of bodily recollection into which the lived past has condensed, as it were, and from which new meanings may unfold." (Fuchs, 2012a, p. 20)

[10] The felt sense is difficult to verbalize, it can be compared to the tip of the tongue phenomenon. As Gendlin writes: "Suppose you have been listening to a discussion and are about to say something relevant and important. The others are still talking. You don't have your words prepared. All you have is a felt sense of what you want to say." (Gendlin, 1981, p. 85) However, a felt sense is not a hidden or suppressed idea that is intuitively given in one way or another. On the contrary, a felt sense is the process of articulating meaning, the felt sense has its internal complexity, it consists of various elements: "Usually, when we are about to say something, we have *the felt sense* of what we want to put across, and the right words come as we speak. The felt sense includes dozens of component parts, perhaps hundreds: the meaning you want to put across, the emotional color you want to give it, the reasons why you want to say it to those particular people, the reaction you hope to elicit from them, and so on." (Gendlin, 1981, p. 85).

behaving or to the process of emerging autobiographical memories. Body memory is a process, an associative process in which bodily sensations (e.g., Gendlin's felt senses) may or may not lead to rememberings and new meaningful content. Therefore, we must abandon the idea that body memory refers only to repressed content that is barred from the sight of consciousness; rather, body memory refers to those unintegrated or unprocessed mental states that can be transformed and processed by implicit as well as explicit memory work.

Fuchs does not state that the meaning cores of body memories are explicit representational contents or veridical episodic memories, but rather under certain conditions (especially in the case of PTSD patients) the body *re-enacts* the past in the present. Here, I would like to suggest that the re-enactments of body memories can also be seen as the manifestations of the affective unconscious. A volatile mood, an unusual behavior, or an outburst of anger in a therapeutic situation are telling examples of the blind spots of the affective unconscious. Fuchs' body memory can be understood as a transitional zone between the non-representational and representational unconscious, or as a bridge between autobiographical memory and unarticulated bodily affects. As we have seen, Fuchs argues for a horizontal unconscious in which the repressed is instantiated in our habitual bodily behavior.

In sum, the notion of body memory weaves the various threads of affective unconscious and personal identity into one holistic network. The conscious comprehension of a bodily "felt sense" suggests the ongoing dynamic between the non-representational and the representational elements of consciousness. Of course, the process of affect transformation is a phenomenological and psychoanalytical conundrum. The concept of body memory provides a way to the deeper layers of the self and also gives us a model for descriptively approach to the process of affect transformation. The task of Chap. 6 will be to explain the intriguing temporal unfolding of repressed affect through the phenomena of *Nachträglichkeit* and retroactive self-understanding.

References

Bortolan, A. (2020). Affectivity and the distinction between minimal and narrative self. *Continental Philosophy Review, 53*, 67–84.

Brudzińska, J. (2012). Depth phenomenology of the emotive dynamics and the psychoanalytic experience. In D. Lohmar & J. Brudzińska (Eds.), *Founding psychoanalysis phenomenologically* (pp. 23–52). Springer.

Casey, E. S. (2000). *Remembering: A phenomenological study* (2nd ed.). Indiana University Press.

Fuchs, T. (2012a). The phenomenology of body memory. In S. C. Koch, T. Fuchs, M. Summa, & C. Müller (Eds.), *Body memory, metaphor and movement* (pp. 9–22). John Benjamins Publishing Company.

Fuchs, T. (2012b). Body memory and the unconscious. In D. Lohmar & J. Brudzińska (Eds.), *Founding psychoanalysis phenomenologically* (pp. 69–82). Springer.

Fuchs, T. (2017). Self across time: The diachronic unity of bodily existence. *Phenomenology and the Cognitive Sciences, 16*(2), 291–315.

Fuchs, T. (2019). The interactive phenomenal field and the life space: A sketch of an ecological concept of psychotherapy. *Psychopathology, 52*(2), 67–74.
Gendlin, E. T. (1981). *Focusing* (2nd ed.). Bantam Books.
Gibson, J. J. (1983). *The senses considered as perceptual systems*. Praeger.
Kozyreva, A. (2017). *Phenomenology of affective subjectivity: Analyses on the pre-reflective unity of subjective experience*. Universitäts Bibliothek.
Levine, P. A. (2015). *Trauma and memory: Brain and body in a search for the living past*. North Atlantic Books.
Merleau-Ponty. (2012). *The phenomenology of perception*. (D. A. Landes, Trans.). Routledge.
Ratcliffe, M. (2008). *Feelings of being: Phenomenology, psychiatry and the sense of reality*. OUP.
Roediger, H. L. (1990). Implicit memory. *Retention without remembering. American Psychologist, 45*(9), 1043–1056.
Schacter, D. L. (1996). *Searching for memory: The brain, the mind, and the past*. Basic Books.
Schacter, D. L., Chiu, C. Y., & Ochsner, K. N. (1993). Implicit memory: A selective review. *Annual Review of Neuroscience, 16*, 159–182.
Schechtman, M. (1996). *The constitution of selves*. Cornell University Press.
Smith, N. (2010). *Towards a phenomenology of repression – A Husserlian reply to the Freudian challenge*. Stockholm University Press.
Stern, D. N. (1985). *The interpersonal world of the infant: A view from psychoanalysis and developmental psychology*. Basic Books.
Summa, M., Koch, S. C., Fuchs, T., & Müller, C. (2012). Body memory: An integration. In S. C. Koch (Ed.), *Body memory, metaphor and movement* (pp. 417–444). John Benjamins Pub.
Ullmann, T. (2015). A narratív, a traumatikus és az affektív szubjektivitás. In I. Bujalos, M. Tóth, & T. Valastyán (Eds.), *Az identitás alakzatai* (pp. 21–37). Kalligram.
Ullmann, T. (2017). Phenomenology of experience and the problem of the unconscious. In M. Gabriel, C. Olay, & S. Ostritsch (Eds.), *Welt und Unendlichkeit: Ein deutsch-ungarischer Dialog in memoriam László Tengelyi* (pp. 141–161). Verlag Karl Alber.

Part III
Psychopathology and the Minimal Self

Chapter 5
Phenomenological Psychiatry of Schizophrenia

Abstract This chapter provides an indirect approach to the problems of the minimal self and self-affection. The chapter examines the disintegration process of the minimal self in schizophrenia. Recently, considerable work has been directed at the ipseity-disturbance model of schizophrenia, which facilitated a thought-provoking dialogue between clinical neuroscience and phenomenologically inspired psychiatry. The main purpose of the following investigations is to draw attention to the intricacies of self-disorders, which may indirectly shed some new light on the nature of the minimal self. Body memory and the horizontal unconscious, discussed in the previous chapters, have explanatory value in understanding the nature and phases of self-disorder. The examination of the disturbances of the minimal self also brings us closer to the account of self-affection and the affective core self.

Keywords Minimal self · Self-disorder · Ipseity · Hiperreflexivity · Ipseity-disturbance model · Schizophrenia · Phenomenological psychiatry

The ipseity-disturbance model of schizophrenia (henceforth IDM) is hotly debated in contemporary interdisciplinary studies and offers a modified view of the minimal self as a target phenomenon for phenomenological psychiatry. At first sight, phenomenological psychiatry seems to be an antipsychiatry movement, since proponents of antipsychiatry (e.g., Thomas Szasz and R. D. Laing)[1] made enormous efforts to criticize biological psychiatry and introduce philosophical frameworks for understanding mental illness. In short, the phenomenon of ipseity-disturbance presupposes that the primarily philosophical notion of *ipseity* can be used as a working

[1] However, for the sake of precision and clarity, it is worth noting that Laing did not regard himself as the advocate of the antipsychiatry movement. Evidently, Foucault's and Sartre's impact on Laing's contribution to phenomenological psychiatry is conspicuous. I contend that he can be also considered the harbinger of the contemporary IDM-model in that respect that he meticulously examined the different stages and appearances of alienation from the intersubjective world. His existentialist, phenomenological ruminations have led to an intricate phenomenology of the life-world, in which the basic relational styles of the family are of central importance.

L. Horváth, *The Affective Core Self*, Contributions to Phenomenology 130,
https://doi.org/10.1007/978-3-031-56920-3_5

hypothesis for explaining the symptoms of schizophrenia spectrum disorders. There is an important question to ponder: how can we transform a philosophical notion to the target phenomenon of psychiatric research? One possible answer to this question is the fact that phenomenological psychiatry is essentially a descriptive undertaking and speaking of *ipseity* does not entail a reified notion of the self. However, Sass (2017) entertains the idea that it is not necessary for phenomenology to be purely descriptive enterprise; rather phenomenology could even contribute to causal explanations of how our experience of the world affects the brain and behavior. The empirical support for this argument stems from modern neuroscience that accepted the role of "downward causation"; that is, functional changes in brain activity can be the result of mental states, or more specifically, psychosocial conflicts and stressors. An extreme example in Sass's list is the case where inactivity or social withdrawal may lead to cerebral atrophy. In short, certain attitudes and behaviors may determine the biological properties of human beings (cf. Sass, 2017, p. 321).

Based on six different approaches to schizophrenia, Lysaker & Lysaker (2010) claim that phenomenologically and psychoanalytically inspired approaches "offer a feel for the illness's first-person dimensions" (p. 333). Sass (2017) criticizes biological psychiatry for its renewed tendency to neglect the empathic and psychological understanding of symptoms in favor of underlying biological disorders. Concern for the subjective side of anomalous experiences is restricted to "superficial checklists of symptoms" (p. 313). Laing (1978), in his seminal work, *The Divided Self*, has already addressed the vicissitudes of reification tendencies is biological psychiatry, observing the dissociative tendencies in schizophrenia, anticipating the phenomenon of hyperreflexivity during symptom formation. Compulsive self-monitoring (or hyper awareness) is a frequently recurring theme in phenomenologically oriented psychiatry. Among others, Klaus Conrad introduced the concept of convulsive reflection (*Reflexionkrampf*), and Borda and Sass reconsidered Firth's (1979) theory[2] of the excessive awareness of the cognitive unconscious as a precursor of the current hyperreflexivity model (Borda & Sass, 2015, p. 470).

Recently, Fuchs & Schlimme (2009) tried to integrate phenomenological psychiatry into the framework of embodiment in their programmatic paper. Since the 1990s onward the notion of the embodied mind has received considerable attention. Varela and Thompson, in their neurophenomenological approach tried to correlate neurobiological findings with first-person experiences (cf. Varela, 1996; Thompson, 2007). More recent studies of phenomenological psychopathology continue to explore the destabilization and disintegration of the self. More specifically, phenomenological psychiatry focuses on the problem of the *embodied self*. The authors

[2] According to the observations of Firth, hyper-awareness occurs when, for example, a subject notices and remembers the colors of ties at a party, i.e., irrelevant information comes into view under the guise of environmentally relevant information. These perceptual abnormalities can lead to delusional systems or even to hallucinations (Firth, 1979, p. 230).

argue that the *body* is the fundamental, implicit layer of experiential reality.[3] In addition, Stanghellini incorporates the notion of pre-reflective self-awareness into psychopathology when he writes:

> I experience myself as the perspectival origin of my experiences (i.e., perceptions or emotions), actions and thoughts. This primordial access to myself, or primitive form of egocentricity, must be distinguished from any explicit and thematic form of I-awareness, since it is tacit and implicit, although experientially present. This primitive experience of myself does not arise in reflection, i.e. from a split between an experienced and an experiencing self, but is a pre-reflexive phenomenon (Stanghellini, 2009, p. 56).

There are several points of intersections between Fuchs' and Stanghellini's interpretations. For example, the tacit, implicit self-awareness is closely tied to the experience of the lived body. One of the prevalent symptoms of schizophrenia is the alteration of bodily experience. Stanghellini and Fuchs speaks of disembodied experience, which frequently occurs even in the pre-psychotic phase. In this context ipseity-disturbance is the pathological modification of the (bodily) minimal self. Therefore, bodily self-awareness plays a crucial role in the IDM-model. Sass (2017) underscores the undisputable contribution of William James in defining and examining the relationship between the self and the body. Sass remarks that for James the self cannot be a purely spiritual element that can be investigated by the process of introspection. Rather, even introspection reveals bodily processes "taking place within the head" – as James has formulated. Introspection may unintentionally transform the "transitive" aspects of the self (such phenomena as feelings of relationship, of processes, and of activity) into "substantive", objectified ones (p. 181). According to Sass' interpretation, James believed that reflection or introspection does not discover the signs of an inner self, but only "a variety of kinesthetic of physiological processes happening all by themselves." (*ibid.*). In general, phenomenological psychopathology scrutinizes such anomalous self-experiences in which the intertwined nature of lived time, lived body, and the stream of experience is fragmented, or at least disturbed by psychosocial stressors, existential crises, or any other kind of affliction.

In the following chapters, two main facets of the ipseity-disturbance model are described: the altered *self-affection* and, as a consequence, the strange phenomenon of *hyperreflexivity*. Stanghellini's notion of *ipseity* is influenced by Michel Henry's philosophy: "Thus, ipseity is the implicit, pre-reflexive, immediate, non-conceptual, non-objectifying and non-observational sense of existing as a subject of aware-ness. It is prior to, and a condition of, all other experience." (Stanghellini, 2009, p. 56) The rationale behind this argument is that there can be cases in which the self, understood here as ipseity, disintegrates, and consequently a special kind of dualism takes place between the person's acute reflection and his or her lived body: "The

[3] "The subject body functions as the medium and background of our experience. Although it is itself not perceived, it operates in every action and interaction with others, without requiring explicit attention. It encompasses those abilities and dispositions that are neither representations nor rules but are actualized in our everyday life in a tacit or implicit mode, before we can reflect upon our experience" (Fuchs & Schlimme, 2009, p. 571).

crisis of ipseity is the clearest expression of the shape schizophrenic life assumes as a deanimated body (i.e., a body deprived of the possibility of living personal experiences – perceptions, thoughts, emotions – as its own) and also as a disembodied spirit (i.e., as a sort of abstract entity which contemplates its own existence from outside – a third-person perspective view, or a view from nowhere)." (Stanghellini, 2009, p. 58).

Besides *ipseity*, another important target phenomenon of contemporary phenomenological psychiatry is the atmospheric alteration of the lifeworld. The progenitor of this approach is Klaus Conrad, who further elaborated the notion of "delusional mood" or "atmosphere" that is often filled with tension and anticipation (cf. Mishara, 2010, p. 9). Jaspers claimed that delusional moods (*Wahnstimmungen*) are without precise or objective content and yet are insufferable for the subject (cf. Jaspers, 1973, p. 82). As we shall see, Sass's description of the stages of prodromal schizophrenia, which is a period preceding the onset of full-blown delusions, relies heavily on Conrad's notions of *Trema* and *Apophany* to (Conrad, 2013 [1959], cf. Sass & Ratcliffe, 2017, p. 95). In general, the following phenomenological investigations are trying to unravel the lived world of patients, who are supposedly in the pre-onset stages of schizophrenia. The priority of endogenous self-disturbance over psychosocial stressors is open to debate in these approaches. There is also a heated debate on the importance of self-disorder in the development of positive and negative symptoms of schizophrenia.[4] This intriguing issue is to be discussed in the following section. It is also worth mentioning that these phenomenological approaches frequently criticize the operational diagnostic systems of clinical psychiatry.[5] In the following section, I will analyze in greater detail the IDM-model, which was developed to serve as an interface between neurobiological and phenomenological studies of schizophrenia.

[4]According to Sass & Parnas (2003) the negative symptoms are poverty of speech, affective flattening, avolition, apathy, anhedonia, and withdrawal from social life. The positive symptoms may involve the lack of boundaries between self and others and diminished self-affection, i.e., patients feel such a way as if they have lost the "sense of inhabiting" their own actions, thoughts, feelings, etc. (p. 431, 433). Ratcliffe (2008), in his discussion of schizophrenia, cites the DSM-IV-TR to summarize the positive and negative symptoms. The positive symptoms are delusions, hallucinations, disorganized speech, self-monitoring of behavior and the negatives are the lessened range and intensity of emotional expression, decrease in fluency and production of thought and speech, and last but not least, the loss of volition (p. 190).

[5]As Stephenson and Parnas critically argue: "It should be remarked, that contemporary psychiatry has lost sight of this Gestalt due to the development and dominance of the psychiatric operational diagnostic systems (i.e., DSM-Vand ICD-10), which gradually have left behind the focus on the actual lived world of the patients and abandoned the original rich psychopathological descriptions in favor of simple questionnaires and checklists." (Stephenson & Parnas, 2018, p. 631)

5.1 The Ipseity-Disturbance Model of Schizophrenia

The inner world of the schizophrenic has been a philosophical and psychological conundrum, but recent phenomenological psychiatry is attempting to explain the prodromal phase of schizophrenia with renewed impetus. Thomas Fuchs, Louis Sass, and Joseph Parnas provided new insights into the strong connection between the minimal self and embodiment. For them, Merleau-Ponty's phenomenal body (*Leib*) and Michael Henry's self-affection are the main building blocks for interpreting social alienation and bizarre delusions. As already argued in the previous chapters, pre-reflective self-awareness and self-affection have priority over any occurrent intentional act and they also serve as conditions of possibility for personhood. In fact, the lived body consists of a multitude of "I can" (primordial self-referentiality) and has an absolute constitutive role in conscious life (Merleau-Ponty, 2012, pp. 133–138). As Fuchs argues: "The self-referentiality that is rooted in the auto-affectivity of the body is indeed imparted to all our perceptions, actions, and thoughts." (Fuchs, 2005a, p. 96) In this respect, Fuchs and Schlimme (2009) describe schizophrenia as a kind of disembodiment of the self, accompanied by the impairment of automatic processing and Gestalt-formation, leading to dissociation between the lived body (*Leib*) and self-awareness. The fragmentation of self-awareness is interpreted as a malfunction of the pre-intentional or passive syntheses. As a result, the patients' phenomenal field is overloaded with insignificant details of sensations.[6] In this way, persons in the prodromal phase of schizophrenia lose their habitual meaning-structures and fall out from the intersubjective lifeworld. According to this model, in melancholia, for example, the so-called "hyperembodiment" sets in, which can also be interpreted as an affective depersonalization. In the first phase of this mode of being the body becomes a heavy obstacle imbued with negative feelings like anxiety, followed by depersonalization or the "nihilistic culmination of melancholia", in which the subject is dissociated from his or her "decaying body" and becomes a detached, alienated observer of his own life. Fuchs & Schlimme (2009) conclude that: "This kind of disembodiment is not restricted to depression but occurs in other depersonalization syndromes as well, in which the affective sense of self is disintegrated and the body is experienced as an object among others." (p. 573). Of course, this skeletal outline above cannot be considered a completely exhaustive analysis of disembodiment.

Before delving into the problems of the ipseity-disturbance model of schizophrenia (IDM) it is necessary to identify the key dilemmas of contemporary schizophrenia-research, which is itself a vast and complex domain. Sass (2017) remarks that the

---

[6]Fuchs (2005a) argues that Husserl and Merleau-Ponty intuitively grasped the phenomena of *knowing-how* and implicit knowledge by the concepts of *passive syntheses* and *operative intentionality* (p. 97). Single elements of perception and action are spontaneously organized into meaningful wholes without any conscious or intentional action. This means that in our day-to-day living we automatically know how to read or how to dance. The exploration of the implicit layers of cognition and action is one of the most fruitful research projects in contemporary cognitive psychology and phenomenology as well.

main concern of schizophrenia research is to legitimize the term "schizophrenia" as a tenable diagnostic category. Fundamentally, schizophrenia is a spectrum disorder and a controversial category in its own right. Sass argues that the boundaries of the term are uncertain and essentially ill-defined, but despite the ongoing criticism it always returns as a diagnostic concept (p. xv). For the sake of simplicity, it is worth distinguishing between the biomedical and phenomenological conceptions of schizophrenia. From the biological perspective, schizophrenia$^{(a)}$ could designate a neuro-cognitive disorder with pathological biomarkers that can be characterized by *in vivo* and *post mortem* clinical studies.[7] The second definition, let us call it schizophrenia$^{(b)}$, has a much wider horizon of meaning: we can include not only the medical considerations but also philosophical, phenomenological, psychoanalytical, and last but not least the anti-psychiatric insights. The concept of schizophrenia discussed in this chapter is intended to serve as an interface between the biomedical and phenomenological standpoints. In light of this interface function, the third schizophrenia concept can be named as schizophrenia$^{(c)}$, which also suggests a phenomenologically oriented research interest. As we will see, Sass and Parnas, and other authors of contemporary approaches focus on holistic and qualitative analysis, and this may be one of the reasons why the research on the prodromal phase of schizophrenia has come to the fore in recent discussions. Obviously, we can find the plethora of diagnostic approaches with respect to schizophrenia spectrum disorder. For example, Sass (2017) highlights that among the classic reductionist movements one can find the so-called *modular* approaches, which distinguish between symptom groups and cognitive dysfunctions, or, one can pay attention to the so-called *dimensional* views, which identify symptoms as points on a scale. These latter approaches and other holistic studies may obfuscate the distinctive characteristics that may lie behind the condition of schizophrenia (p. xvii). Thus, paradoxically, contemporary phenomenological psychiatry, while remaining faithful to the basically anti-reductionist perspective of phenomenology, also seeks to gain a foothold among neurobiological approaches by reintroducing the concept of self-disorder. In this respect, the reference to self-disorder is not without ambiguity: it remains an open question whether self-disorder refers to an endogenous neurobiological disorder, i.e., to the affliction of an ineluctable disease, or whether it has a broader meaning that includes the disorders of interpersonal relationships and existential crises.

Another important issue is the stigmatizing tendency of the term. Sass (2017) remarks that there has been a slight change in the contemporary use of the term. Contrary to the customary phrases of "schizophrenic person" or "schizophrenic patient", the expressions of "patient or person *with* schizophrenia" become prevalent. Sass argues that the former, customary usage implies a strong identification between the person and his or her illness, whereas the new phrasing tries to avoid stigmatization and places more emphasis on the status of the person as opposed to

[7] For example, recent genetic studies (more precisely, "genome-wide association studies" or GWAS) identified several genetic markers, which demonstrate the involvement of the immune system and other peripherical factors in the aetiology and pathophysiology of the disorder (cf. Pardiñas et al., 2018; Zhiqiang et al., 2017).

the illness (p. xvii). Of course, this person-centered, or more precisely *self*-centered, approach is also dominant in contemporary phenomenological psychiatry. Contemporary phenomenological psychiatry, including the IDM-model, was inspired and motivated by earlier, particularly German phenomenological psychopathologist. Eugene Minkowski's idea of the loss of "vital contact" or Wolfgang Balankeburg's loss of "natural self-evidence" are frequently cited theories in Fuchs' and others' contributions. Another progenitor of recent approaches is Klaus Conrad who also dealt with alienating experiences (cf. Sass, 2017, p. xiii). However, the question of the status of phenomenology in these frameworks is not without difficulties. As the short introduction above indicates, contemporary models transformed the problem of *ipseity* into a target property of psychopathological, and even neurobiological, studies. By contrast, Morley (2003) highlighted the fact that the real merit of phenomenological psychiatry is the return to life itself, and the fact that phenomenology could offer a holistic, integrative framework even for biomedical research. Consequently, the meta-scientific and critical attitude of phenomenology must be acknowledged and maintained in contemporary discussions (pp. 89–90). Zahavi (2005) contends that psychopathology, as a "heuristic device", can help to unravel the hidden layers of ordinary experiential reality. At the same time, he warns against the unreflected use of psychopathology. He does not propose a simplified contrastive analysis, but rather accentuates the consequence that the distortions of normal experiential reality indirectly reveal the constitution of normality, which is, he adds, in itself an achievement (p. 133).

Ratcliffe (2008) sees a *hermeneutic* relationship between psychiatry and phenomenology. Descriptions of pathological experiences might inspire phenomenological inquiries to reconsider their own descriptions of experiential reality. The result of questioning phenomenological assumptions might contribute to the new understanding of psychiatric illness (p. 9). It is surprising from the hermeneutical perspective when Ratcliffe goes on to argue that phenomenology could "supply explananda for scientific explanations" (*ibid.*, p. 123). The task of phenomenology is to give detailed and clear descriptions of the phenomena. However, Ratcliffe also shows the drawbacks of the mutual relationship between neuroscience and phenomenology. He describes "existential feelings" as such modes of orientations that cannot be explained by neuroscientific research (*ibid.*). In this context, existential feelings are similar to the Heideggerian *moods*, which are the pre-given structures of being-in-the world, and they also have a latent constitutive role in scientific endeavors. (In the following section I will further discuss the role of existential feelings in psychopathology.) In general, Ratcliffe (2011) argues for the role of feedback loops between phenomenology and the empirical sciences, which could lead to a weak naturalism. In this context, phenomenology aims to unravel and interpret anomalous self-experiences; the alternative modes of being-in-the-world. Ratcliffe is faithful to the fundamental insights of Husserlian phenomenology in arguing for the primacy of phenomenology over the ontology of scientific naturalism. He argues that "to adequately explore alternative ways of being in the world, one must first recognize the contingency of a way of being in the world that the intelligibility of empirical science depends on." (p. 35). The contingency of our being and the contingency of our scientific and pre-scientific endeavors could seriously undermine any psychopathological undertaking. In other words, the question of normativity comes into sight.

Overgaard (2007) observed a vicious circle in phenomenologically oriented psychiatry: on the one hand, a psychiatrist might characterize anomalous experiences based on the phenomenological accounts of the "normal" experience. On the other hand, the search for anomalous experiences and their psychopathological correlates, might confirm the presuppositions of normality offered by the phenomenologist. Overgaard notes that this circularity necessitates that phenomenology, defined as the study of experiential reality, must not be a "closed subject" (p. 40). As a result, the cross-disciplinary approach in question requires constant reflection from both camps of scholars. The task of the following chapter is to examine the role and utilization of the phenomenological concepts in the IDM-model of schizophrenia, keeping in mind the above-mentioned admonitions and qualms.

5.2 Existential Feelings and Self-Affection

Fuchs (2005b) made a distinction between two layers of intentionality in embodied perception: in normal circumstances the (1) *active* or bodily intentionality is in synch with (2) *conscious* intentionality (p. 135).[8] Merleau-Ponty claims that consciousness is grounded in the lived body's own intentional arc; that is, perception, action, and the agency of the subject are interconnected in day-to-day living. The Merleau-Pontian habitual and lived body is not only initiates actions or performs skills, but also provides various subconscious reactions to events in the environment. Fuchs sees the *bodily mimesis* as a kind of interaffectivity between persons on the one hand, and as a *bodily resonance* with the environment on the other: "We can empathically feel objects moving up and down, fastening or slowing down, their 'crescendo' or 'decrescendo'. Things are perceived by the body through an assimilation or *mimesis*, and in handling them, it also appropriates and *incorporates* them." (Fuchs, 2005b, p. 135) Thus, following Merleau-Ponty, Fuchs describes the Husserlian *conscious intentionality* and the Merleau-Pontian *active intentionality* (i.e., the pre-reflective performance of the body) as the two complementary aspects of the same intentional relation.

In addition, Fuchs introduces the notion of "inversion of intentionality", which graphically illustrates the self-referential and paranoid character of schizophrenic

[8] As it was discussed in the previous chapters, *passive synthesis* works automatically under the level of noetic intentionality. There is also a *pre-affective hyletic synthesis* under the surface of active object-consciousness. However, for Merleau-Ponty, passive synthesis is an abstract construction, and he, as a consequence, emphasizes the pre-intentional role of bodily intentionality. Bodily based intentionality is a kind of *mimesis*; a spontaneous process in which our *body schema* adapts to the environment (Merleau-Ponty, 2012). Not surprisingly, Husserl also accentuated the role of the body as kinaesthesis in the pre-intentional hyletic synthesis.

delusional mood and perception. In case of the loss of consensual meaning-structures and habituated perceptual-patterns, patients tend to feel that external objects are staring back at them from the perspective of a "hidden subjectivity". This uncanny feeling may give rise to explicit paranoid delusions. For example, the "passive reception of images" generates paranoid anxieties of being watched or manipulated by unknown forces. In the end, the subject finds him or herself in a completely self-referential, egocentric world (Fuchs, 2005b, p. 137). The result of the dissociation between the intentional and pre-intentional layers of experience is a disengagement from the practical world. The subject is distanced or dislodged from the consensual lifeworld:

> Instead of the common and intersubjective significance of things or situations, e.g. 'this is a table set for the meal', there arise idiosyncratic or archetypal fragments of meaning, always alluding to the patient and his body. The smell of the soup on the table may suddenly take him back to his childhood or to another country; the white colour of the tureen may signify his purity and innocence. But the meal could also be prepared to celebrate a mass in which he shall be sacrificed; the screwed legs of the table indicate that he shall be tortured. In this way, the mimetic faculty of the lived body creates idiosyncratic meanings that seem unmotivated… (Fuchs, 2005b, p. 137)

This passage draws attention not only to the *loss of common sense*[9] but also to the fragmentation of inner time-consciousness brought about by amplified auto-affection. Contemporary scholars reveal the crucial role of affection and mood in mental disorders. This does not mean that the various and diverse symptoms of schizophrenia can be reduced to some kind of mood or affective disorder, but there is a marked, frequently observable, change in the overall existential orientation in persons with schizophrenia. Considering the reports of persons with schizophrenia, Sass speaks of an "ontological unhinging" that means the "fundamental transformation in the coordinates of experience itself" (Sass, 2017, p. 236).

What I intend to underscore here is Sass' observation that patients with a general uneasiness and anxiety are not concerned with some specific content, but rather their fears and delusions show a more pervasive and obscure character. As we shall see, Ratlciffe (2008) also draws attention to the overall existential changes in schizophrenia. Early descriptions of schizophrenia, such as those of Krapelin's and Bleuler's accounts, have already focused on the diminished or even the absence of affect in schizophrenia. But Ratcliffe has serious objections to the presupposition that affective states are internal states of the subject that can be projected onto objects in the environment. In contrast to the internal-external division, a more nuanced and sensitive phenomenology seeks to reveal the *phenomenology of*

[9] Blankenburg argues that psychopathology should be more philosophical and the term "common sense" is useful in describing psychopathological states of alienation. The loss of common sense in a person with schizophrenia is not identical with a loss of IQ or thought disorder (i.e., stammering and struggle for words), but rather the disturbance can be traced back to the *pre-predicative* layer of understanding; for example patients do not know how to behave, they have difficulty with practical actions, and the ability to make contact with other people is missing. As one of the patients eloquently put it: "I find that I no longer have footing in the world. […] It seems I lack a natural understanding." (Blankenburg, 2001, p. 307).

belonging: the altered existential background cannot be shoehorned into the model of the external-internal dualism, but rather begins by examining bodily and wordly changes simultaneously (p. 189).

Despite the non-dualistic attempts in contemporary phenomenological descriptions, internal affective states have their own legitimate use in several frameworks. In particular, the IDM-model focuses on the problem of self-affection, and for Fuchs auto- or self-affection is not only the primal feeling of being alive, but also the second layer of intentionality. Without affection and evaluative emotions, the world would be bleak and meaningless, and, as Husserl argued, attention is modulated by affective forces. In a similar vein, Fuchs speaks of "affective affordances"; through which objects and persons can be perceived as attractive or repulsive factors. In the case of affective intentionality, the self- and world-reference are simultaneously present (Fuchs & Koch, 2014, pp. 2–3). In other words, the pre-reflective attunement of the lived body discloses and evaluates a certain situation and suddenly reveals personal feelings in these circumstances, e.g., relations, interests, conflicts (Fuchs, 2013, p. 626). Fuchs emphasizes that bodily feelings are the *very medium* of affective intentionality. Thus, pre-reflective self-awareness is not an abstract visuo-spatial relation to the world, but rather a complex achievement grounded in bodily/ affective intentionality. Fuchs was inspired by Herman Schmitz, who also argued that the lived body is primarily a *felt body*, and perceiving through the lived body's medium is basically a receptivity to the vibrations of the surroundings (Schmitz et al., 2011, p. 251).

Fuchs draws on Merleau-Ponty's ideas, who argued that the lived body not only "vibrates" and "resonates" with the multisensorial environment, but also has the peculiar ability to spontaneously project a past situation into the present.[10] Merleau-Ponty claims that the living present always carries the past in itself as a kind of "pre-history", and the synthetic activities of consciousness retrieve the past against the background of the unpredictable future.[11] On Fuchs' account, the loss of common sense and the habituated performances of the patients are due to the displacement between *active-bodily* and *conscious* intentionality. The result is a fusion of body and environment based on the mimetic function of the lived body. Inspired by Merleau-Ponty's analysis of the body schema, Fuchs argues: "With increasing weakness of the intentional, objectifying perception, the ecstatic potential of the lived body awakes. Turned inside out, as it were, it may change into a single sensual

[10] Synesthetic sensations and sympathetic feelings are the telling examples of the entanglement of body and environment, in the latter case, for example, we may have a certain feeling in our lived body that "empathically" mirrors surrounding objects moving up and down (Fuchs, 2005b, p. 135).

[11] "It is not a personal act by which I myself would give a new sense to my life. Within sensory exploration, I do not give a past to the present and orient it toward a future as an autonomous subject, but rather insofar as I have a body and insofar as I know how 'to see.' In addition to being a true history, perception also confirms and renews in us a 'pre-history.' [...] There are no connected objects without an act of connecting and without a subject, there is no unity without a unifying, yet every synthesis is simultaneously taken apart and remade by time, which, in a single movement, puts it into question and confirms it because it produces a new present that retains the past." (Merleau-Ponty, 2012, p. 250)

surface that fuses with the environment." (Fuchs, 2005b, p. 138) Following Merleau-Ponty, Fuchs (2005b) accentuates that in schizophrenia the dissolution of ego boundaries can be explained by the exaggeration of the *pathic* or *mimetic* function of perception. Hallucinatory sensations may detach from ordinary objects and obtrude into the gaze of the subject like phantoms or figments of imaginations. Other manifestations of this disoriented state can be the birth of "obtrusive and idiosyncratic meanings" or the spontaneous obtrusion of dream states into consciousness. The crux of the argument is the loss of common sense, the loss of the intersubjective footing in the world and of the intersubjective constitution of reality, which can lead to the egocentric world of psychosis (p. 138).

Another central claim that is of pivotal importance is the underlying affective and hyper-associative schematization process in delusional perception. Fuchs (2005b) says that "Having lost their objective and conceptual unity, things may gain an overwhelming wealth of expressions, charged with emotional qualities. Especially the gaze of others, the quintessence of expression, obtains a captivating and piercing power." (p. 138). Due to the disturbance of the sense of reality the patient expresses "idiosyncratic or archetypal fragments of meaning". The simple smell of a soup on a table may conjure up a childhood memory or the image of a distant country. Paranoid fears are characteristic in delusional perception. Fuchs recounts that the patient may infer from the screwed leg of the table that he shall be tortured (*ibid.*, p. 137). Ratcliffe (2017), who was also fascinated by Merleu-Ponty's description of a hallucinating schizophrenic patient, reformulated the mimetic function of the lived body as the a "conjuring trick" of the body. Determining the nature of hallucinatory experience is a very complicated matter, but according to Ratcliffe, two fundamental, recurring aspects can be discerned: while hallucinations may vary widely in type and complexity, the "*feeling* of perceiving" and "being *affected* by something" are the most common sentiments. Hallucinations may have impoverished sensory content and lack appropriate horizonal structure, but they certainly have an ambiguous feeling component.[12] Ratcliffe quotes from the frequently cited autobiography of the schizophrenic girl, Renee: "I cannot say that I really saw images; they did not represent anything. Rather I Felt them" (Ratcliffe, 2017, p. 198, cf.

[12] The horizonal structure means that we do not experience things in themselves but rather in an appropriate context. Ratcliffe, drawing on Husserl's and Merleau-Ponty's philosophy, regards the horizonal structure of consciousness as the framework of possibilities and the capacity to pre-reflectively discriminate between the modalities of intentionality. With regard to hallucinations, he argues that "One of the things that distinguishes hallucinations from veridical perceptual experiences is that they lack the full horizonal structure of perception. An entity, as hallucinated, does not present itself as amenable to various kinds of perceptual, practical, and interpersonal access. For instance, certain intersensory possibilities are absent." (Ratcliffe, 2017, p. 196).

Sechehaye, 1970, p. 58–59).[13] What is real and what is illusory (or hallucinatory) is not only the result of higher-order cognitive processes, but there is an underlying "sense of reality" owning to socialization.

Ratcliffe (2017) shows that the sense of reality, the experience of our surroundings in general, is shaped by the wider social structures and the habitual trust in perception and other people (p. 148). The social context helps us to recognize what appears to be salient and what kinds of meaning and significance can be attached to things (p. 140). In short, our being-in-the-world is rooted in our habitual certainty. Not surprisingly, our crystallized habitual certainty can only be reflected indirectly through unusual or even traumatic events. It is probable that sooner or later a meteorite will hit the earth and wipe out human and non-human life, but in everyday life no one pays attention to such dreadful scenarios or to the statistical chances of the impact. It is far more likely that the world economy will collapse in the near future for ecological, financial, and political reasons, but in our ordinary circumstances we brush aside these disturbing ruminations without hesitation and do what we have to do to survive. The above examples may illustrate the fact that habitual certainty has multiple layers, including existential, perceptual, and affective.[14] The central question to ponder: how the loss of trust can lead to the fragmentation of the self; a state in which mistrust, anxiety, and, in later stages, delusions or even verbal hallucinations arise.

[13] It is worth to cite the further sentences from the autobiography to demonstrate how horrifying and disturbing can be a hallucinatory experience: "It seemed that my mouth was full of birds which I crunched between my teeth, and their feathers, their blood and broken bones were choking me. Or I saw people whom I had entombed in milk bottles, putrefying, and I was consuming their rotting cadavers. Or I was devouring the head of a cat which meanwhile gnawed at my vitals. It was ghastly, intolerable. In the midst of this horror and turbulence, I nonetheless carried on my work as a secretary. But with what hardship! Adding to the torment, strident noises, piercing cries began to hammer in my head, their unexpectedness made me jump. Nonetheless, I did not hear them as I heard real cries uttered by real people. The noises, localized on the right side, drove me to stop up my ears. But I readily distinguished them from the noises of reality. I heard them without hearing them, and recognized that they arose within me" (Sechehaye, 1970, p. 59). The passage draws attention to the curious phenomenon of "double-bookkeeping". The concept was coined by Jaspers and further described by Sass. The first-person account of schizophrenic subjects may suggest that hallucinatory experiences can be distinguished from perceptual experiences; however, not only the former but also the latter seems to be a unique intentional state. In other words, the patient may find himself or herself in the perplexing state of believing in a fact or situation that is also known to be false or delusional. What complicates matters further is the fact that hallucinations are superimposed on the ordinary world (Ratcliffe, 2013, p. 241; cf. Ratcliffe, 2017, pp. 198–199). For the detailed analysis of double-bookkeeping, see for example Gallagher (2009). and Bortolotti & Broom (2012).

[14] Habitual certainty is inseparable from the anticipation-fulfillment structure of intentionality. As Ratliffe describes: "Many of us anticipate most thing in the guise of habitual certainty. It does not occur to us that we will be deliberately struck by a car as we walk to the shop to buy mil or that we will be assaulted by the stranger we sit next to on the train. This is not to suggest that we naively trust everybody or feel safe all the time, regardless of circumstance. The point is that, when we do feel unsafe in a certain situation or explicitly distrust a particular person, this involves a departure from our default attitude." (Ratcliffe, 2017, p. 121)

One of the major challenges of phenomenological psychiatry is to explain the transition from the delusional atmosphere, which is essentially an affective change, to full-blown psychosis including delusional states and hallucinations. Ratcliffe proposes that the delusional mood or atmosphere, considered ineffable by Jaspers, can be characterized as a "pervasive and unpleasant feeling of uncertainty" that precedes the articulated delusions (Ratcliffe, 2017, p. 111). That is, between the ineffable atmospheric alteration of the sense of reality and acute paranoid delusions lies the dimension of being-in-the-world, which is difficult to represent. Delusional beliefs, and estrangement in general, imply the antecedent change in the sense of belonging to the world; not only the content of thinking and feeling is altered, but also the framework or horizon of the self-world relation (cf. Sass & Ratcliffe, 2017, p. 92). In a very similar vein, Fuchs argues that the sense of reality does not exclusively depend on higher order cognition (i.e., belief or judgement), but rather there is an "affective foundation" of shared reality (Fuchs, 2013, p. 616). Ratcliffe highlights the fact that even emotional regulation depends on the environment and intersubjective relations:

> If one could not feel at ease in the presence of others, one could not become comfortably immersed in a film at the cinema or in the communal atmosphere of a church service. And, to the extent that one enjoys drinking and smoking with other people in a given social environment, the regulatory effects of these pastimes depends on who else is there and how one relates to them. As for the effects of music, these can involve the elicitation of all sorts of interpersonal feelings. Change in the style of interpersonal experience could thus influence which feelings, if any, a piece of music elicits (Ratcliffe, 2017, pp. 150–151).

Delusions arise from the disrupted integrity of perceptual experience and affective attunement. Ratcliffe lists several reasons for the emergence of delusions. The most basic reason can be the disengagement from practical activity and isolation from the public world. The result may be the disruption of the structure of intentionality, namely, one is deprived of the pre-reflectively felt difference between perceiving or imagining. The ensuing ambiguity and indeterminacy of perception can lead to paranoid delusions that grounded in the actual style of "affective anticipation". Anxiety and the dread of the unknown are the most prevalent negative affective anticipations that can lead to full-blown delusions and hallucinations. Ratcliffe emphasizes that anxiety is not only the result of clear verbal hallucinations, but it plays a crucial role in the genesis of hallucinations (Ratcliffe, 2017, p. 71). Ratcliffe (2017) argues that the (social) alienation leads to the erosion of habitual certainty, which also contributes to the disintegration of anticipation-fulfillment structure. The subject suddenly finds himself or herself in a perplexing state in which he or she is unable to clearly sense the marked differences between intentional states. More precisely, the changes in interpersonal experiences may include: "(a) the lack of interpersonal validation, (b) loss of phenomenological coherence, (c) disengagement from practical activity, and (d) wide-ranging disruptions of affect regulation" (p. 151). Another outcome of the estrangement from other people and the consensual world is the diminishment of the distinction between memory and perception. Ratcliffe argues that due to passivity or more severe derealization, distant memories may resurface, and certain memories may also lead to verbal hallucinations (pp. 162–163; cf. Michie et al., 2005).

The idea of existential feeling, mentioned above, is Ratcliffe's other important contribution to the explanation of anomalous self-experiences. As I mentioned earlier, existential feelings are often likened to Heideggerian moods in the sense that they have an enveloping quality; however, Ratcliffe highlighted the role of the simultaneous presence of the lived body and the feelings of belonging to the world in their constitution.

Fuchs (2013) made a subtle distinction between mood and affective atmosphere. The former lacks intentionality but serves evaluative purposes and has a dispositional character (it renders the subject to feel, to perceive, and to act in a certain way), whereas the latter is, by definition, the affective quality of experienced spatial and interpersonal situations (p. 616). Fuchs argues that affective atmospheres are determined by the structural qualities of perception. The idea of affective atmosphere can be distinguished from Heidegger's mood and was inspired by Gestalt psychology and Stern's vitality affects. What are the sources of affective atmospheres? Fuchs (2013) argues that the two main sources of affective atmospheres are the physiognomic or expressive qualities of objects and the intermodal features of perception (i.e. rhythm, intensity, dynamics, and so on). The affective atmospheres are unitary dynamic Gestalts that integrate the various characteristics and features of situations. Suggestive examples of affective atmospheres are the "feeling the hilarity of a party, the sadness of a funeral march, the icy climate of a conference, awe-inspiring aura of an old cathedral, or the uncanniness of a somber wood at night" (p. 616). Affective atmospheres are considered more objective and distinct than existential feelings and moods. Furthermore, Fuchs claims that Jaspers' "delusional mood" can be described as "a characteristic atmosphere of perplexity, uncanniness, and enigmatic significance in which the patient senses an inexplicable change in his environment" (p. 617).

It is fruitful to consider the close connection between Fuchs' notion of auto-affection and Ratcliffe's existential feelings. Existential feelings are feelings *in* the body and feelings of being-in-the-world; that is, as Fuchs accentuates, affectivity is an embodied and extended phenomenon at once (Fuchs, 2013, p. 626).[15] Ratcliffe argues that when we analyze alienated experiences, in which the loss of trust in the world and others is salient, we need to pay attention to the elusiveness and temporality of holistic existential feelings: "Existential feelings unfold in time, arising, changing, and disappearing, but they are also ways of experiencing time, of inhabiting time." (Ratcliffe, 2015, p. 38) Existential feelings are not identical with episodic emotions, nor are they propositional attitudes or beliefs. Ratcliffe argues that the affective changes and pathological beliefs presuppose the "feelings of being" (Ratcliffe, 2008, p. 9). According to Ratcliffe (2008), existential feelings are capable of shaping our "dispositions to act and regulate activities", and they also

[15] It is worth noting that Daniel Stern's notions of "affect attunement" and "interaffectivity" had a serious impact on development of the core self (Stern, 1985).

delineate the space of possibilities of experience and activity (p. 121).[16] On the one hand, existential feelings have a holistic, global character, but, on the other hand, Ratcliffe supposes a discernible structural complexity under the features of homogeneity and elusiveness. The main components of existential feelings are bodily proprioception, kinaesthesia, and interoceptive awareness of visceral feelings (Ratcliffe, 2008, p. 123). These are also the phenomenologically inextricable components of general existential orientations. Another important facet of existential feelings is that they provide the possibility to reconsider or even to dismiss traditional dualisms such as the bodily and the worldly, the internal and the external, the subjective and the objective, and so on (Ratcliffe, 2008, p. 8). Existential feelings bring forth the context or horizon of practical relatedness, and it may be the case in psychopathological circumstances that an altered sense of possibility ensues from the disintegration of intentional relatedness (Ratcliffe, 2008, p. 121).

On my account, owing to the role of the lived body, there is a close connection between the affective relief structure and existential feelings,. Ratcliffe often refers to Husserl and Merleau-Ponty to circumscribe the nature of existential feelings. Both thinkers emphasized the role of bodily dispositions and activities in the constitution of the overall horizonal structure of experience. Ratcliffe (2008) cites Husserl's passages in which he speaks of the "affective pull" of "enticing" objects that has a serious impact on consciousness as well as on the lived body (*Leib*). And, in turn, the dispositions of the lived body are mirrored in perception: "The possibilities that structure perception are felt bodily potentialities, which Husserl refers to as 'kinaesthese'" (p. 132). In sum, Ratcliffe (2008) asserts that existential feelings constitute the world as a "universal horizon" (p. 133). The alternation of existential feelings are often occur indirectly through the mediation of the altered bodily relationship to the world. Because of an altered existential feeling, the world may seem suffocating or overwhelming. Furthermore, the change of the space of practical possibilities may unnoticedly lead to an altered sense of reality (Ratcliffe, 2008, p. 135).

Fuchs' dual-intentionality structure also involves an affective attunement below the level of conscious awareness. Fuchs, following in the footsteps of Merleau-Ponty, speaks of the "ground of becoming" and the "origin of spontaneity"; that is, a passive-automatic infrastructure lurks beneath conscious intentionality and is closely connected to animate life (Fuchs, 2017, p. 11). Alternatively, but in a very similar vein, Schmitz speaks of the "primitive present", which is based on the radiating emotional atmospheres of the lived body, and he argues for the affective

[16] It is noteworthy that existential feelings do not amount to the sum total of possibilities: "Existential feelings do not consist of the sum total of specific possibilities involved in an experience. Rather, they constitute the general *space* of possibilities that shapes ongoing experience and activity. In addition to changes in specific possibilities, the overall possibility space can be heightened, diminished, open or constrained. It can be fairly fixed in structure or changeable. It can involve a different overall balance of active and passive possibilities, taking the form of an arena of things to pursue or a realm of potential happenings. In conjunction with this, one can have an overall feeling of control, of belonging, of helplessness or of vulnerability." (Ratcliffe, 2015, pp. 121–122)

involvement of self-consciousness that is prior to any self-ascription (i.e., self-narrative).

> Therefore, facts of affective involvement contain a self-consciousness prior to any identification and self-ascription, since in their plain factuality, without regard to their propositional content, the subjectivity for the conscious subject already contains a 'mine-ness' [Meinhaftigkeit], as Kurt Schneider has called it (1950, 130). This priority of self-consciousness is only possible if, in the case of affective involvement, the conscious subject is, without identification, aware of itself. (Schmitz et al., 2011, p. 249)

Schmitz, in line with Zahavi, accentuates the role of time and duration in the constitution of pre-reflective self-awareness. Schmitz argues that the *primitive present* discloses itself as a direct self-acquaintance in cases of raging anger, fear, mass ecstasy, melancholy, etc. (Schmitz et al., 2011, p. 248). The primitive present can be conceived as the purest form of self-affection and self-consciousness in which reflective consciousness has halted and the stream of subjective time seemingly stopped; therefore, we live in the intensity and atmosphere of the actual affection. Regardless of the differences in terminology, the above-mentioned authors demonstrate that pre-reflective self-awareness is inseparable from the affective infrastructure of consciousness. In the analysis of retroactivity, I intend to show that self-affection is not merely the emergence or pure presence of primordial affections (such as the feeling of being or the background sense of belonging), but also has its own intentional and evaluative patterns embedded in the horizontal structure of consciousness (i.e., in the affective relief structure).

Up to this point, phenomenological considerations underscored the intertwined nature of the lived body and being-in-the-world; the projective capacity of the body was closely connected to existential feelings and to the sense of reality. Ratcliffe claimed that the disturbance of the modalities of intentionality leads to altered self- and world-experiences. And Fuchs argued that the dissociation, due to modified auto-affection, between the two layers of intentionality (active-bodily and conscious intentionality) could lead to altered sense of reality. Nevertheless, auto-affection is also a hetero-affection, since disorders of self-regulation are due to environmental or relational disturbances. The disruption of self-world relation is often caused by the loss of basic trust. Ratcliffe refers to Biswanger's description of *disappointment* in which the subject "falls from the skyes" or loses the grip on the world. Altered self-world relation may lead to the existential feeling of estrangement (Ratcliffe, 2008, p. 135). Ratcliffe convincingly argues that the role of loss of trust, traumatic experiences, and other intersubjective consequences in general, are significant factors in the origin of self-disorders. In the following chapter I will turn to the intriguing dilemma of disturbed self-affection.

5.3 From Self-Affection to Hyperreflexivity

Louis Sass, in his groundbreaking book, *Madness and Modernism*, provided the detailed account of the altered self-world relation that occurs during the prodromal phase of schizophrenia. The task of mapping the prodromal phase of schizophrenia is enormously difficult, since the subjects often complain about ineffable moods and say that their metaphors are totally inadequate to clearly express their inner turmoil. Not to mention the drawback, that certain patients do not want to, or unable to talk about their mindset. Sass describes the onset of schizophrenia as an alienation; persons with schizophrenia often feel themselves distanced, disconnected from the world as well as their bodily being. The phenomenological significance of the prodromal phase is that it is not in itself a clear-cut psychopathological entity, but leaves room for scholars to utilize and carefully interpret autobiographical accounts of the changing self-world experiences.

Sass (2017) contends that the schizophrenic breaks, like epileptic seizures, often begin with an aura. He borrows the term *Trema* from Klaus Conrad to depict this preliminary stage. Trema refers to the actor's stage fright that precedes the act of performance; in this frightened state the subject is" filled with anticipation and dread" (pp. 25–26). Sass (2017) argues that in the early phases, the onset of pre-psychotic period can be likened to a kind of mood disorder; that is, a strange, uncanny mood hangs over mundane objects and activities, and the typical emotions such as joy and sadness are suddenly gone. It is noteworthy that the phase of *Trema* poses a challenge even to the descriptive phenomenological accounts, since its volatile and elusive nature makes it compatible with the above-mentioned moods, existential feelings, and affective atmospheres. It is difficult to pin down or categorize it as an affective state, which is why Sass tends to see it as an enveloping, all-encompassing ontological change, perhaps reminiscent of the ontological insecurity discussed by R. D. Laing (1978).

The outcome of the *Trema* could be the phase of "delusional percept," in which a normal or insignificant act of perception is seemingly imbued with a special kind of meaning and significance. However, there is a special kind of mood that precedes the fully developed delusions. Sass calls this stage the "truth-taking-stare", which is the translation of the German expression of "*die Wahrnehmungstarre*" and denotes a rigid, transfixed state of perception (Sass, 2017, p. 26). The truth-taking-stare is accompanied by the dialectical notion of *Apophany* or anti-epiphany. Subjects often intuitively feel as if they could comprehend the hidden meaning of life, but in the end they are unable to adequately express their sentiments. The Greek word *Apophany* means "to become manifest," and the subject in this state feels a strange dialectic between meaningfulness and meaninglessness. Sass contends that Giorgio de Chirico's journal mirrors this uncanny stage in which the familiar becomes unfamiliar and vice versa. Furthermore, the feeling of *déjà vu* is often integral to Apophany (Sass, 2017, p. 26, 32). Sass attempts to illustrate the onset of psychosis with a surrealist mood. Through de Chirico, Sass arrives at Nietzsche's term of *Stimmung,* and he regards it as the main symptom of the onset of schizophrenia. The

Stimmung is a compound of different experiences, and it persists or reoccurs throughout the course of schizophrenia. In addition, it can also occur in various types of the schizophrenia spectrum. According to Sass, the *Stimmung* is a mood or a state of mind that accompanies the truth-taking-stare. Sass gives a succint definition for both *Stimmung* and the truth-taking-stare: the former is essentially a perceptual or emotional experience, while the latter refers to a particular mode of action (Sass, 2017, p. 27). These intertwined states have their own internal structure and complexity which has also been explored and elaborated by Sass.

Sass utilizes the autobiography of the schizophrenic girl, Renee, to illustrate the stages and the progression of *Stimmung*. He distinguishes three closely related stages, the "visions" of Unreality, Mere Being, and Fragmentation. The above-mentioned stage of Apophany constitutes the fourth stage which may appear in the subsequent phases of the psychotic break (Sass, 2017, p. 28). The three interrelated stages are integral parts of the *Trema*, and the fourth stage of Apophany is somewhat independent. In the state of *Unreality,* the world becomes an aimless, emotionless, nebulous sphere where objects have lost their practical meaning-structures; that is, they are not suitable means for familiar intentional actions anymore. Sass claims that the world transforms into a kind of simulacrum or second world, but there is no sign of this radical transformation in the subject's behavior. The state of unreality radically alters the experience of space, but in comparison to neurological patients, the subject does not stumble over objects (Sass, 2017, p. 29). In the stage of unreality, everything is distant and unreal.

Renee recounts the story that she was visited by her friend. Sass underscores that in the altered experience of Unreality Renee was unable to recognize her friend as a living being who tries to empathize with her: "I knew her name and everything about her, yet she appeared strange, unreal, like a statue. I saw her eyes her nose, her lips moving, heard her voice and understood what she said perfectly, yet I was in the presence of a stranger." (quoted by Sass, 2017, p. 29, cf. Sechehaye, 1970, p. 36) In the second phase of the *Trema*, called *Mere Being*, everyday objects like a chair or a jug, obtrude into the focus of awareness and their sheer existence becomes fascinating for the subject. For Sass, the stage of Mere Being can be approached and interpreted through Sartre's and Heidegger's philosophy. The patient's attention focuses on the "mute thereness of the world" (Sass, 2017, p. 30). The fundamental question of metaphysics can be used as an analogy for the stage of Mere Being. On the one hand, Sass cites the Heideggerian question: Why is there something rather than nothing? On the other hand, he uses Sartre's terminology when he writes that in this stage "*essences* recede while sheer *existence* obtrudes" (Sass, 2017, p. 32).

In the third stage of *Fragmentation*, the perceived objects are shattered into pieces of decontextualized sensory data. One patient complained that everything disintegrates like a photograph torn into pieces. Another patient told that there is no way to create a coherent Gestalt of a watch from perceived sensory elements without exhaustive concentration.[17] Fragmentation may occur in various forms and intensities. Perceived objects may seem disconnected and devoid of ordinary context. Furthermore, the situation in which Renee was unable to experience her therapist's face as a coherent Gestalt may indicate the stage at which the perception of a single object is shattered into pieces. Another important facet of fragmentation is the reification of words. To put it simply, the direct referential use of language is thwarted by a felt fissure between words and objects on the one hand, and by the reification of words on the other (Sass, 2017, pp. 31–32). The final stage of the aforementioned *Apophany* is the "exhausting symbolic thinking". Apophany is a mood or vision in which the world resonates with figurative significance. The frightening aspects of Apophany are the ensuing paranoid delusions: for example, random events are interpreted as inevitable, or paranoid ruminations obtrude into consciousness (e.g., the glance or behavior of people are suspicious, unnatural, it may be that they are impostors, etc.). Every single event is imbued with meaning, and due to a "delusional tension" the feeling arises that something is bound to happen, but the subject is unable to explicate what exactly will happen (Sass, 2017, p. 33).

The stages of *Trema* symbolize the global changes of subjectivity. According to the current trend, the "global alterations of subjectivity" and self-disorders are combined and complementary phenomena. In addition, empirical studies demonstrate that self-disturbance is a phenotypic marker of schizophrenia spectrum disorders and it may have a predictive value in marking the onset of psychosis (Nelson et al., 2009, p. 808).[18] According to Nelson and his colleagues (2009), neurobiological researches have already differentiated several levels of selfhood, including Damasio's core self or the narrative and autobiographical self and so on. The minimal self also appears in this context as the *pole* of reflective self-awareness. These studies have something important to add to the diachronic continuity and the synchronic unity of the self. In this framework, ipseity (i.e., the minimal self) is understood as "self-presence of the self-as-subject" (p. 808). In the following section, I

[17] Sass refers to Case 22 from Chapman's work: „I have to put things together in my head. If I look at my watch I see the watch, the watchstrap, face, and so on, then I have got to put them together to get it into one piece." (Chapman, 1966, p. 229) Chapman argues that the visual perceptual constancy is shattered in schizophrenia. He further adds that, based on the testimonies of the same patient, not only perceptual constancy, but also the visual imagery of memories can be disturbed. The patient strenuously constructs a mental image during a conversation, but he or she is unable to hold onto it because of the incessant flow of new information (Chapman, 1966, p. 230).

[18] In a similar vein, Parnas and Sass (2011) argue: „Self-disorders, considered as a core of schizophrenia, may also constitute essential 'phenotypic' markers that can help to define the boundaries of the schizophrenia spectrum disorders and thereby improve our classificatory systems. [...] Another important point is that self-disorders seem to be persisting features of schizophrenia that occur well before the onset of psychosis. This makes them highly relevant for ongoing projects of early-illness detection and intervention in populations considered to be at increased risk for psychosis. A final point is that self-disorders may be crucial for biological research in providing phenotypes that are closely associated with the basic neural-disease correlates." (p. 543)

will briefly examine the hyper-reflexive disorder of schizophrenia and then return to the question of operationalizing the minimal self.

When the sense-bestowing implicit processes are endangered, exaggerated consciousness arises to compensate for the ensuing fragmentation. Sass (2017) characterizes this heightened awareness as "exigent introspection with hyperintentional quality". The exigent introspection can be approached from different angles. It is a possible explanation that the subject begins to pay more attention to the cognitive unconscious, i.e., habitual stimuli, because of neurobiological abnormalities. It may well be that the hyperreflexivity in question is brought about by the "exacerbation of introverted temperamental disposition", or by an "overfocused cognitive style" (p. 186). At this stage the subject is unable to perform his or her usual practical activities without the accompanying reflection. The minute details of perception and language production obtrudes into the focus of consciousness; what was previously at the periphery of attention now springs forth to the sphere of focal attention. The lived body, as the medium of being-in-the world, is objectified, and the subject defines himself or herself as a robot or a puppet, often directed by invisible external forces. Sass (2017) argues that hyper-reflexive awareness breaks through in an affliction-like fashion, leading to the fragmentation of the self. However, it is not an easy task to identify the main causes of the fragmentation process. Sass demonstrates that there is a certain ambivalence in the explanatory role of hyperreflexivity. That is, the fragmentation of the self can be both the *cause* and the *consequence* of the hyperfocused introspection (cf. p. 186).

The "hyper-reflexive configuration of awareness" can be described as a "constant reflexive scrutiny", or a "simultaneous introspection", coupled with the trap or enclosure of interiority, where the normally latent stimuli come into the domain of focal awareness (e.g., automatic popping-up, kinaesthetic bodily sensations, inner speech, etc.) (Parnas & Sass, 2011, p. 537). Parnas & Sass (2011) claim that the notion of hyperreflexivity can be distinguished from hyper-reflectivity. The latter term alludes to a conscious, deliberate process, while the former refers to a more profound change: the "structural alteration of the process of phenomenalization" (p. 540). The authors claim that hyperreflexivity indirectly reveals the pre-intentional, passive accomplishments of consciousness. Thus, echoing Merleau-Ponty, it is reasonable to speak of "operative hyper-reflexivity" that covers the perplexing perceptual and psychosocial anomalies of the prodromal stage, such as the loss of the sense of spontaneity, easiness, and lenience in life (p. 540.). The subject is spontaneously begins to reflect on the automatic popping-up and popping-out phenomena occurring in his or her phenomenal field, or involuntarily focuses on kinaesthetic bodily sensations, inner speech, and the presuppositions of his own thinking (p. 537).[19]

[19] Hyperreflexivity leads to the reification or objectivation of implicit processes: "Phenomena that would normally function as the very medium of self-awareness come to be experienced, rather, as reified, spatial-like entities (spatialization of experience) or even as externalized objects. Trapped in the enclosure in his interiority, the normal world directedness becomes disrupted and impeded, and immersion in the world is impossible." (Parnas & Sass, 2011, p. 536)

The authors conclude that the heightening of consciousness engenders the diminishment of self-presence and self-affection. In other words, the diminishment of self-affection and the ensuing hyperreflexivity are the two sides of the same coin. The authors claim that the person with schizophrenia has too little self-affection on the one hand, and too much self-consciousness on the other hand (Parnas & Sass, 2011, p. 537). Diminished self-presence could be the cause of reports in which patients characterize themselves as automatons, cyborgs, or simply as machines. When Parnas & Sass (2011) characterize the disturbance of the minimal self as diminished self-presence and self-affection, they are also attempting to solve the mystery of phenomenal consciousness. For the authors, self-affection corresponds not only to vital energy (i.e., the brute fact of being sentient or alive), but also to an underlying implicit realm from which desiderative and meaning-bestowing tendencies spring forth:

> The feelings of diminished self-presence and waning first-person perspective are, on this view, the expression of a self-disorder, best described as a disorder of self-affection. And this self-disorder is at the root of the disturbed relationship to objects and the world that is typical of schizophrenia. Self-affection is, after all, the very condition for the experience of appetite, vital energy, and point of orientation. It is what grounds human motivation and organizes our experiential world in accordance with needs and wishes, thereby giving objects their significance for us as obstacles, tools, objects of desire, and the like… (Parnas & Sass, 2011, p. 536)

The above quate can be interpreted as the introduction of an implicit unconscious which consists of the network of pre-intentional and non-representational processes that exert their influence in a self-affective way. On the same page, Parnas and Sass also argue that without the established constitutive rules of self-affection the structure of thought and perception dissolves. The "hyper-reflexive mutation" entails a constant introspection that erodes the passive infrastructure of ipseity or self-affection; and, as a result, mental contents become objectified and spatialized (Parnas & Sass, 2011, p. 535). The intimate connection between thinking and the self may also dissociate and in severe psychotic states inner monologues may transform into hallucinatory voices (*thoughts aloud*), accompanied by diminished self-presence.

The brief analysis of schizophrenic self-disorder provides a good example for the interdisciplinary application of the minimal self. However, the concept of self- or auto-affection raises is not unambiguous. Self-affection turns out to be an umbrella term that designates the pre-intentional, passive accomplishments of consciousness. Here I would like to propose an extended concept of self-affection that includes the dispositions and habits of the lived body (body memory) and the desiderative aspects of drives and instincts. Self-affection may also prefigure the modal structure of intentional states. As we have seen, the disturbance of self-affection may lead to delusional mood and other pre-psychotic episodes in which anomalous or unusual intentional states emerge. In these unusual states, the subject's normal modal structure of intentionality is disrupted, and anomalous states such as verbal hallucinations and inserted thoughts may occur (cf. Ratcliffe, 2017, p. 51). According to my proposal, the immanent self-manifestation of phenomenality, i.e., the feeling of being alive and pre-reflective self-awareness, can be enriched with this broadly understood self-affection.

This broad sense of self-affection is also compatible with Husserl's affective relief structure. If we accept the hypothesis that self-affective occurrences of affective and mental states are not all-or-nothing phenomena, but rather rely on the passive infrastructure of consciousness, we can establish a close connection between the affective relief structure of perception and self-affection understood as the immanent self-manifestation of phenomenality. The idea of the affective relief is compatible with the "diminishment of self-affection" in schizophrenia.

From a phenomenological point of view, we can see a close correlation between the fragmentation of ordinary perception and the disruption of self-affection. Unusual or anomalous affective states and experiences may emerge such as the "*praecox*-feeling" and the "Kretchmerian paradox". According to Sass (2004) these phrases refer to a strange, ineffable, and contradictory qualities of schizophrenic emotionality, which contribute to the process of alienation from the consensual lifeworld. A person with schizophrenia often displays the feeling of strangeness, remoteness and can be over-sensitive and cold at the same time (Sass, 2004, p. 130). In their recent article, Sass and Feyaerts critically remarks that the concept of the *diminished* self-presence (i.e., diminished self-affection) cannot adequately explain the paradoxical experiential combinations of schizophrenia. For instance, the loss of ownership of subjectivity may coexist with a grandiose sense of being at the center of the "All". In this delusion self-presence seems diminished and expanded in an alternating mode (Feyaerts & Sass, 2023, p. 6). In contrast to diminished self-affection or diminished minimal self, hiperreflexivity seems compatible with the diversity of symptoms that occur in schizophrenia spectrum disorder. Sass and Feyaerts argue that the idea of hyperreflexivity accounts for the seemingly antithetical symptoms and for the variability of symptoms. In this context, variability means that symptoms may wax and wane over time, imposing themselves forcefully or fading away. By contrast, the "diminished minimal self" suggests a persistent condition despite its changeable ways of manifestation. In addition, it refers to an "underlying state of being" without explaining the process that might have brought it about. It also leaves open the question of why the sense of self remains relatively stable prior to the illness onset. Furthermore, it is also unanswered how the self can be "regained or reconstructed" during recovery (Sass & Feyaerts, 2023, p. 8). In contrast to diminished self-affection, the advantage of hyperreflexivity is that it may allow empirical confirmation and can be used to predict the development of psychotic symptoms. The increasing or decreasing patterns of hyperreflexive may indicate the onset, diminishment, or remission of psychotic symptoms (Sass & Feyaerts, 2023, p. 8).

As we have seen, it is open to debate what is at the core of the ipseitiy disturbance model of schizophrenia. Recent developments offered the hypothesis that hyperreflexivity can account for the paradoxes of symptoms in schizophrenia. A distance has also emerged between hyperreflexive awareness and the diminished minimal self. As Sass and Feyaerts argue, while diminished self-affection or minimal self leads to the loss of self-consciousness, hyperreflexivity, that seems unique to our species, can be understood as exaggerated self-consciousness and endless self-questioning (Sass & Feyaerts, 2023, p. 10).

Fazakas and his colleagues reformulated the idea of ipseity with the help of Sartre's concept of coenesthesia. They argue that coenesthesia is the bodily element of ipseity and it can shed a new light on the anomalies of self-experience. In contrast to the analysis of hyperreflexive awaraness, the authors try to explore the affective depth of selfhood and revitalize the philosophical concept of self-affection. The author's starting point is to give "thickness to the self" because the minimal self loses its explanatory value and descriptive force if it cannot account for the disorders of the self. Coenesthetic affectivity is closely related to Ratcliffe's existential feelings, it is a "tinted self-affection" that also reveals the lived body as the foundation of the minimal self. The *mineness* of experiences is not a mere analytical feature of the self, but rather the feeling of being an embodied subject. Coenesthesia also constitutes a link between the three main features of the IDM-model, such as diminished self-affection, hyperreflexivity, and disturbed hold or grip on the world. Coenesthetic affectivity may function as a "trouble générateur" in Schizophrenia Spectrum Disorders. The authors argue that in schizophrenia aberrant coenesthetic elements may invade consciousness and cause changes in the translucent matter of consciousness (i.e., in the experience of *Leiblichkeit*). The philosophical outcome of their proposal is that the *for-me-ness* of experiential reality is always coenesthetically tinted (Fazakas et al., 2023, p. 12). If we compare the latest developments around the IDM-model, then we can see that Sass and Feyaerts pushed into the background the constitutive role off self-affection and underscored the role of hiperreflexivity in symptom formation. However, the phenomenological significance of self-affection remained central to the ide of coenesthetic affectivity. In other words, coenesthesia provides a phenomenological "thickness" to the minimal self and self-affection. However, we should not neglect the problem of intersubjectivity in the search for the "trouble générateur" of Schizophrenia Spectrum Disorders. Both the model of hyperreflexivity and coenesthetic affectivity emphasizes the altered world-experiences during the onset of schizophrenia.[20]

The self-disorder model of schizophrenia also places emphasis on the role of intersubjectivity in self-constitution. However, it must be admitted that the metaphorical and figurative self-descriptions of patients, or persons in prodromal phase, are open to further interpretation. To put it simply, patients often complain about the ineffable, inchoate, or atmospheric quality of the above-mentioned *Stimmung* and delusional mood, which undermines the attempt of operationalization and opens the way to the hermeneutic stance. It is important to point out that the limitations of language, empathy, and understanding of first-person accounts of schizophrenia can lead us to a kind of "phenomenological underdetermination". Ratcliffe (2017) argues that phenomenology cannot be the read off of first-person reports. In his analysis of hallucinations accounts, he shows the ambiguity of first-person reports

[20] For example, Fazakas and his colleagues argue: "Coenesthesia thus plays a role in every process of sense-making, in the process of making sense of the world or of ourselves." (Fazakas, et al., p. 12).

by highlighting the fact that several "nonphenomenological factors" influence and modify the content of the accounts. For example, education, linguistic abilities, age, different contexts, or the subject's willingness to communicate can cause difficulties in the process of interpretation. Another frequently observed perplexity of first-person reports is that there is no clear line between metaphorical and literal understanding of the accounts (pp. 54–55). It is a difficult task to untangle the perplexing descriptions and make general inferences about the person's self-experience and being-in-the-world.

Varga (2012) has critically noted that scholars have paid less attention to the "world-immersedness-side" of the reports and have shifted toward the indicators of self-disorder (p. 183). Parnas & Sass (2011) view the disorder of the minimal self as a generative disorder (*trouble générateur*). The authors claim, based on the contributions of Eugene Minkowki, that the ipseity disorder may be a core feature of schizophrenia that give rise to positive and negative symptoms, or conversely, the symptoms may lead us back to the generative disorder ipseity disturbance (p. 541). At first glance, the IDM-model seems to pull us toward the operationalized definition of the minimal self, including the process of self-affection. In defense of Parnas and Sass, it has to be recognized that the loss of common sense suggested by the patients' accounts always presupposes world-immersedness.

Pienkos and his colleagues (2017) conducted qualitative research and introduced the concept of "anomalous world-experience". The increasing amount of analysis of delusional mood suggest that the disorder of self-affection (i.e., ipseity or minimal self) is paired with anomalous world-experience.[21] Varga's contribution to this dilemma is to provide a situated account of non-reflective self-awareness.[22] Varga (2012) utilizes the previously discussed *trema*-phenomenon and argues for the priority of "trema-like general alienation from the environment", which leads to the disruption of the implicit sense of presence and familiarity of being-in-the-world (p. 184). He offers at least two interpretations to the cases when patients complain that "I don't feel myself' or "I am not myself anymore": (1) the reports may allude

---

[21] See for example, Ratcliffe (2013, 2017), Sass & Ratcliffe (2017).

[22] The concept of non-reflective self-awareness (NRSA) comprises pre-reflective self-awareness, but it has also a wider territory of meaning. First of all, Varga (2012) relies on Sartre's non-reflective self-consciousness that cuts across the duality of the knower and the known, and designates an immediate, non-cognitive relation. In addition, he mentions the syndrome of "transient global amnesia" in which a global memory loss takes place, and the abilities to form new memories is impaired as well. The most obvious symptom of the disease is the repetitive questioning. However, empirical studies hypothesized that in spite of the memory impairment and the decline of cognitive capacities, the NRSA could be preserved (p. 173, 176). The concept of NRSA also refers to the bodily, non-conceptual consciousness. By referring to Bermudez and Gallagher, Varga states that a self-referential somatic proprioception gives us the pre-reflective sense of being in our bodies. The proprioceptive body schema leads to the differentiation between the outer and the inner, between our own and others' experiences (Varga, 2012, p. 179). Contrary to Zahavi's pre-reflective self-awareness, the thesis of NRSA puts more emphasis on the bodily foundation of self-awareness. Nonetheless both conceptions claim that the "mineness" of experiences cannot be explained by higher-order monitoring theories because of the ensuing infinite regress of reflective mental states.

to the disturbance of non-reflective self-awareness; (2) there is a disturbance in the immersion of the self in the world (p. 187). However, as I see it, one can also find several "situated accounts" in the IDM-model. For instance, Nelson & Sass (2009) came up with the remarkable case study of Adam, in which the social stressors significantly contributed to the development of hyper-reflexive awareness. Adam's activities were controlled and monitored by his girlfriend, Fiona. The resulting anxiety and entrapment may have contributed to Adam's persistent self-monitoring behavior.[23] That is, hyperreflexivity can occur not only for endogenous reasons, but also for relational reasons (p. 494).

Carruthers and Musholt (2018) also underscore the lack of the role of "world-directed consciousness" in the self-disorder theory. In addition, the authors point to a serious methodological dilemma, when they question the reliability of measuring disturbances of pre-reflective self-awareness. By definition, this implicit awareness is non-representational, but the task of measuring the changes in self-experience inadvertently utilizes reflective self-evaluations. Patients need to examine the general sense or character of their own *ipseity* by focusing on specific anomalous bodily experiences, thoughts, voices, activities, and so on (pp. 697–698). As a result, empirical studies indirectly investigate explicit self-representations, leavening behind the philosophical idea of the self. That is, the initial goal of the interrogation is to describe the non-representational character of self-experience, but the very act of identifying and locating subjective, bodily experiences can radically alter the target phenomenon.

Ratcliffe (2017) argues that diagnosis of schizophrenia might be compatible with other types of anomalous experiences. He revisits the autobiography of Renee and draws attention to the distressing life events that occur in the biography (e.g., Renee's father had a mistress, her mother became depressed and said that she would kill herself and so on). Ratcliffe argues for the priority of loss of trust in the environment and other people due to traumatizing life events (p. 183). Renee felt extreme anxiety and alienation from the consensual world, and her only refuge was her therapist. Ratcliffe concluded that the structure of experience, the sense of reality, and the sense of self are intersubjectively constituted. Meaningful contact with others establishes the possibility that we live in a familiar world rather than in the estranged, delusional "sub-universe" of a person with schizophrenia (Ratcliffe, 2017, p. 184). He is highly skeptical about the postulation of a presocial sense of self: "self-experience is equally interpersonal experience" – he asserts (Ratcliffe, 2017, p. 181).

[23] It is worth quoting a short passage from the conceptualization of the case study: „Adam frequently spoke about his year-long relationship with his girlfriend, 'Fiona.' It transpired that a central dynamic in this relationship involved feelings of exposure, vulnerability, and subservience. Adam was prone to experience Fiona as controlling and intrusive in her attitude toward him; also, he felt that she did not trust him. For example, she would check his Internet usage, leave him feeling guilty if he wanted to spend time with friends, and question him closely ('she badgers me') regarding his activities and plans. Although Adam tolerated this relationship dynamic, it created in him a sense of anxiety and entrapment (e.g., 'It makes me feel tight, like I'm caged…')." (Nelson & Sass, 2009, p. 494)

At the end of our current discussion of self-disorders, we have returned to Fuchs' considerations. Following Merleau-Ponty, Fuchs repeatedly asserted that it is almost impossible to overcome the inherent dualism in the discussions of the self-world relation. Self-disturbances can be approached from the minimalist framework as well as from the intersubjective framework. These approaches can be complementary but not exhaustive or exclusive. In general, the features of self-disorder demonstrate the embeddedness and vulnerability of the self. The brief survey of the field shows that there is a debate between the proponents of the minimal self and of the social self. However, the multidimensional or multifactorial analysis may lead to new ways of clarification. Despite the different, often controversial, strategies for defining the minimal self, the role of affectivity has received enormous attention. In this section, I intended to show that the strange cases of alienated self-experiences can also draw attention to the affective unconscious, defined as the passive-affective infrastructure of the self. Thus, the minimal self implies the horizon of the non-representational affective unconscious. In Chap. 7, I will try to reformulate the problem of the minimal self in terms of affectivity and to introduce the affective core self, which can be conceived as the constellation of the affective unconscious. I propose the idea that the affective core self is not an elementary self, akin to the formal minimal self, but rather a malleable, constantly evolving self, constituted by sedimented traits, dispositions, habits and modified or schematized by traumatic experiences. The problem of time-consciousness and its relation to the minimal self will be addressed again in more detail in the analysis of retroactivity.

References

Blankenburg, W. (2001). First steps toward a psychopathology of "common sense". *(Trans. Aaroon L. Mishara) Philosophy, Psychiatry, and Psychology, 8*(4), 303–315.

Borda, J. P., & Sass, L. A. (2015). Phenomenology and neurobiology of self disorder in schizophrenia: Primary factors. *Schizophrenia Research, 169*(1–3), 464–473.

Bortolotti, L., & Broome, M. R. (2012). Affective dimensions of the phenomenon of double bookkeeping in delusions. *Emotion Review, 4*(2), 187–191.

Carruthers, G., & Musholt, K. (2018). Ipseity at the intersection of phenomenology, psychiatry and philosophy of mind: Are we talking about the same thing? *Review of Philosophy and Psychology, 9*, 689–701.

Chapman, J. (1966). The early symptoms of schizophrenia. *The British Journal of Psychiatry, 112*(484), 225–251.

Conrad, K. (2013 [1959]). *Die beginnende Schizophrenie. Versuch einer Gestaltanalyse des Wahns*. Psychiatrie-Verlag.

Fazakas, I., Bois, M., & Gozé, T. (2023). Giving thickness to the minimal self: Coenesthetic depth and the materiality of consciousness. *Phenomenology and the Cognitive Sciences*. https://doi.org/10.1007/s11097-023-09951-w

Feyaerts, J., & Sass, L. (2023). Self-disorder in schizophrenia: A revised view (1. Comprehensive review: Dualities of self- and world-experience). *Schizophrenia Bulletin, sbad169*. https://doi.org/10.1093/schbul/sbad169

Firth, C. D. (1979). Consciousness, information processing, and schizophrenia. *British Journal of Psychiatry, 134*(3), 225–235.

Fuchs, T. (2005a). Corporealized and disembodied minds: A phenomenological view of the body in melancholia and schizophrenia. *Philosophy, Psychiatry, and Psychology, 12*(2), 95–107.
Fuchs, T. (2005b). Delusional mood and delusional perception – A phenomenological analysis. *Psychopathology, 38*(3), 133–139.
Fuchs, T. (2013). The phenomenology of affectivity. In K. W. M. Fulford, M. Davies, R. G. T. Gipps, G. Graham, J. Z. Sadler, G. Stanghellini, & T. Thornton (Eds.), *The Oxford handbook of philosophy and psychiatry* (pp. 612–631). Oxford University Press.
Fuchs, T. (2017). Self across time: The diachronic unity of bodily existence. *Phenomenology and the Cognitive Sciences, 16*(2), 291–315.
Fuchs, T., & Koch, S. (2014). Embodied affectivity: On moving and being moved. *Frontiers in Psychology, 5*(508), 1–12.
Fuchs, T., & Schlimme, J. E. (2009). Embodiment and psychopathology: A phenomenological perspective. *Current Opinion in Psychiatry, 22*(6), 570–575.
Gallagher, S. (2009). Delusional realities. In M. R. Broome & L. Bortolotti (Eds.), *Psychiatry as cognitive neuroscience: Philosophical perspectives* (pp. 245–266). OUP.
Jaspers, K. (1973). *Allgemeine Psychopathologie (Neunte, unveränderte Auflage)*. Springer.
Laing, R. D. (1978). *The divided self*. The Penguin Force.
Lysaker, P. H., & Lysaker, J. T. (2010). Schizophrenia and alterations in self-experience: A comparison of 6 perspectives. *Schizophrenia Bulletin, 36*(2), 331–340.
Merleau-Ponty. (2012). *The phenomenology of perception*. (D. A. Landes, Trans.). Routledge.
Michie, P. T., Badrock, J. C., Waters, F. A. V., & Maybery, M. T. (2005). Auditory hallucinations: Failure to inhibit irrelevant memories. *Cognitive Neuropsychiatry, 10*(2), 125–136.
Mishara, A. (2010). Klaus Conrad (1905–1961): Delusional mood, psychosis, and beginning of schizophrenia. *Schizophrenia Bulletin, 36*(1), 9–13.
Morley, J. (2003). Phenomenological and biological psychiatry: Complementary or mutual. *Philosophy, Psychiatry, and Psychology, 9*(1), 87–90.
Nelson, B., & Sass, L. S. (2009). Medusa's stare: A case study of working with self-disturbance in early phase of schizophrenia. *Clinical Case Studies, 8*(6), 489–504.
Nelson, B., Fornito, A., Harrison, B. J., Yücel, M., Sass, L. A., Yung, A. R., Thompson, A., Wood, S. J., Pantelis, C., & McGorry, P. D. (2009). A disturbed sense of self in the psychosis prodrome: Linking phenomenology and neurobiology. *Neuroscience and Biobehavioral Reviews, 33*(6), 807–817.
Overgaard, S. (2007). What, if anything, can phenomenology teach psychopathology (and vice versa)? *Husserl Studies in Japan, 4*(5), 32–42.
Pardiñas, A., & Holmans, P. (2018). Common schizophrenia alleles are enriched in mutation-intolerant genes and in regions under strong background selection. *Nature Genetics, 50*, 381–389.
Parnas, J., & Sass, L. A. (2011). The structure of self-consciousness in schizophrenia. In S. Gallagher (Ed.), *The Oxford handbook of the self* (pp. 524–546). OUP.
Pienkos, E., Silverstein, S., & Sass, L. S. (2017). The phenomenology of anomalous world experience in schizophrenia: A qualitative study. *Journal of Phenomenological Study, 48*(2), 188–213.
Ratcliffe, M. (2008). *Feelings of being: Phenomenology, psychiatry and the sense of reality*. OUP.
Ratcliffe, M. (2011). Phenomenology is not a servant of science. *Philosophy, Psychiatry, & Psychology, 18*(1), 33–36.
Ratcliffe, M. (2013). Delusional atmosphere and the sense of unreality. In G. Stanghellini & T. Fuchs (Eds.), *One century of Karl Jaspers' general psychopathology* (pp. 229–244). OUP.
Ratcliffe, M. (2015). *Experience of depression: A study in phenomenology*. OUP.
Ratcliffe, M. (2017). *Real hallucinations: Psychiatric illness, intentionality, and the interpersonal world*. MIT.
Sass, L. A. (2004). Affectivity and schizophrenia: A phenomenological view. In D. Zahavi (Ed.), *Hidden resources: Classical resources on subjectivity* (pp. 127–147). Imprint Academic.
Sass, L. A. (2017). *Madness and modernism: Insanity in the light of modern art, literature, and thought* (revised ed.). Oxford University Press.

Sass, L. A., & Feyearts, J. (2023). Self-disorder in schizophrenia: A revised view (2. Theoretical revision: hyperreflexivity). *Schizophrenia Bulletin*, sbad170. https://doi.org/10.1093/schbul/sbad170

Sass, L. A., & Parnas, J. (2003). Schizophrenia, consciousness, and the self. *Schizophrenia Bulletin, 29*(3), 427–444.

Sass, L. A., & Ratcliffe, M. (2017). Atmosphere: On the phenomenology of "atmospheric" alterations in schizophrenia – Overall sense of reality, familiarity, vitality, meaning, or relevance (ancillary article to EAWE domain 5). *Psychopathology, 50*(1), 90–97.

Schmitz, H., Müllan, R. O., & Slaby, J. (2011). Emotions outside the box – The new phenomenology of feeling and corporeality. *Phenomenology and the Cognitive Sciences, 10*(2), 241–259.

Sechehaye, M. (1970). *Autobiography of a schizophrenic girl*. American Library.

Stanghellini, G. (2009). Embodiment and schizophrenia. *World Psychiatry, 8*, 56–59.

Stephenson, H., & Parnas, J. (2018). What can self-disorders in schizophrenia tell us about the nature of subjectivity? A psychopathological investigation. *Phenomenology and the Cognitive Sciences, 17*, 629–642.

Stern, D. N. (1985). *The interpersonal world of the infant: A view from psychoanalysis and developmental psychology*. Basic Books.

Thompson, E. (2007). *Mind in life: Biology, phenomenology, and the sciences of mind*. HUP.

Varela, F. (1996). Neurophenomenology: A methodological remedy for the hard problem. *Journal of Consciousness Studies, 3*(4), 330–349.

Varga, S. (2012). Non-reflective self-awareness: Towards a 'situated' account. *Journal of Consciousness Studies, 19*(3–4), 164–193.

Zahavi, D. (2005). *Subjectivity and selfhood: Investigating the first-person perspective*. MIT Press.

Zhiqiang, L., Chen, J., & Shi, Y. (2017). Genome-wide association analysis identifies 30 new susceptibility loci for schizophrenia. *Nature Genetics, 49*, 1576–1583.

Part IV
The Unconscious and the Minimal Self

Chapter 6
Time-Consciousness and Affective Identity

Abstract This chapter combines the problems of the affective unconscious and affective identity with the minimal self, more precisely, with the notion of the (less) minimal self. In order to accomplish this theoretical step, the following discussions take into account the dilemma of the retroactive constitution of the near and distant past. Retroactive constitution is embedded in the larger problem of time-consciousness. The chapter examines the phenomenological notion of the affective retroactive awakening of the past and tries to broaden the view of retroactivity by including the phenomenon of intrusive memories. The chapter also provides extensive analyses of the Freudian idea of afterwardsness (*Nachträglichkeit*) and compares it with the intentional accomplishment of retroactive sense-bestowing. The general aim of these analyses is to show the importance of retroactivity in the constitution of the affective unconscious and affective identity. In addition, the chapter expands the view of retroactivity to include traumatic retroactivity. Traumatic retroactivity includes intrusive memories and also incorporates the temporal profile of Freud's afterwardsness.

Keywords Intrusive memory · Affective unconscious · Trauma · Retroactivity · Afterwardsness · Affective awakening

In the previous chapters, my aim was to show that the affective relief and self-affection broadens the notion of the minimal self. The excursion towards the IDM-model has shown that affective disturbances, including altered moods, existential feelings, and affective atmospheres can be seen as the indicators of self-disorder. While we have been focusing on the role of self-affection in self-constitution and self-disorder, we should not forget the equally fundamental role of narrative self-understanding. Phenomenologically speaking, the representational, thematic and the non-representational, unthematic layers of self-constitution are only artificially separated. The multidimensional approach to the self does not neglect the first in favor of the second. By adhering to the layered structure of the self, the present work does not attempt to eliminate or overshadow the role of narrative self-understanding in self-constitution.

L. Horváth, *The Affective Core Self*, Contributions to Phenomenology 130,
https://doi.org/10.1007/978-3-031-56920-3_6

The aim of this chapter is to uncover the affective unconscious that horizontally schematizes our perceptual field and contributes to the diachronic and synchronic continuity of the self. The *affective relief* and *body memory* were the two most important building blocks for the introduction of the affective unconscious. In order to introduce the notion of affective identity, which is based on the affective unconscious, the different alternatives of the phenomenological unconscious were discussed and comparative analysis tried has shown the zones of convergences and divergences between the phenomenological and psychoanalytical unconscious. The psychopathological consequences of altered self-affection were addressed by the IDM-model of schizophrenia. Up to this point, the role of time-consciousness was only marginally present in explaining the process of self-constitution and the phenomenological unconscious.

In the following sections, I will focus on the role of time-consciousness in the constitution of the affective identity, i.e., the affective core self. The discussion of retroactive self-understanding attempts to link the passive-associative network of the affective unconscious with implicit body memory. Or, to put in other words, implicit body memory can be embedded in the passive-associative processes of the phenomenological unconscious. This chapter also discusses the various intentional relations to the past. As we will see, the various forms of remembering are both cognitive and affective reactivations of the past. At first glance, it seems difficult to observe the differences between the recollections of the past, because in the streaming living present the different types of past-relations can be present and seamlessly intertwined. However, Husserl and recent phenomenological inquires distinguished between explicit memory, involuntary memory, and affective awakening of the past (i.e., the process of retroactive awakening). What is important in this fine-grained analysis is not just the content of memory, but also the mode of their givenness, i.e., they are different intentional relations to the past. These phenomenological investigations are also consistent with the basic tenet of the narrative self-constitution view, according to which the retrospective and prospective intentions play an important structural role in self-constitution.

There is a remarkable affinity between body memory and retroactivity: each type of recollection involves an implicit, pre-reflective meaning-bestowing process that operates at the boundary between representative and non-representative consciousness. The central aim of the following chapters is to show that the various threads of the phenomenological unconscious can be woven into the problem of affective identity. And, from the perspective of affective identity, we can extrapolate an affective core self that extends the scope and completes the *formal* notion of the minimal self.

6.1 The Fragility of the Phenomenological Unity

As previously demonstrated, the problem of pre-reflective self-awareness is inseparable from the problem of diachronic continuity of the self. Zahavi's notion of the minimal self and Fuchs' body memory stresses the importance of explaining the continuity of self-awareness and personal identity. Husserl, and more recently

Zahavi, put forward the idea that the synthesis of intentional states occurs spontaneously in time-consciousness. As a result, the sense self is self-affectively constituted in the experiential flow of time.

Interestingly, modern memory research is also concerned with the interrelated aspects of subjective time and self-constitution. Endel Tulving introduced the notion of "autonoetic awareness (or autonoesis)" as the condition of possibility for remembering. Tulving argues that remembering, perceiving, imaging, and dreaming are all characterized by a special kind of consciousness and a feeling component. Mental time travel, i.e. the act of recollection, has three interrrelated aspects: (1) the sense of subjective time that evokes the felt distance of the past; (2) the difference between the above-mentioned intentional states; and (3) the mental time travel requires the traveler, the self. The self in question cannot be reduced to the self of the mirror test; it is rather a self that exists in subjective time or, in other words, is extended to the past and the future (Tulving, 2002, p. 2). According to Tulving (2002), episodic memory "is the only memory system that allows people to consciously re-experience past experiences" (p. 6).

Schacter (1996) turns attention to a significant change in memory research, when he says that memories cannot be compared with simple computer files, or snapshots. We do not store pure memory files, rather the remembering self is attached to the meaning, sense, and emotions generated by past experiences (p. 5). Furthermore, the act of remembering has a significant affective impact. Schacter mentions observations in which remembering to sad episodes brought people to tears, and the recollection of joyful events led almost immediately to the sense of elation (p. 4). And, of course, there are the curious cases of Freud's hysterics who suffered from reminiscences. The various observations of the relationship between the self and time converge to the view that the past exerts its power at different levels. We can speak of the distant, forgotten past, which manifests itself implicitly in repetition or other kinds of behavioral signs, or we can voluntarily recall past events, or we can be astonished by involuntary memories[1]. And how does the self fits in this picture? For example, Williams and his colleagues argue that autobiographical memories are the "database from which the self is constructed" and the interaction between the self and memory is the most important function of autobiographical memory and autobiographical knowledge. However, memories are not faithful representations of past events, but are altered, distorted, and fabricated in order to reflect the "central aspects of the self". The self-memory system (SMS) designates the reciprocal relationship between autobiographical memory and the self, which consists of currently acquired beliefs, traits, and habits (Williams et al., 2008, p. 25).

---

[1] Involuntary memories can be seen as the representational manifestation of the affective unconscious. Their vivid, emotional aura provides an affective connection with the near or distant past: "Involuntary memories are conscious and unintentional recollections of personal experiences and have been described as being peculiarly vivid and emotional and having a strong feeling of immediacy. Involuntary memory retrieval appears to be quite a common experience in everyday life…" (Williams et al., 2008, p. 48) The authors also highlight that involuntary memories are often vivid recollections and in contrast to voluntary memories they have a greater effect on the subject's moods.

There are several representational and non-representational relations to the past. For present purposes, the central problem is to get insight into the process of articulating meaning. It is reasonable to suppose a twilight zone between the non-representational and the representational relation to the past, where meaning gradually emerges. According to Kozyreva (2017), this grey area is the right place where the dialogue between memory research and phenomenology can be established (p. 179). A phenomenologically oriented study, therefore, attempts to approach and describe the schematization process that occurs between the implicit and explicit modes of remembering. How can we take into account the newly found, articulating past in the here and now? Psychoanalysis has always been concerned with the difficult problems of the resurrected past. The false memory debate of the 1990s and the constructive theory of memory retrieval are clear examples of this. The overall aim of Chap. 6 is to discuss how the implicit meaning structures are transformed into explicit representations. The phenomenological idea of retroactivity and Freud's idea of *Nachträglichkeit* allow us to address the problem of the indeterminate (implicit) past that nonetheless exerts an influence on the present and reconfigures or re-schematizes the underlying affective unconscious of the self. The modified or reconfigured affective unconscious has a strong influence on affective identity. In this regard, I will use the already introduced notion of the affective core self which is the sedimented but also malleable and renewable affective-associative network of the affective unconscious. For preliminary purposes, let us turn back to the problems of consciousness and the self.

Consciousness is not an empty transcendental ego of object constitution or the simple ray of attention, but is consist of the stream of experiences and of the modalities of intentionality (i.e., perception, memory, belief, and imagination). As Ullmann puts it: "Consciousness is like a swarming chamber of a huge railway station, in which crowds of phantasmatic images, phantasmatic sensations mix with thoughts, memories, and perceptions." (Ullmann, 2012b, p. 126) Goldie also illuminates the random associative process that lurks beneath the organizational principles of an external, narrative perspective:

> … there will very likely be all sorts of traces of thought and imagination and fragments of stream-of-consciousness or 'flashbulb' memory that flit through one's mind from time to time, many of them I dare say hardly reaching a level of conscious awareness. Moreover, many episodes of one's life are really so insignificant, boring and humdrum that these fragments and traces with be all there is – there will be no narrative thinking about them (although perhaps there could be) (Goldie, 2003, p. 312).

To put it more drastically: beneath the superficial layer of consciousness, a subconscious associative network works tirelessly. It is quite a miracle that there are instances of coherent conscious states against the backdrop of the potential psychosis; that is, without the passive organizing principles of association and the affect-regulating process of the intersubjective world it is easy to fall prey to paranoid, or any other kind of pathological, thinking. Zahavi claimed that the (more) minimal self, understood as the for-me-ness of experiences, gives us a sense of integrity, coherence, and identity. Moreover, the for-me-ness character of experiences is inseparable from the qualia of what-it-is-like-ness; that is, the ownership of mental, bodily, and affective states entail phenomenal consciousness and vice versa. The ultimate composition, namely the what-it-is-like-*for*-me-ness character of intentional states, has been called the pre-reflective level of self-awareness (see Sect. 2.4). However, Ullmann considered the concept of the minimal self to be too minimal, without a real explanatory value. Ratcliffe has also called into question the pre-social minimal self: "The first step is to ask just how minimal the minimal self can be: does it incorporate only the sense that you are having some kind of experience, or does it also include the sense that you are having an experience of one or another type, such as perceiving, imagining, engaging in inner speech, or remembering?" (Ratcliffe, 2017a, p. 18) Ratcliffe goes on to say that schizophrenia is basically a relational disturbance; his statement indirectly supports the idea that even the minimal self is interpersonally constituted (Ratcliffe, 2017b, p. 150).

In general, the crux of Ratcliffe's (2017a) argument is that the minimal self directly involves the pre-reflectively felt distinctions between the modalities of intentionality. Under normal circumstances, for example, we know and feel the differences between remembering and perceiving; as Tulving's aforementioned observations also suggest. However, the minimal self cannot be reduced to the modalities of intentionality alone, since they do not entail "the sense of being a singular locus of experience". Ultimately, Ratcliffe (2017a) adheres to a weaker claim: the felt difference between the modalities of intentionality is a necessary but not sufficient condition for the constitution of the minimal self (p. 18). Another important thesis is that the modalities of intentionality are configured by a unified anticipation-fulfillment structure and this latter structure presupposes "a cohesive set of felt bodily dispositions" (p. 19). That is, what we can affectively or cognitively anticipate is highly dependent on our ingrained bodily dispositions – this kind of body-environment cohesion is inherent in Fuchs' body memory. Moreover, the self is developmentally dependent and environmentally maintained in adulthood. As also mentioned in Chap. 2, the critical remarks of Ratcliffe and others motivated Zahavi to introduce the distinction between the (less) minimal self of the adult and the (more) minimal self of the child (cf. Zahavi, 2017). For Ratcliffe, the minimal self's experiential aspect is not the *for-me-ness* of experiences, but the "a sense of which type of intentional state one is" or the ability to discriminate between intentional states:

> If your experience did not respect the distinction between perceiving and remembering, we would have no sense of temporal location. And, if we could not distinguish imagining from perceiving, experienced boundaries between self and environment would break down, to the extent that we would lack any sense of spatial location. [...] Without some sense of spatiotemporal location, it is difficult to see how any experience of being in a singular, coherent locus of experience could be sustained. Hence minimal-self experience has to discriminate between type of intentional state (Ratcliffe, 2017a, p. 27).

According to Ratcliffe, the idea of the disruption of the modalities of intentionality has more explanatory value for psychopathology than the abstract *for-me-ness character* of ginennesses. Zahavi (2020) reconsiders the role of for-me-ness in psychopathological cases. For-me-ness, in this context, refers to the stability of first-person perspective or to the sheer fact that even hallucinated voices are given for the first-person perspective of the subject, and not publicly available occurrences[2]. But how can dissociation and disownership symptoms be explained? What is missing or what is altered in anomalous self-experiences is the "sense of endorsement and self-familiarity" of mental states; the quality of warmth and intimacy of thoughts, as William James suggested (p. 649). Zahavi's distinction between the *familiarity* and the *ownership* of mental states can be fruitful, but what is intriguing is the fact that the process of *disownership* runs parallel with estrangement and the ensuing delusions and hallucinations. It seems that the disturbance of the ownership of experiences implies both the emergence of anomalous intentional states and the disintegration of the unity of the self (including synchronic and diachronic unity).

As for the disintegration of the self, a horrifying but eloquent self-description can be found in Elyn R. Saks's autobiography. She calls her set of symptoms disorganization:

> Consciousness gradually loses its coherence. One's center gives way. The center cannot hold. The "me" becomes a haze, and the solid center from which one experiences reality breaks up like a bad radio signal. There is no longer a sturdy vantage point from which to look out, take things in, asses what's happening. No core holds things together, providing the lens through which to see the world, to make judgement and comprehend risk. Random moments of time follow one another. Sights, sounds, thoughts, and feelings don't go together. No organizing principle takes successive moments in time and puts them together in a coherent way from which sense can be made. And it's all taking place slow motion (Saks, 2015, p. 13).

The frequently quoted passage is understood by Ratcliffe as the loss of *phenomenological unity* (Ratcliffe, 2017a, p. 16). The quotation also sheds some light on the shattered temporal unity of self-experience and world-experience. The breakdown of unity goes hand in hand with the disrupted anticipatory structure. Anticipation means not only the protentional structure of time-consciousness, but also the interaction and relation of different intentional states:

> My imagining might take my writing in a new direction; something I remember might prompt me to do something else; and an incident in my environment might distract me from the task. Like the temporal profiles of the intentional states concerned, these interactions

[2]It is difficult, if not impossible, to give the clear phenomenological account of verbal hallucinations. Ratcliffe (2017a) argues that hallucinatory voices can be approached by the phenomena of inserted thought; that is, verbal hallucinations (VHs) and thought insertion (TI) may have a common ground and they are most probably perception like experiences with a thought like content. What is conspicuous here, according to Ratcliffe, is the affliction like character of the hallucination. It is hard to determinate their acoustic, quasi-acoustic, or thought-like character even from first-person perspective, but what is most certain is that the suffering subject is "being-affected" by them (p. 63). And Ratcliffe adds that these disturbing internal experiences are anteceded by the "all-enveloping experience of passivity and disengagement from the social world" (*ibid.*, p. 69). Thus, it may be plausible to suppose that the voice hearer has first-person access to his or her voices, though patients are often uncertain about the source of the "voice", but Ratcliffe, unlike Zahavi, adds weight to the disintegration of the modalities of intentionality. The for-me-ness character of experiences may remain intact (i.e., the hallucination occurs in the patient's phenomenal field), yet a new, unfamiliar kind of intentionality may lead to the complete disorganization or disintegration of the self; as the uote from Saks also demonstrates.

> structured in characteristic ways. I experience the influence of my imaginings on the flow of my thoughts as distinct from the influence of my memories. The way in which a type of intentional state interacts with others, such as the way in which memory can drift into an imagining and vice versa, is also integral to its anticipation-fulfillment profile (Ratcliffe, 2017a, p. 167–168).

Ratcliffe focuses on the habitual, affective, nonpropositional *style of anticipation* that manifests itself in the cohesive modal structure of intentionality, and it is intersubjectively created, supported, and normalized. The anticipation structure suggests that even the modal structure of intentionality, i.e., the implicit differences and similarities between perception, imagination, and memory, are constituted, sustained, and regulated by the intersubjective lifeworld. Ratcliffe also shows that in the case of traumatic experiences, the sedimented "habitual certainty" is disrupted and our entire anticipatory structure is disintegrated. Traumatic events can lead to a shattered self, meaning that the experience of intersubjective and instrumental relations is radically altered:

> It does not occur to us that we will deliberately struck by a car as we walk to the shop to buy a milk or that we will be assaulted by the stranger we sit next to on the train. This is not to suggest that we naively trust everybody or feel safe all the time, regardless of circumstances. The point is that, when we do feel unsafe in a certain situation or explicitly distrust a particular person, this involves a departure from our default attitude (Ratcliffe, 2017a, p. 121).

From the perspective of affective identity traumatic events are of central importance since they not only give rise to the loss of trust and certainty, but they also jeopardize the integrity of intentionality and involve "a pervasive alteration in the sense of time" (Ratcliffe, 2017a, p. 113). In short, the modal structure and the temporal profiles of intentionality can be disturbed by traumatic events and lead to delusional perception.

Traumatic events and subsequent traumatic experiences give rise to profound phenomenological changes. Ratcliffe (2017a) draws a close analogy between the experiential changes caused by trauma and delusional mood. However, he adds that speaking of a causal relation between trauma and phenomenological change may sound dubious. There could be several other causes behind the changes of existential orientations. It is also difficult to find a clear distinction between a specific traumatic event and the effects of that event. For Ratcliffe, the "experience of trauma" is a broad category that encompasses the possible outcomes of a (1) discrete traumatic event, (2) the chronic exposure to repeated traumatic events or even (3) the impact of wider social and cultural upheavals. What seems to be certain is that both shocking interpersonal events (e.g., torture) and subtler forms of alienations lead to the disruption of the "habitual style of interpersonal anticipation" (pp. 115–116).

In the following section, I will examine the phenomenology of affective retroactive awakening which could contribute to the study of constructive memory processes, and conversely, the retroactive awakening of the past could be understood as one of the phenomenological sides of constructive memory retrieval. The following chapters will focus on the mutual relationship between the present and the past. The remembered past is not a neutral reservoir of accumulated experience, but rather a fluctuating structure that often bears the marks of reconstructive processes initiated from the present. The problem of trauma and the phenomenological concept of traumatic subjectivity will also be elaborated in order to illustrate the role of retroactive self-understanding in the constitution of the self. On my account, both retroactive self-understanding and narrative self-understanding preserves and maintains the coherence and stability of the self and the former process is integral to the latter. How does retroactivity fit into the picture of trauma? If we accept that trauma can shatter the habitual certainty of the subject and entails the loss of trust and the disintegration of the habitual anticipatory structure of the subject, then the function of retroactivity might be to preserve the integrity of the self, including the habitualized (modal) structure of intentionality. Retroactive self-understanding may preserve the integrity of the self by re-schematizing the past in the light of the present circumstances, but if both retroactive and narrative self-understanding fail, then the self may disintegrate.

It is noteworthy that the phenomenon of affective retroactive awakening is a general idea that refers to the retroactive understanding of the past. The process of retrospection, as we have seen, is incorporated into narrative thinking and into the constitution of the narrative self. However, Husserl's concept of retroactivity is more specific than pure retrospection. Retroactivity is not only a conscious, deliberate intentional act, but also a subconscious, passive associative process that can proceed in an automatic or spontaneous manner. The second important characteristic is that Husserl observes the constitutive role of retroactivity in the stream of consciousness. The affective retroactive awakening of the past is not necessarily a metacognition, or reflexive intentional act, but the spontaneous activity of passive genesis. The third characteristic, which is closely related to the second, is that retroactivity is bound to affectivity; that is, articulated affections or even the pre-affective hyletic synthesis may give rise to retroactive genesis of meaning. As recently Ullmann, Kozyreva, and Summa has claimed, the process of retroactive awakening can be a generic intentional activity embedded into the more usual and conspicuous modalities of intentionality (i.e., perception, imagination, belief, and imagination). Retroactive processes may also play a crucial role in the constitution of the affective relief structure (Mishara, 1990), in the genesis of truth (Marosán, 2020), in the experience of personal change (Ullmann, 2010), in the constitution of our affective identity (Kozyreva, 2017, 2018), and in the occurrence of involuntary memories (Summa, 2014)[3].

The task of the following chapter is to narrow the scope of retroactivity. The present work will focus on the possible relationship between trauma and affective retroactive awakening. For start, let us turn back to the dilemmas of implicit memory, and reconsider the implicit and affective presence of the past from phenomenological point of view.

---

[3] For the detailed account of affective awakening, see Chap. 3. For the systematic overview of the Husserlian concepts of passivity and retroactivity see Steinbock (2004), Mishara (1990), Marosán (2020).

6.2 The Passive-Implicit Past and Emotional Memory

As we have seen in Chaps. 2 and 3, implicit memory is one of the main research interests of contemporary psychology and social psychology. According to Schacter's definition we can speak of implicit memory when past experiences exert an influence on the subject without conscious awareness or conscious recollection (Schacter, 1996, p. 161). The case of psychogenic amnesia and other clinical observations suggest the presence of hidden emotions and affects after an incident (i.e. traumatic event); the person may bear the mark of the physical or emotional shock without explicit recollection. Schacter's observations indicate that the implicit effect of the past plays a serious role in altering the structure of the personality:

> Though the clinical observations I considered earlier are largely anecdotal, they remind us that implicit effects of past experiences may shape our emotional reactions, preferences, and dispositions – key elements of what we call personality. Freud, of course, realized something similar when he postulated a dynamic unconscious and emphasized the important role of early experiences. With the tools available to modern cognitive neuroscience, we can now explore the memory processes and systems that contribute to our likes and dislikes and our *habitual ways of responding to the world.* While our sense of self and identity is highly dependent on explicit memory for past episodes and autobiographical facts, our personalities may be more closely tied to implicit memory processes (Schacter, 1996, p. 233, *emphasis added*).

In my view, the key element in the above citation is "habitual ways of responding to the world". The reactive or habitual patterns of behavior and thinking might be the psychological formulation of the affective unconscious and affective identity. Recently, in contemporary phenomenology, Fuchs, Kozyreva, and Ullmann highlighted the importance of implicit memory. Kozyreva (2017) pointed out that cognitive psychology studies the impact of implicit processes in experimental circumstances and tries to explore the relations between the different memory-systems. The role of phenomenological inquires could be to complement memory research with the investigation of the pre-thematic states at the intersection of implicit and explicit cognition. Kozyreva mentions that in cognitive psychology special test-conditions measure the difference between explicit and implicit recall. However, several intriguing mental phenomena can be found in the intermediate zone between implicit and explicit memory; for example, varieties of recognition, involuntary memory, emotional and traumatic memory. There is a common "experiential structure" in these phenomena: they point to a non-representational, pre-thematic relation to the past that is most noticeably appears in approaches to body memory (p. 183).

It has been emphasized in this study that the problem of pre-reflective self-awareness can be reformulated through the phenomenological understanding of implicit memory. Kozyreva argues that implicit memory can be imagined in phenomenological research as a non-representational relation to the past. In my view, in contemporary approaches, *implicit body memory* and Husserl's *affective relief* can be seen as the two phenomenological interpretations of implicit processes. In addition, in Chap. 4, I attempted to show that body memory is integral to the affective relief structure. Research on implicit memory can also draw attention to the underlying affective and cognitive schematism. As Ullmann (2017) has noted, the (affective) unconscious is not only a reservoir of repressed affects and representations, but also an "affective sphere" that is implicitly present in da-to-day living and often inadvertently schematizes, influences our interpersonal relations[4].

Merleau-Ponty's contribution to the current discourse on body memory and affectivity is undeniable. The non-representational operative intentionality lays the foundations of higher-level object-intentionality. Fuchs claims that the term of body memory cannot be found in Melreau-Ponty's oeuvre, but his notion of "sedimented practical schemas" reminds us of procedural memory. But how do we arrive from skills, habits, and capacities to the problem of the repressed? How can we recognize the scars of the past in bodily expressions and dispositions? As was discussed in Chap. 4, the repressed affections are not representations in a barred compartment of the psyche, but rather manifestations in the style of being. Now it is time to formulate and define the notion of traumatic subjectivity more precisely. Following Fuchs, Kozyreva, and Ullmann, a composite picture begins to take shape in which traumatic subjectivity refers to a non-representational past-relation, instantiated and occasionally enacted by the lived body. In addition, the projective capacity of the lived body gives rise to the holistic presence of existential feelings in the lifeworld.

Inspired by Merleau-Ponty, Fuchs claims that the lived body has the peculiar capacity to actualize the past and constructing virtual presents (Fuchs, 2012, p. 11). The most glaring examples of these projective or constructing processes can be the flashbacks of patients with PTSD, or the bodily enactments in hysterical symptoms[5]. The complexity and intricacy of conscious life may leave room for various experiences that are hard to explain and categorize. Merleau-Ponty abandoned the representational account of memory and attempted to capture the primordial relation between the lived body, time, and space:

> The part played by the body in memory is comprehensible only if memory is, not only the constituting consciousness of the past, but an effort to reopen time on the basis of the implications contained in the present, and if the body, as our permanent means of "taking up attitudes" and thus constructing pseudo-presents, is the medium of our communication with time as well as with space (Merleau-Ponty, 2012, p. 210).

It is frequently mentioned in contemporary phenomenology that bodily dispositions disclose or delimit structures of possibilities in the lived world. However, the body is not only the bedrock of the living present, but also enables the above-mentioned

[4] "Unconscious affectivity or the affective unconscious is an »invisible« system of affective structures, schemes and Gestalts and as such it is opposed to phenomenally appearing feelings and emotions. In phenomenological terms: the affective unconscious is an emotional or affective horizon (Spielraum der Möglichkeiten) or a prefigured implicit sense-structure." (Ullmann, 2017, p. 160)

[5] The flashbacks are often the mixture of real and feared or imagined events. The term flashback did not appear until the late 1960s when it referred to the experiences of LSD users. Sometimes LSD users reexperienced certain aspects of their drug-induced hallucinations. Later the term was applied to the involuntary recollections of Vietnam war veterans (Schacter, 1996, p. 207).

"mental time-travel". The above citation highlights the role of "implications" in the living present, which can be likened to the retrieval *cues* used in the process of remembering (cf. Schacter, 1996, p. 62, 70).

Section 4.2 has introduced the concept of the horizontal unconscious, which refers to the subjectively lived world that is permeated by the implicit influence of the past. Drawing on Merleau-Ponty and Fuchs, Kozyreva defines the (implicit) past as a "modus of one's bodily existence" that stays unnoticed and effective through habits and dispositions (Kozyreva, 2017, p. 155). Kozyreva, based on Merleau-Ponty's and Fuchs' considerations, defines the past as incorporated dispositions that exist in the mode of oblivion. The expression of the "past in the mode of oblivion" designates not only the expressive quality of the lived body, but also the special cases of emotional memory. Kozyreva claims that emotional memories are closely related, but independent elements of the implicit realm. She mentions several psychological experiments in which people suffering from Alzheimer's disease are able to preserve their *emotional attitudes* toward certain people or life events. Thus, Kozyreva puts forward the argument that "not all off emotional memory is necessarily implicit, but part of the implicit memory clearly relates to the retention of affection and feelings" (Kozyreva, 2017, p. 177).

Before the detailed analysis of traumatic subjectivity and other kinds of past-relation, it is worth stressing the interrelated aspects of the previously discussed themes. The non-representational past-relation pervades the living present and based on the medium of the lived body. Or, conversely, the implicit habit-structure is actuated by the projective capacities of the lived body. The implicit body memory gives the living present a feeling toned, atmospheric quality and, without conscious deliberation, demarcates the fields of interests and repulsions. Ratcliffe's existential feeling and Fuchs' implicit body memory are thematically closely related[6]. Furthermore, body memory concretizes the seemingly abstract concept of the affec-

[6]The definition of body memory is not restricted to the interiority of the lived body, rather Fuchs argues that acquired predispositions and capacities to perceive, act, and desire extend out to the environment and constitute an invisible relational network. Body memory comprises practical skills, for example the ability of typewriting (Fuchs, 2019, p. 2). The phenomena of bodily resonance, intercorporeality, and interaffectivity allows for Fuchs to reconsider the status of mental disorders. He argues that mental disorders are basically disturbances of bodily and interbodily existence. In this respect, the aim of therapeutic relationship is to develop more adaptive and beneficial interactions with others. Therefore, the focus is shifted from the inside of the psyche to the outside, i.e. to the in-between region of patient-therapist relationship (cf. Fuchs, 2019, p. 2). Therapeutic relationships uncover the existential feelings of patients including the loss of trust, that is a recurrent theme in the traumatized patients' narratives. Existential feelings are closely related to body memory since they both "feelings of the body" and "ways of finding oneself in the world", by definition (Ratcliffe, 2008, p. 2). In short, in Ratcliffe's notion of existential feelings one can find the synthesis of body memory and the Heideggerian *Befindlichkeit*. Furthermore, existential feelings point toward the affective foundations of higher cognition. In Fuchs' own words: "Much of what is currently conceived as belonging to higher domains of cognition, belief, or judgement – for example, the sense of reality, the sense of being oneself, paranoid ideation, even delusions – is actually based on an affective foundation, namely on the tacit background feelings of the lived body in relation to others." (Fuchs, 2015, p. 616)

tive relief structure. Drawing on Merleau-Ponty's work, Ullmann characterized the non-representational level of affective schemas as an "invisible nervation" that permeates our normal, habitual ways of thinking and acting. Here I would like to suggest that that both implicit body memory and the affective unconscious (i.e., the "invisible nervation" of affective schemas) can be incorporated into the affective relief structure, constituting a condensed, habituated non-representatial sublayer of consciousness. However, the characterization of the affective unconscious as a network of habits, traits, skills, capacities, and affective-reactive schemas etc., suggests a static image of the affective core self. Without inner time-consciousness, the affective relief structure would lose its ability to bring the past into the present or to project the self into the future. The implicit layers of the self can be scrutinized and made explicit because the intentional life of consciousness unfolds in time. It is my contention that retroactive awakening and *Nachträglcihkeit* gives us the opportunity to dynamize the invisible nervation of the affective unconscious. Ullmann argues that the process of retroactivity is not only an ordinary experience in our intentional life, but also a means of gaining insight into the process of change. Retroactivity is both an intentional state of consciousness and a means to observe the phenomenological unconscious. Thus, retroactive meaning-bestowing is particularly important from genetic phenomenological point of view. Ullmann argues that retroactive understanding takes place both on the level of the intentional objects (e.g., we recognize a melody) and on the level of life history (e.g., we experience a turning point in our life but recognize it belatedly) (Ullmann, 2010, p. 281).

Here, I would like to add to this methodological observation that the process of retroactive meaning-constitution enables us to observe the reconfiguration of the affective relief structure in life history. And the affective relief is incorporated into the implicit body memory. Merleau-Ponty argued that the lived body constitutes the "style of being" and also the "locus of being" from which the orientations to the past and the future arise. As we have seen, the affective retroactive awakening of sense is a complex process with intentional and pre-intentional levels. The analysis of traumatic subjectivity will give us phenomenological insights into the pre-reflective sense-bestowing process of the lived body. The passive-associative processes of the lived body can reconfigure and maintain the affective relief structure (i.e., the affective core self).

6.3 The Affective Schematism and Traumatic Subjectivity

Following Fuchs, Kozyreva regards traumatic memory as a special kind of emotional memory. Emotional memory is also the part of the implicit unconscious in which everlasting habits may reside owing to traumatic experiences. Kozyreva claims that the impact of trauma is mirrored in the alterations of the subject's dispositions, which manifest themselves in the social world. After a traumatic encounter, a formerly harmless situation may elicit anxiety and despair because it acts as a retrieval cue that reminds the subject to the abuse or any other kind of trauma. These considerations show that the traumatic memory in question cannot be identical with the Freudian repressed. Kozyreva tries to avoid the step of placing traumatic experiences in the hidden registers of the psyche, instead, she describes the influence of the past as a change in one's "implicit dispositions" (Kozyreva, 2017, p. 181).

Regarding the issue of affective identity, Ullmann (2017) proposed the idea that, in a broad sense, we are all traumatic subjects. His claim is based on the psychoanalytic notion of separation anxiety. The primordial trauma of separation from the mother remains in a non-representational form in the unconscious. He argues, however, that this kind of traumatic subjectivity leads to theoretical difficulties: we cannot know for sure whether everyone has a traumatic subjectivity (over and above narrative identity), or whether the precondition for traumatic subjectivity is an encounter with traumatic situations in one way or another (p. 155).

Despite his cautionary remark that the claim that *everyone* is a traumatic subject is too general and basically dissolves the concept of trauma in a vague generality, I am inclined to accept the *strong claim* that we are all, to some degree, traumatizable. It is plausible to suppose that we are all traumatic subjects to a certain extent, since it is impossible to live a life without subversive experiences (e.g., physical shock) or milder emotional traumas[7]. In addition to Otto Ranks' "birth trauma" or Freud"s "oedipal conflict," there can be several other reasons for using the concept of traumatic subjectivity: psychosocial conflicts, existential crises, major disappointments, and the phenomenology of illness or life-threatening situations may play a central role in the constitution of identity. I do not intend here to consider the role of evolution or the individual vulnerability in the genesis of trauma, or to discuss the cultural and institutional embeddedness of the concept (cf. Hacking, 1995, Golden and Bergo, 2009). Traumatic experiences disrupt the continuation of the life history and induce a fragmentation process in the self.

Kozyreva's and Ullmann's reflections on trauma crystallize and complete the concepts of the affective unconscious and affective identity. Traumatic events can be engraved in the affective relief structure that constitutes the affective unconscious of the self. However, the problem of traumatic subjectivity and its relation to time-consciousness poses a challenge to the idea of affective identity. Traumatic experiences can be seen as enigmas or blind spots in the affective unconscious, waiting for conscious integration and narrative elaboration.

Both Ullmann and Kozyreva draw attention to the affective-associative processes that establish a meaningful connection to the past. In addition to explicit remembering and active recollection, there is the underlying phenomenon of "affective awakening" that, according to Kozyreva, sustains the process of conscious recollection. Let us pay attention to the associative complexity of the experiential life:

> Our field of conscious experience is not uniformly organized: there are objects and groups of objects which stand out against the background; there are sounds which attract more attention than others; there are thoughts and feelings that are more salient, while others are less prominent and yet constantly present; there are memories which suddenly occur and

[7] Ullmann's considerations are mainly based on Levinas's, Bernet's, and Tengelyi"s ideas, and he also relies heavily on psychoanalytical insights (cf. Ullmann, 2015; 2017).

> others already incorporated within our way of being. Nevertheless, the fact that our experience is multifaceted and variable does not mean it is chaotic or disorganized (Kozyreva, 2017, p. 119)

Kozyreva argues that an adequate phenomenological undertaking must take into account not only the principles of Gestalt formations, but also the affective dimension of associative synthesis. Her argument is based on the Husserlian notions of "affective force", that surfaced in the lectures on passive syntheses. In a sense, the principle of similarity and unity-formation provides the opportunity to connect everything with everything else, but despite the lurking danger of associative chaos, we constitute a well-organized and coherent phenomenal field. Kozyreva (2017) underscores the underlying principle of the "degree of vivacity"; that is, every associative awakening (whether impressional or retroactive) requires an affective force that sets the chains of associations in motion and organizes them (p. 120). In Husserl's word: "An actual connection of unity always and necessarily presupposes affective force or affective differentiation." (Husserl, 2001, p. 221). As a result, conscious life is bombarded by spontaneous associations that not only obey the rules of similarity but are also brought about in self-affective way. As we have previously seen, Mishara stated that the ego is not the master of its own phenomenal field. In other words, the ego is not the master of its own associative processes. Here, I do not intend to follow the path of Smith's (2010) comparative analysis of the Husserlian and Freudian notions of association, but it is worth mentioning that the Freudian *free association* and the Jungian *association test* demonstrated the possibility of the emergence and the repression of affectively charged associations (cf. Szabó, 2016, p. 105).

Ullmann highlighted the fact that the problem of retroactivity is marginal in Husserl's analyses concerning passive syntheses. Nonetheless, the introduction of the role of affective awakening may reformulate and compete with the initial conception of time-consciousness. The temporal modification of primal impression and retention is, at the same time, an affective depletion; the initial novelty or vivacity of experience slowly fades away into the retentional oblivion of the distant past. While the temporal modification can be understood by Husserl as the momentary preservation of the impressions, the retentional phase is also the diminishment of affective intensity. That is, as Kozyreva and others has accentuated, there is the twofold mechanism of preservation of sense and diminishment of affective strength in the spacious present of time-consciousness. The implicit phenomenon of retroactivity is capable of awakening empty presentations from the structure of sedimentation. However, as it has been mentioned in Sect. 3.2, the clouding over of retentional consciousness is not identical with complete affective depletion. The Husserlian *marginal consciousness*, at first sight, directly enables the annihilation of impressions in the retentional flow but at the same time it tends to retain the affective properties of the previously unnoticed but subliminally perceived impressions. As a result, the periphery of consciousness is full of hyletic including latent affective forces which are waiting for retroactive integration. Affective awakening serves as a tool for the integration of the subliminal content into conscious awareness. The possible integration of subliminal impressions is not only the result of retroactive intentional activity, but is also closely related to the anticipatory structure of consciousness. Due to the anticipatory nature of consciousness, in *fresh* retentions one can find retentionally overclouded but affectively still living impressions.

Let us examine this phenomenon from the point of view of the ego, that is, from the *noetic side* of intentionality. The anticipatory structure of consciousness is not limited to the protentional structure. The just-past retentional flow contains the sphere of the unfulfilled anticipations that can potentially be integrated belatedly under the right circumstances. To put it more dramatically, we can speak of anticipations within the retentional flow that are implicitly present at the periphery of the ego's attentional sphere. This kind of anticipatory process is not identical with constructive perceptual processes. The gist of the argument is not to point to the belated synthesis of the previously unnoticed hyletic elements, or to the process of filling-in in the case of ambiguous perception, but rather to the reawakening of the just-past or even the distant past from the periphery of the ego. As was also discussed in Chap. 3, Husserl introduced the metaphor of affective force in order to provide a tangible description of the oscillating movement of consciousness between the present and the past.

The central concern here is to show that the process of retroactive sense-bestowing, as a special kind of intentional relation, can contribute to the reconsideration or even the expansion of the minimal self. The for-me-ness of experiences is the basic characteristic of the (more) minimal self. However, I argue that the for-meness or simply the mineness of experiences is a more complex question in the case of the (less) minimal self. If we leave behind the formal notion of the (more) minimal self, then the problem of ownership or mineness can also be connected with the dilemmas of explicit remembering and retroactive awakening.

What I intend to show is the fact that the construction of autobiography is highly dependent on the actual traits and preferences of the subject. And in different time periods, different memories come to one's mind as relevant or irrelevant ones. These constructive and combinative processes have been demonstrated by contemporary memory research and also by the proponents of narrative self-understanding (a good example is Schechtman's empathic access). In short, the actual self relates selectively and constructively to his or her involuntary and voluntary memories, often appropriating them without reflection and clear reasons. The most significant factor in altering the acknowledgment or appropriation of memories is lived time itself. However, the dividing line between constructive memory processes and the phenomenology of retroactivity cannot be obliterated. The retroactive awakening of the past can be phenomenologically categorized as an affective and cognitive schematization process that can awaken the just-past and also the distant past. Its main purpose is to recognize hidden or unnoticed meaning-structures in the light of new knowledge acquired in the present.

And how does all of this relate to the minimal self? I argue that the core of personality (i.e., the affective cores self) can be altered by milder subversive or more dramatic traumatic experiences in lived time. According to my proposal, disintegration and reintegration processes may occur in the affective core self (i.e., in the (less) minimal self) with the contribution of explicit (conscious) and implicit (subconscious) retroactive processes. In this context, the term "disintegration" refers to the wide spectrum of unusual intentional states studied by the advocates of the self-disorder theory, including the disengagement from practical concerns, delusional mood, and delusional perceptions (i.e., concrete delusions and hallucinations)[8]. The "reintegration" refers to the return to the consensual world and the recovery of perceptual cohesion. In this regard, Merleau-Ponty and recently Ratcliffe have pointed out that perception plays a constitutive role in stabilizing identity and self-world relations; when perceptual cohesion is disrupted, the integrity of the modal structure of intentionality is also endangered, which, as Ratcliffe argues, is identical with the disruption of the minimal self (Ratcliffe, 2017a, pp. 164–165). Retroactivity is also closely connected to perception, since the most common forms of retroactive awakenings are based on the unnoticed elements in the just-past retentional flow. As noted earlier, retroactivity could also play a pivotal role in the reconfiguration of the self through the integration of traumatic experiences into the fabric of autobiographical narrative.

Kozyreva discussed the problem of retroactivity in the broader context of associative synthesis. In particular, she argued for the crucial role of affectivity in the genesis of association. It is not an exaggeration to say that the retroactive awakening of the past complements the phenomenological insights about the organization of the self and recollection. In a similar vein, Ullmann has convincingly argued that retroactivity is not the exceptional mechanism of consciousness, but is also of central importance from the standpoint of genetic phenomenology. The retroactive constitution of sense unravels the ways in which consciousness recreates the past and alters the self. It also provides an opportunity to examine the phenomenological unconscious (Ullmann, 2010, pp. 275–276). Retroactivity, on the one hand, sheds light on the mechanism of *becoming conscious* of something, and, on the other hand, leads to the question of unarticulated affections, which can be compared to the Leibniz's *petites perceptions* (Ullmann, 2012a, p. 98). The problem of unarticulated affections refer to the fact that what is awakening in retroactive understanding is not an empty presentation, but rather an unnoticed affection. One can find a particular non-linear or transverse movement (i.e., counterflow) in the linear rentential stream of consciousness. Our general intuition suggests that the present is derived from the past and the future is projected from the present. The present, schematized by past experiences, is the springboard for our projects in the future. However, retroactivity is a special associative process, because it constitutes a past at the core of the living present by means of the *deferred articulation* of intentional objects (Ullmann, 2012a, p. 99). The result of this special past-relation is that the past is inconceivable as a timeless reservoir of experiences. In other words, the phenomenological unconscious that is retroactively constituted is incompatible with the Freudian timeless unconscious (i.e., the id). The retroactively constituted past seems incompatible with the Freudian unconscious. However, Freud's ruminations over

[8] For the discussion of self-disorder, and especially schizophrenic self-disorder see Chap. 5.

the process of memory construction offers interesting points of theoretical convergence between phenomenology and psychoanalysis. As we shall see in Sect. 6.4, the dilemmas of early trauma and *Nahcträglichkeit* clearly show that Freud himself reckoned with and struggled with the problem of constructive memory processes. And there is another element of special significance in affective retroactive awakening: the affective force that awakens the past always emerges from the relief structure of the living present. An affectively neutral phenomenal field could not provide the opportunity to reconfigure the past. In short, retroactive understanding is embedded in the affective relief structure of the living present, which is, of course, also applies to the other modalities of intentionality (remembering, imagining, etc.).

From here, at least two divergent arguments can be offered. The first argument goes back to the problem of narrative self-understanding and offers everyday scenarios in which we usually reconsider or reevaluate past events in an explicit way. First, it can be critically argued that the retroactive understanding of the past is the basic ingredient of narrative self-understanding and in this case, we only open up and then close an ironic gap between the past self and the present self and between the past event and the reevaluated past in the present. These objections follow directly from Goldie's considerations (see Sect. 2.2). However, even Goldie's concept of narrative self-understanding reckons with the importance of affective states, i.e., the emotional and evaluative import of past experiences are the key ingredients of narrative self-understanding. One of Goldie's examples is particularly illuminating: imagine a scenario in which the protagonist sees himself from the perspective of observer memory. As Goldie recounts a story: "I remember from last night: I can *see* myself now, shamefully making a ridiculous fool of myself in front of all those people, getting up on the table and gleefully singing some stupid song." (Goldie, 2012, p. 52) The significant aspect of this thought experiment is that feelings originating from the present infuse the memory of the party. Goldie (2012) contends that "I remember it as I now feel about it." (p. 52) Moreover, the feeling in the moment can be infused into the content of *what* is remembered and also into the *way* of remembering (p. 53). However, it is important to note that the scenario cannot be interpreted as a direct analogy of retroactive understanding in a narrative context. Goldie is not talking about the complete affective resurrection of the past; on the contrary, in this thought experiment, the shame that arises in the present moment makes the awakening of the glee of last night's party impossible. We do not travel back in time and recapture the past, but rather we inhabit the position of "omniscience" from which we consciously or unconsciously alter the way of remembering (Goldie, 2012, p. 52.). As we shall see, one of the most important factors of retroactivity, at least according to Ullmann's interpretation, is that a *renewed past* can arise because of the backward radiating affective force. In Goldie's scenario, however, the emphasis is only on the reevaluation of the past because of the new emotional and evaluative import.

The second argument emphasizes the imaginable scenarios of surprise, disappointment, physical or emotional shock, and thus leads to the problem of traumatic subjectivity. The most radical manifestation of traumatic retroactive awakening occurs when new information in the phenomenal field triggers an associative reaction and the traumatic past is powerfully enacted in the present. The flashbacks and enactments of PTSD patients may represent the most extreme case of this process. Involuntary memories may also be the result of a milder form of retroactive awakening. Retroactivity offers a new way of describing the complex phenomenology of the resurrected past. As we haves seen, however, it is highly contestable to speak of the faithful representation of the past in retroactive reconstructions.

The studies of constructive memory processes suggest that the authentic reproduction of the past is literally impossible. Even the PTSD flashbacks, which are rigid, decontextualized, and dissociated states, are not entirely immutable. De Sousa (2017) argues that in cases of PTSD the decrease in attention is coupled with increased arousal, and PTSD patients are unable to tell a coherent story about the traumatic events. Memories and associated emotions lose their narrative structures. As a result, *somatosensory flashbacks* can occur in various forms and modalities. While the non-traumatic experiences are automatically organized and integrated into personal narratives, the main characteristic of somatosensory flashbacks is the lack of integration: the synthetizing effect of a coherent episodic memory or narrative is missing, and thus the sensory experiences of visual, olfactory, affective, auditory, and kinaesthetic information might occur in a relatively independent, dissociated manner (p. 158). This particular type of somatosensory flashback and other type of somatoform symptom may represent the pathological indeterminacy of the past. Pathological in the sense that the subject is unable to narratively characterize the traumatic event. As has been argued in relation to body memory, bodily enactments could be the signs of subpersonal meaning-bestowing processes. De Sousa (2017) argues that, in PTSD, there is a dissociation between the personal and subpersonal elements of memory and feeling: "Dissociated traumatic memories sometimes appear to be transmuted into 'somatoform' or bodily symptoms. In these cases, explicit memory and normal emotional response are replaced by apparently organic symptoms such as back pain, fatigue, headaches, or other unexplained pains." (p. 158) Admittedly, from clinical or neuroscientific perspective, it is highly speculative to speak of implicit or subpersonal meaning-bestowing processes. The system approach of cognitive psychology and neuroscience sees decoupled modules and disturbances of information processing, whereas the phenomenologist tries to unravel the active and passive genesis of personal meaning.

However, there is a thematic convergence between phenomenology and psychoanalysis regarding the relationship between affection and cognition. That is, the affective intensity of a memory does not contribute to its validity. The process of memory retrieval is filled with distortion, displacement, repression and suppression. Despite of the convincing affective power of awakened memories, we have to accept the indeterminacy of the past. Comparing Husserl's and Freud's ideas on the unconscious Tengelyi (2007a) argued that the consequence of *retroactive constitution* is the image of a partially inaccessible and uncontrollable past. Everything that can be retrieved from the past is embedded in the *horizon* of the actual present, which is sustained and opened up by the actual concerns of the present. Tengelyi (2007a) also draws attention to the late Husserl's definition of the ego. In his late manuscripts, Husserl described the ego as an unstable identity; the ego is constituted as an identity of inconsistencies and inner self-contradictions. The ego is therefore a vague unity (*"vagen" Einheit*). The process of retroactive meaning constitution plays an important role in the establishment of the ego's insecure unity. The notion of lived time, in which identity is constituted and updated, requires the complementary role of retroactive constitution (pp. 193–194).

From the above considerations, we can draw an intriguing phenomenological observation. Not only the future is the open horizon of possibilities, but also the past is in a state of constant fluctuation. As a result, the self finds itself at the intersection of the indeterminacy of the past and the incomprehensible openness of the future. Indeterminacy and unfathomable openness may give rise to otherness and alterity at the core of the self. To put it into the context of traumatic subjectivity, we can say that the radical experience of otherness may appear in our most intimate memories due to the retroactive reevaluation of the past. The experiential quality of the inconceivably alien is not confined to the open horizon of the future, but it can also appear in our personal memories. Our most cherished memories may give rise to shocking recognitions or bewilderment owing to the process of reevaluation. Using the Husserlian metaphor of "backwards-radiating affective awakening", it may be plausible to speak of *affective shockwaves* of subversive and traumatic experiences that flow backward in time from the present and inscribe their affective qualities in our identity constituting memories. The retroactive awakening and evaluating of the past does not lead necesseraly to the disintegration of the self, on the contrary, contributes to the preservation of the coherence and integrity of the self.

From the perspective of the Husserlian framework, it is important to note that retroactivity is essentially an affective-associative connection between the past and present on the level of *content*. However, it has to be admitted that that not only the *noematic* but also the *noetic* side of consciousness is involved in the affective awakening of the past. Both phenomenological and cognitive psychological approaches highlight the constructive elements in remembering. In this respect, let us remind ourselves to the counterintuitive observation of memory research: the reconstruction of the past mirrors our present concerns, purposes, and habits. From a phenomenological viewpoint, Kozyreva (2017) argues that active recollection, at its roots, is always also a process of reconstruction. Kozyreva was inspired by research on implicit memory. Schacter provides the description of "emotionally driven retrospective bias", which is a general feature in the reconstruction of traumatic events. Researchers have found that after a sniper attack at an elementary school "children who were at school during the attack tended to remember themselves as being in a situation of greater safety than they actually were" (Schacter, 1996, p. 206). Children reconstructed the event according to their current emotional needs in order to reduce anxiety. Another interview of an incident in 1988 shows that the declination of post-traumatic stress influenced the memory of the shooting, the subjects tended to remember the shooting as less threatening than it was. The studies indicate that the subjects reconstructed the past through the "filter of their later emotional states" (Schacter, 1996, p. 207). Another intriguing memory bias is the so-called mood-congruent retrieval: depressive patients are tend to remember negative experiences when they are in a sad mood, which eventually intensifies their sadness and leads them to the negative feedback cycle of clinical depression (Schacter, 1996, p. 211; cf. Blaney, 1986, p. 231). In other words, a negative emotional filter can be observed in depression.

Bower (1981) made a distinction between mood-congruity effect and mood-state-dependent retention. The former suggests that people tend to learn more about events that are in synch with their current emotional state, the latter refers to the phenomenon that people are able to recall an event more precisely if they somehow reinstate the emotional tone of the learning period (p. 147). Experiments in cognitive psychology has shown that people in a pleasant mood recall a greater percentage of pleasant experiences than ones, while in an unpleasant mood they recall the higher percentage of negative, unpleasant experiences. In addition, there is an obvious shift in the appraisal of the emotional intensity of events. For example, when the subjects were in a positive mood, the recollected incidents were rated as more pleasant and less unpleasant, and feeling bad in the present can transform the recalled experience into unpleasant. That is, there was an apparent shift in the affective-emotional ratings of events that reflected the actual mood of the subject (Bower, 1981, p. 133). These were the most common instances of mood-state-dependent retrieval. In case of depression, the affect-state-dependent memory means that "the sadder the patient is, the quicker he or she can call up sad experiences relative to happy experiences." (Bower, 1981, p. 134)[9].

In my view, these studies can strengthen the phenomenological intuitions about the affective schematization processes that operate under the level of conscious intentional activity. Kozyreva widens the scope of the affective retroactive awakening from the perspective of implicit memory research. For her, it is conceivable that the starting point of every active recollection is a passive affective awakening (Kozyreva, 2017, p. 189). How can we bolster the underlying affective motivations in case of remembering? Several studies demonstrate the importance of *emotional arousal* in remembering. For example, according to Schacter, an event with high-arousal is easier to recall than an event with low-arousal, and the accuracy of memory is related to the emotional arousal elicited by the experience. In addition,

[9] Bower proposes the intriguing idea of semantic-network approach to integrate the role of emotion in cognition. The hypothesis supposes that each distinct emotion is a node or unit in memory that fuses together other aspects of the emotion in question (e.g., one specific node of emotion is associated with expressive behavior, autonomic patterns, and appraisals, etc.). The blending or mixing of emotions is also conceivable according to this model, but one specific emotion may inhibit the other, for example, fear as a contradictory emotion could suppress joy and sexual arousal (Bower, 1981, p. 135). Bower introduces the analogy of the electrical network, that demonstrates the activation process of associative clusters in the case of mode-state-dependent effects: "The contents of consciousness are the sensations, concepts, and propositions whose current activation level exceeds some threshold. Activation presumably spreads from one concept to another, or from one proposition to another, by associative linkages between them. A relevant analogy is an electrical network in which terminals correspond to concepts or event nodes (units), connecting wires correspond to associative relations with more or less resistance, and electrical energy corresponds to activation that is injected into one or more nodes (units) in the network. Activation of a node can be accomplished either by presentation of the corresponding stimulus pattern or by prior activation of an associated thought." (Bower, 1981, p. 135)

Schacter argues that: "Arousal may also influence what is remembered from an emotional experience by focusing our attention on specific aspects of an experience." (Schacter, 1996, p. 209) The following finding can be another example of the affective motivational processes: people who were exposed to slides of traumatizing images (e.g., images of car accidents) were able to remember more central elements and fewer peripheral details of the episode. Under traumatic conditions, attention focuses on "emotionally arousing parts of the episode" (Schacter, 1996, p. 210). There are, of course, theoretical and methodological contrasts between the experimental settings of psychology and the first-person, phenomenological accounts of phenomenologists.

The above-mentioned findings and studies do not verify the Husserlian affective awakening, but only support it; that is, my intention was not to transform the notion of affective awakening into a target phenomenon for cognitive psychological experiments. However, I suggest that the experimental findings can contribute to the development and enrichment of the Husserlian concept of affective retroactive awakening. In short, interdisciplinary research on the problem of affective schematization is a highly conceivable and valuable endeavor. Nevertheless, Kozyreva's general claim that every association is incited by underlying affective awakenings or motivations is essentially a phenomenological claim. Moreover, not only contemporary memory research, but also Freudian psychoanalysis has indirectly contributed to the study of affective awakening. For present purposes I side with Ullmann (2015) in arguing that we can gain a clearer picture of affective identity through the lens of traumatic subjectivity. From a psychoanalytic angle, the dynamics of affective forces, especially the fluctuations of pleasure and pain, have a powerful impact on the constitution of the living present and on the recollection of the (indeterminate) past. (This effect has already been demonstrated by the projective capacity of the lived body.) Obviously, the retroactive reconfiguration of the past can be examined from the perspective of narrative self-understanding and self-constitution.

The concept of retroactive meaning-constitution can be extended to include traumatic retroactivity. I propose that traumatic retroactivity is the subtype of retroactive self-understanding. In the following chapters I will distinguish between two possible forms of traumatic retroactivity. Most importantly, retroactive trauma (i.e., Freud's *Nachträglichkeit*) will be introduced as a particular mode of affective retroactive awakening. Before diving into the problem of the Freudian afterwardness, however, the problem of intrusive memory will be examined. From a phenomenological point of view, intrusions and enactments can also be seen as manifestations of affective retractive awakening.

6.4 Intrusive Traumatic Memory and PTSD

The previous discussions have shed some light on the role of affective awakening in memory retrieval and have shown that retroactivity is an crucial function in the process of self-constitution. When we try to analyze the role of pre-thematic associations inside the experiential flow, we can abandon the concept of the unconscious reservoir. However, as we will see, Husserl himself suggests a more or less objectified notion of the unconscious when he speaks of the sedimentation of the distant past. The conclusion so far has been that the past is not an unchanging representational system, but rather the past is reconstructed in the present by means of scattered fragments and remnants of diverse experiences. In this respect, there is a convergence between contemporary phenomenology and cognitive psychology.

Neisser (2014 [1967]) points out that the constructive view of memory has a long history. For example, Bartlett showed that reorganization and change are the basic rules of memory retrieval. And what about the state of stored information? Or, phenomenologically speaking, is there any chance to presuppose the stratified layer of sedimentation in Husserlian terms? The answer from cognitive psychology is not too reassuring: "Stored information consists of traces of earlier constructive acts, organized in ways that correspond to the structure of those acts" (p. 265). Neisser's succinct statement needs further elaboration. What is of pivotal importance here is the fact that there are no "dormant copies of earlier experiences". For Neisser (2014 [1964]), the reconstruction of a dinosaur serves as an instructive parallel for memory construction. A paleontologist can find all of the bones of the dinosaur and reconstruct its skeleton quite accurately. However, the reconstruction of the past is not identical with the retrieval of diminished sensory data. In this respect, the metaphor of the dinosaur's skeleton is quiet misleading, because in memory retrieval the bone fragments (i.e., the sensory elements of memory) of the dinosaur do not appear in the model of the dinosaur (i.e., in the end result of memory construction). The bones are only "sources of information about how to reconstruct it" (p. 271). According to Schacter's (1996) interpretation: "…only bits and pieces of incoming data are represented in memory" (p. 40). Furthermore, Schacter contends that if we accept that memory is the basic ingredient of the sense of self, then the self is based on fragmentary remnants of experiences (*ibid.*).

At this point, it is a crucial to consider how the malleable past comes back to life again in the course of affective awakening? The analyses of passivity and the affective relief structure have shown that both the living present and the indeterminate past have a peculiar kind of pre-thematic and pre-intentional organization. A pre-intentional activity takes place in cases of voluntary and involuntary reconstructions of past events or in cases of construing an autobiography. But what happens when the past exerts an unbearable influence on the present? Intrusive traumatic memories, repetitions, or the enactments of body memory may shed light on the anomalous wasy of memory retrieval. Before addressing the problem of intrusive memories, it is important to see that "traumatic memory" is a controversial concept. As Loftus and Kaufman (1992) have pointed out, traumatic memories can appear as quite accurate flashbulb memories, or, in other cases, as repressed memories, and sometimes the "return of the repressed" is easily recognizable (p., 222). What I want to explore are the scenarios in which the past does not remain in a state of oblivion but obtrudes into consciousness in a spontaneous and self-organizing way. These phenomena may help to support the claim that an ongoing, implicit (and bodily mediated) retroactive self-understanding lurking behind the back of consciousness. By implicit or bodily retractive awakening I refer to the emergence of body memories, which Fuchs described through the analysis of Proust's madeleine experience and Appelfeld's postwar memories.

In this chapter I consider the phenomenologically significant aspect of post-traumatic stress disorder. In the previous section, I examined the malleable past from the perspective of the actual present, and now I will consider the cases of intrusive memories. From a phenomenological perspective, the intrusive memories can be interpreted as a special form of affective retroactive awakening in which the traumatic past is dissociated from the self and seems to live its own independent life. In my view, the re-experiencing of the past in the present and the diminishment of the clear boundaries between the present and the past are of special importance for phenomenology. According to this proposal intrusive memory is a special kind of affective awakening, and it can also help to integrate Thalia Welsh's notion of *active repressed retentions* into the framework of the affective unconscious. As I have indicated in Sect. 3.3, Welsh introduced the notion of active repressed retentions in order to reconcile the Freudian repressed unconscious with the Husserlian phenomenological unconscious (i.e., the sphere of sedimentation).

Post-traumatic stress disorder (PTSD) appeared as a diagnostic category appeared in the third volume of *The Diagnostic and Statistical Manual of Mental Disorders* (DSM III), and its validity continues to be hotly debated. In the history of the DSM, first the term of "traumatic neurosis" and then "anxiety disorder" were the precursors of PTSD, which was finally introduced in DSM-III. It is obvious that PTSD is not an completely coherent or consistent construct. For example, there is an obvious problem with the so-called *stressor criterion*: a presumed traumatic event does not necessarily lead to the typical PTSD symptoms; there are traumatic situations from which the subject develops a more positive attitude toward life (cf. Hunt, 2010, pp. 51–56). For present purposes, what is important about PTSD is the modification of the temporal structure of intentionality and its relationship with trauma and implicit memory.

Larrabee points out that veterans suffering from vivid remembrances would have been labeled as psychotics a hundred years ago (Larrabee, 1995, p. 352). The historical antecedent of PTSD is the phenomenon of shell shock, first observed in World War I. Soldiers exposed to life-threatening situations developed recurring nightmares and intrusive memories of the shocking events[10]. Many people, who

[10] Ian Hacking argues that in the late nineteenth century the curious cases of railway spine, which were caused by railway accidents, and Charcot's hysterics set the stage for the psychologization of trauma. Pierre Janet made psychological trauma to the central aspect of clinical practice and Freud was committed to sexual aetiology. Hacking accentuates that in case of Janet the cause of trauma was an event or state; that is, the world was at the center in the genesis of trauma. For Freud, on the other hand, trauma almost always involved an intentional action. And these human actions called for the processes of reconstruction in memory and reinterpretation in therapy (Hacking, 1995, pp. 191–192). Hacking further argues that being in shock or acute stress disorder, as the DSM-IV called it, seems to be a human universal (Hacking, 1995, p. 188). Another important historical observation comes from Judith Herman, who argues that the psychologization of trauma is inseparable from political movements. For example, the transformation of shell shock into PTSD is the direct consequence of the political context of anti-war movement, which pushed the cult of war into the background (cf. Hacking, 1995, p. 212).

were executed for cowardice, had actually been suffering from shell shock. After World War II, and especially after the end of Vietnam War, PTSD were acknowledged by medical experts (Schacter, 2001, p. 173). In DSM-IV-TR, the most important diagnostic factors are not the characteristics of the person, but rather the exposure to traumatic events. Owing to this new formulation, the diagnosis of PTSD has had a dramatic impact on forensic psychiatry. However, there is no empirical evidence to support the idea that traumatic pathology is strictly limited to specific traumatic events (Bisteon et al., 2014, p. 670). Bisteon and his colleagues (2014) argue that a strong analogy can be drawn between the onset of PTSD and the Freudian afterwardness (*Nachträglichkeit*): both phenomena describe a mechanism by which a subjective interpretation is incorporated into the aetiology of the trauma (p. 671). (Further elaboration ensues in Sect. 6.4.) Almost all trauma survivors tend to experience disturbing intrusive memories after the traumatic event, but PTSD patients are tend to experience intrusive memories even a long time after the events. Schacter portrays them as if they literally stuck in the past. Individual vulnerability and the subject's relation to the past is also important in the development of the PTSD syndrome:

> Studies of Vietnam veterans and victims of sexual abuse indicate that individuals who remain focused on the past for years after a traumatic event exhibit higher levels of psychological distress than those who focus on the present and future. High levels of psychological distress in turn stimulate even greater focus on the past, thus setting up a destructive self-perpetuating cycle of persistent remembering like that observed in cases of depression (Schacter, 2001, p. 175).

The phenomena of "temporal disintegration" and "avoidance" pose a challenge to both phenomenology and psychoanalysis. It is a natural tendency to try to avoid negative, unpleasant memories; memories aroused by the feelings of shame and guilt are common in this regard. Patients with PTSD often try to avoid recalling trauma-related memories and thoughts due to the overwhelming pain and distress associated with these memories. As a result, a "rebound affect" occurs; that is, the suppressed trauma-related memories will cause even more psychological distress for the subject who is unable to cope with them. Schacter argues that the working through process of trauma (through therapeutic techniques)[11] facilitates the habituation process that reduces trauma-related anxiety and distress:

> Repetition of just about any stimulus or experience will result in what researchers call habituation – a reduced physiological response to the stimulus. If I play a loud sound for you at regular intervals and record physiological activity, at first you will show a strong response to the sound, followed by a gradual drop-off. The same goes for traumatic

[11] Schacter puts heavy emphasis on *imaginal exposure therapy,* in which patients are exposed to trauma-related stimuli and experience vivid images of the event. The results showed reduced levels of anxiety and intrusive memories in Vietnam veterans and survivors of sexual abuse (Schacter, 2001, p. 177). The findings of modern neuropsychoanalysis specified the traumatic-affective reactivations in the disturbance of the communication between the amygdala and the hippocampus (traumatic episodes can lead to a drastic shrinking of the area of the hippocampus responsible for processing and integrating memories; currently this can be the neurobiological basis of PTSD (cf. Fischmann et al., 2013).

> memories: repeated reexperiencing of a traumatic memory in a safe setting can dampen the initial physiological response to the trauma. Attempts to suppress memories of upsetting experiences prevent this normal process of habituation. Suppressed recollections thus retain an extra charge that eventually augments persistence (Schacter, 2001, p. 177).

Phenomenologically speaking, the suppressed memories, endowed with an extra affective charge, can be regarded as associative awakenings without the active participation of the ego. A pre-egoic affective tendency gives rise to the repeating intrusion of the traumatic past and distorts the normal retentional flow of consciousness. It was discussed in Sect. 4.3 that Fuchs defines trauma memory as a foreign body that withdraws from conscious recollection, but the smallest cues can evoke and reactualize the traumatic situation in the present (Fuchs, 2012, p. 17)[12]. Schacter considers the role of flashbacks of war veterans. It is important to note that even PTSD flashbacks are composite formations: they contain both real and imagined elements, most certainly derived from the worst fears of the subject. Schacter recounts the case of a veteran who suffered from a returning flashback in which the corpse of a murdered villager was reanimated again like a zombie from a horror movie (Schacter, 1996, p. 207). Schacter claims that flashback visions revolve around the worst fears and constitute an incoherent mixture of reality and imagination.

The disconcerting experiences elicited by flashbacks can be reconsidered from the perspective of the affective unconscious. A crucial consequence of the affectively charged intrusive traumatic memories can be a disturbance in the affective relief structure, amounting to the disintegration of the self. Veteran soldiers often exhibit the ingrained behavioral patterns of the past or suffer from the incessant return of the horrific scenes of war. As scholars often remark, they literally remain and live in the past. Of course, this conclusion is too simplisctic. To broaden the scope of investigation, I will turn to Ratcliffe's particularly illuminating interpretation. The main question to consider is the reason why these flashbacks are interpreted as *relieved* rather than *recalled* experiences? Ratcliffe (2017a) emphasizes that one of the most important characteristics of traumatic memory is the lack of contextualization. While ordinary autobiographical memories are woven into the person's narrative, traumatic memories lack the proper contextualization due to their shock value. And the suffering subject finds himself or herself in a vicious circle owing to the altered anticipation structure of intentionality. One can feel estranged from the social world due to the traumatic events and, consequently, traumatic memories are anticipated differently than the well-integrated autobiographical memories. The incoming traumatic memory is characterized by an *anxious*

[12] One of Fuchs's illuminating examples comes from the autobiography of Aharon Appelfeld. It is worth quoting from the autobiography to demonstrate the role of retrieval cues in the return of old behavioral patterns: "Sometimes the smell of scroungy straw or the cry of a bird suffices to throw me far away and deeply into myself. – All that has happened then has been imprinted into the cells of my body. Not into my memory. The cells of the body seem to remember better than memory although it is assigned for that. Even years after the war I did not walk in the middle of the pavement or lane, but always close to the wall, always in a hurry, like someone who flees" (Fuchs, 2012, p. 18).

anticipation that differentiates it from other kinds of rememberings and further deepens the experience of isolation from the social world (p. 162). That is, for Ratcliffe the crucial issue is not the veracity of intrusive memories, but rather the altered anticipatory structure that literally causes suffering to the subject; not only the content is disturbing but also the "how" of remembering. He further suggests that regardless of the suffering subject's coping mechanism, there are three main characteristics of intrusive traumatic memories: (1) they are decontextualized; (2) they are anxiously anticipated; (3) they are experienced as variably perception-like (Ratcliffe, 2017a, p. 163).

As we have seen, Ratcliffe places emphasis on the lack of understanding and the erosion of the modal structure of intentionality in cases of intrusive memory. If I understand Ratcliffe correctly, the erosion or disturbance of the modal structure of intentionality can be blamed for the occurrence of "unfamiliar" kinds of intentionality such as misperception, thought insertion, verbal or visual hallucination, etc. Strikingly, Ratcliffe also points to the underlying affective changes that contribute to the disintegration of the minimal self, especially in case of trauma memories. The change in the modalities of intentionality goes hand in hand with the disruption of the sense of time. For example, Schacter argues that the second important factor in the development of PTSD is the "temporal disintegration": survivors of the 1993 Southern California firestorm reported severe distortions in their experience of time:

> …some survivors reported disturbances in their sense of orientation in time: they felt that time had stopped or that the present was no longer continuous with the past or the future. People who experienced high levels of such "temporal disintegration" immediately after the firestorms were especially likely to focus on and ruminate about the event six months later. A year after the firestorms, these same individuals experienced more distress than did people who were able to focus more on the present or future in the intervening months. Temporal disintegration in response to a trauma thus foreshadowed later troubles in people who remained stuck in the past, prisoners of persistent memories (Schacter, 2001, p. 175).

In these situations, the distant past always remains at the forefront of the present; the past does not fade away due to its overwhelming affective force, but rather it disrupts the normal way of temporalization. It is a phenomenological challenge to provide a detailed and subtle interpretation of persistent memories that somehow "live" over and above the habitual temporal order of time-consciousness. They are not empty retentions, lurking in the oblivion of the system of sedimentation, because they are filled with affective force and have a particular content. Persistent traumatic memories not only superimpose themselves on the living present, but at the same time limit the possibility of opening up to the otherness of the future. This brings us back to the argument that PTSD patients sometimes literally live in the past[13].

[13] „One veteran arrived in the laboratory with a loaded gun. When asked to remember events from anytime in their pasts, these regalia-wearing veterans retrieved far more memories from their time in Vietnam than did well-adjusted veterans. They also had difficulties coming up with specific memories of pleasant evens from any period in their lives. They are so emotionally tied to their Vietnam experiences that they attend to and care about little else." (Schacter, 1996, p. 211).

According to Larrabee's (1995) overview, the consciousness of war veterans is flooded with memories. It is an enormously difficult task to clearly characterize the intentional states of PTSD patients. They can be defined as memories or vivid flashbacks of former traumatic events. But, as we have seen, the contextual and behavioral elements are also integral to the experience. Larrabee (1995) suggests that patients often hear noises of the traumatic incident within the actual present and "reacting physically and emotionally as they would have in the actual situation in the war arena" (p. 352). The flashback component of the experience alludes to quasi-perceptual contents and action: the veteran hears mortars exploding, guns firing, soldiers yelling, or running for cover, etc. These paradoxical memories are often actively evoked, but they also have their own autonomous activity and, as a result, they are able to override the physiological triggers of bodily and emotive activity. While not all subjects with PTSD exhibit such behavioral outcomes, patients frequently suffer from hallucination-like representations of the past, vivid nightmares, and affective disturbances (p. 352). According to Larrabee, such trauma-related memories cannot be interpreted in terms of the *narrative* and *pictorial* theories of memory, since, in certain cases, the recollection of the experience is too overwhelming to the subject. Larrabee (1995) argues that "the original trauma experience itself is thus both past (as having been originally experienced) and present (as now re-experienced)." (p. 357) In conclusion, I suggest that the trauma memory in question is an unfamiliar and also *hybrid* intentional state: it is partly representational and partly non-representational; it bears some resemblance to perception, but it is also a mixture of imagination and memory, often accompanied by reenactments. From the representational point of view, the subject can mark the memory as a trace of a past event; from the non-representational point of view, however, there are disturbing cases of enactment and acting out.

From the perspective of behavioral symptoms, Larrabee (1995) reconsiders the problem of intrusive traumatic memories. She proposes the notion of "flash-forward experience" is based on the features of re-enactments. The flash-forward element simply means that the subject may experience uncontrollable *activities*, and the loss of control eventuates the feeling of "craziness" (p. 353). As a result, a mild dissociation ensues. On the one hand, the patient knows that he or she is living through a past experience, but on the other hand, the patient is aware of the overwhelming power and autonomy of the flash-forward experience. In Larrabee's (1995) own words: "trauma memory 'flashes forward' in the person and takes it over" (p 353). Larrabee further speculates that acting out and somatic manifestations are *passive associative processes* in the Husserlian sense.

The crucial question is how can consciousness retain and "regurgitate" past experiences in an unexpected, unwanted, and unacceptable ways? The eruption of trauma memory may be a sign of initiating a conscious meaning-bestowing process; a means of working through. In other words, the inherent motivations of flash-forward memories can be interpreted as preconscious motivations for understanding and integration. For Larrabee (1995), the key to understanding these unusual experiences lies in the phenomenon of "retroactive cancellation". Here, cancellation means the negation of something incomprehensible: the former intangibility of the experience is crossed over and new meaning emerges (p. 360). From a clinical point of view, Hunt (2010) argues that exposure to intrusive memories can cause distress, but can also change the meaning of the traumatic incident. The process is quite complicated, as in case of high arousal and exposure to the traumatic memory the intrusion can be dysfunctional and facilitates avoidance rather than integration (p. 35). Levine (2015) warns against the utilization of prolonged exposure therapy immediately after the traumatizing event. After the shocking event, the subject can be in an emotionally disturbed state, and repeating it in therapy may lead to further traumatization or to the compulsion to repeat the stressful mindset (pp. 146–147). Hunt refers to studies that highlight the observation that trauma-processing cannot be reduced to intrusion; rather, a more subtle and adaptive processing takes place through discussing the trauma with family and friends; i.e., the creation of a narrative is crucial for integration (Hunt, 2010, pp. 65–66). It is worth asking how repression may occur in case of PTSD? Hunt (2010) suggests that due to traumatic events can create an implicit memory network, but he has reservations about the status of the traumatic event *per se*; that is, the emergence of the trauma is highly dependent on subjective vulnerability. In accordance with Ratcliffe's considerations, Hunt agrees with the widely held assumption that a traumatic event can undermine our fundamental beliefs and assumptions about the world (p. 62). Both Hunt and Ratcliffe emphasoize that avoidance, i.e., living as if nothing had happened, is one form of adaptation, and narrative self-understanding of the traumatic event is another.

Hunt explains the process of integration with different memory systems. He argues that the conditioned responses of trauma – including emotional and behavioral responses – are stored in implicit memory networks, from which the conditioned responses can be triggered by situational reminder cues. That is, intrusive memories or reenactments are defined as triggered implicit memories. Several persons with PTSD are able to successfully avoid reminder cues and live an ordinary life, however, there are victims who suffer from the incessant activation of conditioned responses (Hunt, 2010, p. 69). It is another widely held therapeutic technic to try to act on the rigid, repetitive implicit remnants of the traumatic events. But Hunt introduces another pair of memories over and above the implicit-explicit distinction, namely he refers to Brewin and Andrews' dual-representation theory (cf. Brewin and Andrews, 1998). The two memory systems in question are the "verbally accessible memory" (VAM) and "situationally accessible memory" (SAM). The systems operate in parallel, or in certain cases one of them system dominates the other. In the case of traumatization, a dual encoding process takes place. The VAM system stores the information that the person consciously collected before, during, and after the traumatic event. Most importantly, the same system contains narrative elements of the situation. These narratives are created by the person, consciously or unconsciously. The peculiar forms of intrusive memories discussed above, including flashbacks, enactments, film- or perception-like repetitions, etc., are stored in the SAM system. This lower-level system contains more perceptual information about the traumatic event without "verbal codes" and it is difficult to communicate this pre-verbal information to others. According to Hunt, this may account for the fragmented and uncontrolled nature of trauma memories. Recovery from PTSD can work in two ways: first, traumatized persons have to deal with the loss of trust, the shattered belief-structure; second, the classically conditioned flashbacks have to be integrated as well (Hunt, 2010, pp. 70–71).

What can we say about intrusive or persistent memories on PTSD from a phenomenological point of view? On the one hand, intrusive memories may shed some light on the transitional zone between the implicit and the explicit, which can be regarded as a "constitutive dimension of pre-reflective self-experience" – as Kozyreva has formulated (Kozyreva, 2017, p. 173). Due to the high affective charge of intrusive memories, PTSD may indirectly point to the underlying affective-emotional infrastructure of the self. Kozyreva (2017) contends that emotional memory cannot be seamlessly shoehorned into the category of implicit memory, as not all emotional memories are implicit but they are definitely related to the retention of affections and feelings (p. 177). Of course, the emotional memories of amnesic patients – which manifest themselves in retained aversive behaviors – are certainly not identical to the stirred-up emotions of intrusive traumatic memories. Nonetheless, both phenomena have something to do with implicit memory, and, in phenomenological terms, they may contribute to the further elaboration of the affective relief structure defined as a special, unconscious level of the self.

In addition, we can think of intrusive memories as "active repressed retentions". As we saw in Sect. 3.3, Welsh (2002) has suggested that there are good reasons to supplement the Husserlian notion of passive unconscious with an underlying unconscious activity. Welsh argued that the Freud's insights into the repressed unconscious can be incorporated into the phenomenological unconscious. According to her, the psyche is composed of near retentions, far inactive retentions, and active repressed retentions that explain seemingly inexplicable behavior (Welsh, 2002, p. 181). I would like to add another item into Welsh's model of the phenomenological unconscious. How can we phenomenologically describe the intrusive symptoms of PTSD? Intrusive or persistent memories may represent the hidden layer of hyperactive repressed retentions: *repressed* in the sense that avoidance behavior can suppress the intrusion and repetition of traumatic scenarios, but *hyperactive* in the sense that not every person with PTSD is able to perform the suppression, and traumatic memories may have their own (implicit) intentional directedness in the process of narrative integration. Beyond these speculative interpretations only one thing is sure: the intrusive memories are almost certainly cannot be typified as explicit rememberings or as reproductive presentifications of the past. Traumatic memories have their own peculiar non-conscious, repetitive, disturbing, and involuntary aspects that challenges phenomenological categorization. To recapitulate, let me try to make a clear distinction between active repressed and hyperactive repressed retentions. Both of them can be described as passive pre-egoic tendencies waiting for the conscious appreciation at the periphery of consciousness. Welsh's active repressed retention was inspired by Freudian insights and refers to barred instincts, drives, and motivations that can be brought into consciousness through therapeutic techniques. By contrast, the hyperactive repressed retentions are fragmentary in nature and easily reactivated by reminder cues. Moreover, they can be even more devastating and debilitating than the hysterical symptoms.

Matthew Yaw offered an intriguing phenomenological interpretation of PTSD. Drawing on his own war experiences, he argued that the most striking consequence of the reenactment of the past is the radical transformation of the present. He places less emphasis on the notion of intrusive memories and focuses on the drastically changed horizonal structure of the living present:

> For example, when one is having a "flash-forward," all of the features of the environment may seem familiar. But, the situation is frighteningly real. While the underlying source may be a memory, living through a flash-forward is not experienced as remembering or as "being reminded" of the primal traumatic event. The shop-lined Main Street does not remind one of the dilapidated streets of Fallujah, but becomes those very avenues of fear and violence. The pothole does not remind one of the time a bomb detonated, but *is* the site of a bomb about to detonate. The entire present, lived world becomes a different one, transforming into something that may be familiar in memory, but experienced as real enough to instantiate extreme protective behavior, physiological reactions, and highly tangible misery (Yaw, 2015, p. 223).

In Yaw's (2015) description the holistic, experiential transformation of the living present comes to the fore: past and present merge in a familiar yet still frightening constellation. The above description stresses the importance of an underlying sense-meaning process. According to Yaw, in order to live through a flash-forward experience, an object has to be "charged" with a particular meaning (p. 223). Thus, affectivity also plays a pivotal role in this interpretation. Yaw emphasizes the holistic character of intrusive memory: when he looks at a pothole on the road, it immediately reminds him of the detonation of a bomb. What is relevant here is not the associative process, based on the similarity of the circumstances, but rather the avoidance reactions and anxiety triggered by the familiar sight (Yaw, 2015, p. 223.). In this description the pothole is not only a kind of anxiety-provoking trigger, but it has also been transformed into a floating (almost unconscious) meaning-structure that oscillates between past and present.

According to Yaw's own interpretation, "the noetic bestowal of danger" is automatically triggered because of the traumatic event. In other words, he is literally unable to perceive the pothole as a neutral object in the road because of the fear and anxiety that has been ingrained into his consciousness following the detonation that occurred under his vehicle (Yaw, 2015, pp. 209–210). He claims that the noetic judgement (i.e., the pothole is dangerous) constitutes a "habituated fixation" in the ego. Yaw (2015) further argues that there is a distinct difference between cognitive awareness and experiential sense-bestowal. In Yaw's own words: "… I 'know' that it is only a pothole, and yet can't 'shake the feeling' that it is dangerous." (p. 215) The traumatized consciousness resists any antithetical reason. From the angle of the emotional processing theory, Yaw argus that the greater the *affective value* of the fear, the harder to transform the fearful memory through reasoning and other kind of interpretative strategies (p. 219). Yaws illuminating analysis, which is at once a first-person account and a phenomenological description, harmonizes well with other interpretations of traumatic subjectivity. For example, the atmospheric quality of anxiety and the holistic change of the entire living present were conspicuous in Yaw's account. To recapitulate, the symptom is not specifically an intrusive memory, but rather a holistic transformation of one's way of being-in-the-world. In Ratcliffe's formulation, he has *found* himself in the world in a drastically different way because of the situational cue.

In conclusion, I propose that a special kind of affective awakening occurs in case of intrusive memories. Due to the diversity of PTSD symptoms and the above-mentioned interpretations of intrusive memories, it is important to keep in mind that there is no exclusive explanation regarding intrusive memories and PTSD. Nonetheless, I argue that the general idea of intrusive or persistent memory can be approached from a phenomenological point of view. What can we say about intrusive memories in the context of affective awakening? I suggest that we can categorize the tormenting intrusive memories as one special type of associative-affective awakening: let me introduce the notion of "subversive retroactive awakening" in order to phenomenologically reformulate the spontaneous recollection of intrusive memories. Intrusive traumatic memory is subversive in the sense that it destabilizes the self by generataing an unusual intentional state that re-schematizes the living present according to the context of the traumatic past and constitutes an anomalous intentional state, which comprises the mixture and combination of perception, memory, imagination, and bodily enactment. The affectively charged fragmentary experiences can lay claim to conscious, narrative understanding. At first sight, intrusive memory may appear to be the pure form of retroactive awakening, as the past and the present collide in an unusual intentional state. Nonetheless, there is a radical shift between the pure affective awakening of fragmentary past experiences and the conscious-retroactive understanding of the past. This shift is due to the decontextualized nature of the flashbacks: the role of retroactivity here lies in the process of meaning-bestowing, the retroactive cancellation of meaninglessness (as Larrabee has formulated) gives rise to the stabilization of the temporal profiles, to the clear demarcation of the past and the present, including the openness to the future.

According to Husserl, Kozyreva, and Ullmann, retroactive awakening is based on implicit affective phenomena. Therefore, in this context, it seems reasonable to utilize the Husserlian notion of the affective force that radiates backward in time, bringing the past into the horizon of the living present. Intrusive memories insert themselves into the living present, which entails the disruption of the modalities of intentionality, i.e., past and present may exist in a superimposed way, surrounded by the threatening vibe of anxiety and dread. It is important to note that, the present is not only painted, projected, or colored by traumatic memories, but rather the very (affective) core of the self may undergo a dramatic change. Yaw developed the idea that in cases of flash-forward experiences, an entirely new living present emerges due to the ego's attachment to affections such as fear, anxiety, and danger.

The phenomenology of intrusive memories draws attention to an implicit-passive meaning-bestowing process that was initiated in a pre-egoic manner. The traumatic past may run forward into the present owing to the hypervigilance of the traumatized subject. In phenomenological terms, trauma memory may motivate the retroactive schematization process of the past; a past that was retained in the retentional flow of time-consciousness and constituted the wound of incomprehensible otherness in the affective unconscious. In the next chapter, a comparative analysis between retroactivity and *Nachträglichkeit* will be presented in order to analyze the more subtle way of affective retroactive awakening.

6.5 Retroactivity and Nachträglichkeit

The problem of deferred action, revision, or afterwardsness (i.e., *Nachträglichkeit*) is of crucial importance in explaining the reconstruction of the past. The temporal schema of afterwardsness is one of the most complicated issues in the Freudian oeuvre. The motif appears mainly in adjectival-adverbial form (*nachträglich*) and rarely as an abstract term (*Nachträglichkeit*) (Fletcher, 2013, p. 251). In this chapter, I will show that Nachträglichkeit may offer a fruitful comparative analysis between psychoanalysis and phenomenology. Freud did not write a specific work on the problem of afterwardsness, but as Eickhoff (2006) has shown, it was a constantly recurring theme in the entire oeuvre. The first appearance of the term can be found in the famous letter to Fliess (14 Nov. 1897) in which Freud explicitly abandoned his seduction theory (p. 1454). Noel-Smith (2016) traces back the problem of *Nachträglickeit* to Freud's *Project for a scientific psychology* (1895) that contains the case study of Emma. He defines the phenomenon in the following way: "an experience is given retrospective meaning after a later experience" (p. 68). Freud observes that the typical instances of hysterical repression are those cases in which a memory becomes traumatic after the traumatic event. In these strange cases of *deferred action*, the memories have been retrospectively filled with affective force (i.e., sexual excitement), due to sexual maturation that takes place in puberty.

Emma had a compulsion of not being able to go into a shop alone. Her explanation for this aversive behavior was based on a memory stemming from her period of puberty; she recounted that when she was twelve years old, two shop assistants laughed at her because of the dress she was wearing, and one of them had pleased her sexually. During the therapeutic sessions, Freud discloses another memory that dates back to Emma's childhood (she was eight years old at the time). The second remembering depicted a shopkeeper who "grabbed her at her genitals", and Freud traced back her "oppressive bad conscience" to the fact that she has gone back to the shop despite the incident. Freud contends that in case of the *real event* of the first assault there was no way for sexual release, but after the recollection of the scenario, the arising memory eventuated an affective shock. Freud concludes that the repressed memory has only become a trauma through a *deferred action* (Freud, 1966 [1895], pp. 355–56). In short, the case study suggests that a repressed or split off memory becomes a trauma after the event. Moreover, Emma's accounts are quite inconsistent, and Freud focuses on the combination of scenes for etiological purposes. The intriguing aspect of the case study is that *repression* occurs in the second scene in which the shop assistants laughed at Emma. Fletcher (2013) argues that

Emma realizes that she was the victim of sexual harassment in her pre-puberty period and a deferred sexual release is triggered by the sight of the grinning shop assistant. Emma denied that she could clearly remember the first scene, but it is reasonable to assume that an associative link was established due to the similarity of the situations (p. 70). That is, the repressed sexual affect is constituted in the present owing to the reinterpretation of the past experience. On the one hand, there is the "belated precipitation of the sexual affect", and, on the other hand, the retrospective understanding of the primary scene (p. 71). Laplanche (1976) summarizes the case in the following way: "Here we have an instance of a memory exciting an affect which it had not excited as an experience, because in the meantime the changes produced by puberty had made possible a new understanding of what was remembered. Now this case is typical of repression in hysteria." (p. 41) The associative logic behind the awakening of sexual release or excitement is that the laughing shop assistants reminded Emma of the grinning shopkeeper who sexually assaulted her in early childhood. Fletcher (2013) points to the development of Freud's trauma theory: whereas in *The Neuro-Psychoses of Defence* (1962a [1894]) Freud believes in a specific "traumatic moment", in which repression and symptom formation occurs together, in the *Studies on Hysteria* (1955 [1895]) he introduces the auxiliary moment in which dissociated or repudiated representations return to consciousness (p. 61).

From the phenomenological point of view, Bernet (2000) has also drawn attention to the strange temporal structure of *Nachträglichkeit*. His interpretation rests on the role of association. He speaks of "a-subjective" or "anonymous" association between the two events that allows the "primitive shock" to manifest itself in the present and, in general, to run "against the course of time" (p. 163). An appropriate "signification" only retroactively (*nacthräglich*) takes place at the time of the second event. The first event, as a "shock without signification", is associated with the second event, which is a "signification without shock" (p. 163). An important factor in Bernet's argument is the concept of "trace". The trace of the first traumatic event does not simply refer to the act of repression or deferred symptom formation. Bernet raises the question of the status of the subject, who has suffered premature seduction. The subject in question finds himself or herself in a special indeterminate state: "...the one who undergoes this traumatic seduction is not yet – in relation to this event in any case – a subject. But neither is it nothing, since this event touches the subject without it knowing how to speak of it and since it leads indelible traces. 'Trace' is really the right word here, since it is a matter of something that refers to an event, the absence of which is so deep as to have never been truly present for the child." (Bernet, 2000, p. 162) Due to the mechanism of repression and symptom formation, the subject does not define itself as "traumatized subject", but is haunted throughout its life by a "foreign alterity", which lives in the form of the "first shock" and in the "incomprehensible symptom" (Bernet, 2000, p. 164).

The linear temporal structure of traumatization is obfuscated by the new way of understanding the past. From the phenomenological point of view, the upshot is that the traumatic affection "radiated backwards" in time and reconstructed the formerly neutral event as a traumatic event. The developmental stage of puberty retrospectively re-schematized the repressed memory[14]. In short, the revision of a past event may reconfigure the experience in question and "upload" it with affective force. From a phenomenological point of view, we can assume that the revision initiated a backward-radiating affective force that endowed a formerly natural event with traumatic force.

One objection to this simplified view of the re-schematization or reinterpretation of a past event is that even the Husserlian concept of backward-radiating affective force is an ambiguous metaphor. Ullmann has critically remarked that we cannot radiate anything into the past since it would be the arbitrary alteration of the past. Therefore, what is implicitly alluded in the notion of backward radiating affective force are the phenomena of change and genesis. That is, the retroactive revision of the past represents more precisely the unnoticed radical changes of the actual *self* rather than the faithful reconstruction of the past (Ullmann, 2010, pp. 283–284). What was happened in the past is irreversible, therefore the recollected past reflects our current emotional responses in a more pronounced way and does not count as an authentic revival of past emotions (cf. Summa, 2015, p. 177).

The second objection concerns that Freud's concept of *Nachtraglichkeit* is basically insignificant. Freud retroactive trauma is one of his early attempts to explain the traumatic origin of hysterical symptoms and quickly replaced by the theory of seduction and Oedipus-complex. In addition, it is incompatible with the phenomenology of affective retroactive awakening. However, as we shall see, the concept of afterwardsness occurs before and after the refutation of the seduction theory and is also Freud's major contribution to the problem of lived time. A for the incompatibility, it must be admitted that the phenomenological problem of retroactivity, at least in its original Husserlian form, does not deal with the problem of repression. Repression, in turn, poses a challenge to the phenomenological interpretation of a psychoanalytical idea.

The main task of this chapter is to show that that the problem of traumatic subjectivity may provide a general framework for comparing retroactivity and *Nachträglichkeit*. It has to be readily admitted, and it has now became clear, that the problem of traumatization per se is a multifaceted phenomenon that poses a challenge for both psychoanalysis and phenomenology. Despite the challenges and theoretical conundrums, the passive-implicit infrastructure of the self provides a ground for comparing the two phenomena in question. Whereas the idea of the

[14] In *Further remarks on the neuro-psychoses of defence* (1896) Freud argues that not the experiences themselves act traumatically, but rather their revival as memory after the subject has entered the stage of sexual maturity (Freud, 1962c [1896], p. 165.) In the same year, in *Heredity and the aetiology of the neuroses* (1896) Freud stresses his own sexual aetiology in order to reject Charcot's heredity conception. In this paper he also describes the revival of traumatic memory-traces, which have received their affective power thanks to the change in puberty: "Thanks to the change due to puberty, the memory will display a power which was completely lacking from the event itself. *The memory will operate as though it were a contemporary event.* What happens is, as it were, *a posthumous action by a sexual trauma*." (Freud, 1962b [1896], p. 154, *emphasis in the original*) What is relevant for Freud in the specific aetiology of hysteria is a "*passive sexual experience before puberty*". (Freud, 1962b [1896], p. 152, *emphasis in the original*).

affective relief structure mainly focused on the synchronic unity of the self, the problem of retroactivity will highlight the temporal dimension of the phenomenological unconscious and indirectly address the question of the diachronic unity of the self.

For start, let us confront the problem of the retrospective reconstruction of the past, which is an intriguing dilemma in memory research as well as in the contemporary phenomenology of affectivity (cf. Summa, 2015). From psychoanalytical angle, it is worth emphasizing that Freud considers the retrospective process to be a relevant aetiological factor in hysteria:

> Although it does not usually happen in psychical life that a memory arouses an affect which it did not give rise to as an experience, this is nevertheless something quite usual in the case of a sexual idea, precisely because the retardation of puberty is a general characteristic of the organization. Every adolescent individual has memory-traces which can only be understood with the emergence of sexual feelings of his own; and accordingly every adolescent must carry the germ of hysteria within him (Freud, 1966 [1895], p. 356).

It is worth keeping in mind that Freud addresses the problem of affectivity from a clinical perspective, but we can still find productive analogies regarding the phenomenon of affective awakening. In the case study of Emma, the metaphor of the "backward-radiating affective force" may be an apt metaphor for the process of traumatization[15]. As we have seen previously, Emma's memory of the second scene (laughing shop assistants) triggered or released the sexual response. Laplanche contends that this is a sexual reaction in a "double form," since it stems from both the physiological excitation and the new ideas about sexuality generated by the changes of puberty (Laplanche, 1976, p. 40). Fletcher observes a contradiction in Freud's analysis. On the one hand, Freud argued that sexual release was attached to a memory, that is, it occurred in a deferred way; on the other hand, he Freud claims that sexual release occurred prematurely, implying a stage prior to puberty. Moreover, if we accept the first version of "deferred release," then the role of *repression* in symptom formation must be taken into account, which implies that the first scene of the sexual assault persisted in one way or another. Nonetheless, Fletcher claims that Freud is inclined to accept the first scenario of deferred release in this case study and places less emphasis on prematureness (Fletcher, 2013, p. 72, 74).

The phenomenological translation of Emma's traumatic experience may sound like as if the second scene initiated a backward-radiating affective force that, due to the similarity of the retrieval cues, reactivated and loaded the memory of the first scene with affect. It is important to note, however, that the memory with a fresh affective charge is an experience constituted in the living present. That is, an aspect of retroactivity is realized: the affective force, initiated from the present, reanimates or brings back a past experience with new significance and meaning. Furthermore, the self has been radically altered by the affective awakening of the traumatic past.

[15] Of course, the problem of traumatization cannot be exhaustively explained either a phenomenological or a psychoanalytical framework. As was mentioned earlier, individual vulnerability and external stressors delineate a complex set of factors that may play major role in the emergence of traumatization.

What is missing from the picture is the clarified status of the traumatic scenes. Freud suggests that traumatization is a deferred process, since repression and symptom formation occur belatedly, after the event. The retroactive understanding of the past does not necessarily call into question the status of the remembered scenes, but for Freud, the status and veracity of the original traumatic scenes is a vexing problem. These dilemmas are crucial. To put it crudely, in contrast to Freud's efforts, the retroactive awakening of the past is not an "archeological" effort; the phenomenon of retroactivity produces reconstructed past experiences in the new light of the present. Of course, the process of re-evaluation itself can occasionally be traumatic, but the traumatic effect is constituted in the living present. Let us take a fictional example. When one of my old friends told me the secret that he had been a successive bank robber for several year, I could suddenly see his strange behavior in a new light. Now I understand why he kept changing his appearance and underwent several plastic surgeries, or why he kept traveling abroad. In this scenario, because of the staggering new information I may live through a retroactive trauma: I can no longer see my friend's strange behavior in a neutral way anymore. As the example shows, the traumatic impact of the new knowledge may re-schematize the past, but I was traumatized in the present by new, surprising information, and the past events remained intact in the process of retrospection. However, as has been suggested, *Nachträglichkeit* is more than simple retrospection and reevaluation.

In addition, even the utilization of *Nachträglichkeit* is ambiguous in psychoanalytic literature. For example, Birksted-Breen (2003) in her review article showed that Freud distinguished at least three different meanings of *Nachträglichkeit,* and the English and French translations are also bear the mark of uncertainty. Contrary to British developmental models, French psychoanalysis puts heavy emphasis on the non-linear form of temporality. Therefore, according to Birksted-Breen, the first definition simply means "later". The second refers to a "movement from past to future", which is implicit in the two-stage model of trauma. The third meaning, proposed by French psychoanalysis, alludes to a retrospective meaning-bestowing process (Birksted-Breen, pp. 1501–1502).

The second meaning, the movement from past to future, sounds like a counterintuitive idea from a phenomenological perspective, but can be illuminated by the processes of repression and *Nachträglichkeit*. The repressed is prone to be activated later, and this mechanism has been called a "time bomb" by Laplanche (1999). Laplanche also considered three meanings of *Nachträglichkeit*. According to his overview, the first is the temporal meaning of "later"; the second meaning is connected to the seduction theory and designates the direction of time from the past to the future (i.e., it covers the delayed action of the unconscious material); and last but not least, the third meaning conveys the sense of a retroactive meaning-bestowing process. Laplanche adds that Freud never tries to reconcile the deterministic and the hermeneutic meanings of the term. The deterministic interpretation refers to a "time bomb", and the hermeneutic interpretation refers to the meaningful impact of the present on the past. The retroactive meaning-bestowing process was thematized by Jung as "retrospective fantasizing" (*Zurückphantasieren*), but Freud insisted on the deterministic approach, arguing for the organic determinism of repression (pp. 265–266). In contrast to Freud, Jung argues in the following way:

> Hence, in tracing the libido regression, the analysis does not always follow the exact path marked out by the historical development, but often that of a subsequently formed fantasy, based only in part on former realities. In our case, too, the events were only partly real, and they got their enormous significance only afterwards, when the libido regressed. Whenever the libido seizes upon a certain reminiscence, we may expect it to be elaborated and transformed, for everything that is touched by the libido revives, takes on dramatic form, and becomes systematized. We have to admit that by far the greater part of the material became significant only later, when the regressing libido, seizing hold of anything suitable that lay in its path, had turned all this into a fantasy. Then that fantasy, keeping pace with the regressive movement of libido, came back at last to the father and put upon him all the infantile sexual wishes. Even so has it ever been thought that the golden age of Paradise lay in the past (Jung, 1961 [1913], p. 175; par. 394)!

The above quotation suggests that the regression of the libido is capable of reviving the past, but not without phantasmatic distortions. Of course, as the abandonment of the seduction theory shows, Freud himself was confronted with the distorting role of fantasy during the archeology of traumatic memories. Jung's concept of libido-regression can be compared to Husserl's metaphor of backward-radiating affective force; however, naturally, there are paradigmatic differences. For example, Husserl would have regarded Jung's libido theory as a naturalistic conception that should have been eliminated by phenomenological reduction. Laplanche argues that Freud insists on the deterministic approach and was consistent when he rejected the "reversibility of temporal direction in later texts." (Laplanche, 1999, p. 266). In contrast to Freud's ambiguity, Jung's retrospective fantasy (*Zurückphantasieren*) convincingly reflects the phenomenon of affective schematization and colorization of the past[16]. However, I do not intend relegate the problem of retroactive sense-making process to the background by reducing it only to a peculiar form of fantasy activity; things are getting more complicated. Let us now further discuss the Freudian insights into the phenomenon of *Nachträglichkeit*.

In the *Standard Edition* of Freud's legacy, Strachey refers to "deferred action" in order to translate the problem of *Nachträglichkeit*. It is rather confusing that this translation refers to the above-mentioned third meaning of *Nachträglichkeit*, namely the "retrospective attribution of meaning" (*après-coup*) popularized by Lacan and French psychoanalysts in general (cf. Birksted-Breen, 2003, pp. 1501–1502). Furthermore, in 1989 Laplanche introduced "afterwardsness" as the English translation of *Nachträglichkeit*, and in 1993 Otto Kernberg proposed the phrase of "retrospective modification" (House, 2017, 774). According to House, the common translations suffer from imperfections. The tension is most noticeable between *deferred action* and *retrospective modification*. These translations suggest entirely different temporal structures, i.e., determinism is central in the former, and the latter

[16] Freud in *Civilization and its Discontents* (1930) speaks of indestructible residues of memories awaiting the cathexis of instinctual energy, and surprisingly, he also ruminates on the possibility of the obliteration of mental content (Freud, 1961 [1930], pp. 69–71; see also Yiassemides, 2014, p. 106)

term suggests the re-schematization of the past via the actual concerns and needs of the present. The latter meaning also bears a superficial resemblance to the phenomenology of retroactivity. House highlighted the stark contrast between the translations with suggestive metaphors:

> Deferred action is a mental process with the same temporal structure as fireworks or land mines. The results, the deferred actions – beautiful lights in the night sky or destructive explosions – have been determined in the past by what was desired when the device was constructed. Retrospective modification has the opposite temporal structure. Its results are determined in the present on the basis of current needs (House, 2017, p. 776).

The retrospective modification and the Lacanian "weak translation" of *après-coup* is of immense importance to this comparative study. Lacan discussed Freud's case history of the Wolf Man in 1953 and coined the term *après-coup* to translate the adverb *nachträglich*. Regarding the problem of the status of the primal scene, Lacan speaks of a "turning point at which the subject restructures himself" after the fact (*après-coup*) (Lacan, 2006 [1953], p. 213). As House notes, Lacan was not satisfied with the translation, but he explicitly used the term in the sense of "retroactivity" or "retrospective modification" (House, 2017, p. 777). Following in the footsteps of Lacan, Laplanche and Pontalis (1973 [1967]) accentuated the delayed meaning-bestowing process and the role of affectivity: "…experiences, impressions and memory-traces may be revised at a later date to fit in with fresh experiences or with the attainment of a new stage of development. They may in that event be endowed not only with a new meaning but also with psychical effectiveness." (p. 111, cf. House, 2017, p. 779).

Recently, Browning (2018) underscored a phenomenologically significant implicit meaning in *Nachträglichkeit*: there is a bidirectional causality or movement from the past to the present and from the present to the past. Laplanche introduced the metaphor of a house that was defectively built and later collapsed. The collapse is an after-effect that was deferred in time, but the cause of the collapse can be clarified in retrospect. How does this example illustrate the problem of psychic trauma? The traumatic event – as a repressed germ – may give rise to predispositions and vulnerability, but the traumatic effect (i.e., symptom formation) occurs after the event, retrospectively. In terms of affectivity, we migt say that the traumatic excitement was "implanted" in the psyche and was subsequently released by a situation that bore a resemblance to the original traumatic event. In Freud's case studies the advent of puberty brings about the release of affections of anxiety and excitement. Browing argues that *après-coup* designates a complex time structure that encompasses both the role of deferral and retrospection (cf. Browning, 2018, p. 781; 783). The synthesis of deferral and retrospection gives rise to a peculiar temporal structure that may add new insights to the problem of the diachronic continuity of the self. Laplanche introduces the metaphor of the spiral or helix in order to reckon with the non-linear associative connections in lived time:

> The work of a thought does not travel in a straight line like an arrow. It follows a movement I call "spiral" but that, in reality, is a helix. This means that the life of human beings, and especially the movement of their thought, is neither linear, constantly leaving one point to pass on to another, nor circular, constrained to repeat the same sequences. The movement

> of a spiral constantly distances itself from one pole, but at the same time, it is led to return again and again to points along the same vertical. Returning to a vertical, one is led to reconsider the same themes at another and higher level (Laplanche, 2017 [1989–1990], p. 7).

In the psychoanalytic context, the non-linear movement of associations opens the way to rethinking the analytic process. In Birkstes-Breen's words: "Modern conceptions of psychoanalysis no longer aim at unearthing an objective reality but enable a continual reappraisal and modification of the past as internal world. The analytic process is a process of continual reinterpretation in which every new point arrived at encompasses and restructures what has been so far." (Birksted-Breen, 2003, p. 1508) In addition, she highlights that the continuous retrospective evaluation of the past is the result of the subject's openness to the future. Thus, retroactive or reconstructive processes are impeded by the subject's fixations. When the patient gets stuck in a hopeless present situation, then she is not only living through the sudden eclipse of the future but also suffers from a one-sided past-relation. Birksted-Breen (2003) claims that the "retroactive resignification" of the past is essentially a developmental progress, therefore, the "hatred of progressive time produces an attack on retroactive time" (p. 1509). Birksted-Breen also argues that patients with developmental issues often find themselves in obsessive imaginations, or rigid, unchangeable memories come to the surface in the therapeutic sessions. The previous observations bring us closer to the intricate nature of lived time. Openness to the future is a precondition for the retroactive appraisal of the past, and vice versa, a modified past-experience may indirectly re-schematize the anticipatory-structure of the imminent future.

Several interpretations of the altered experience of time can be found in the psychoanalytic context. For example, Birksted-Breen (2003) argues that schizoid patients exhibit the curious mental attitude of "slicing". These patients use sadistic fantasies for protective purposes: one of Birksted-Breen's patient, B, had the recurrent fantasy of slicing a brain like a piece of ham. The aim of this imagination is nothing mor than control. The outside world is apparently too overwhelming; therefore, the subject tries to stop the experiential flow of time and slices it into distinguishable inputs. There are no fruitful associations in the therapeutic sessions, and "The future is eradicated because, like Zeno's arrow, you never get there." (p. 1509) Later the therapy has produced an emotional experience: on the one hand, the patient waited for a woman who never appeared, on the other hand, this fixation led him to a disappointment in which he felt awful because a girl was unable to reciprocate his amorous affections. Paradoxically, this painful memory conveyed for the patient the sense of time, and he recognized how terrible it would be to live imprisoned in a frozen time-slice. For the schizoid patient, the experiential flow of life come to halt, and lived time is transformed into a slideshow with disconnected elements and without affective value. Consciousness can easily find itself in a dissociated state due to traumas, but the successful therapeutic intervention can facilitate self-reflection. In the process of working through dissociation or fixation, the non-linear, oscillatory relationship between the present and the past could encourage the subject to open up to the indeterminate future and reconsider the past from a new angle. As Birksted-Breen (2003) asserts: "The ability to symbolize and for self-reflection necessitates a relationship to time which can allow for the double movement forward and backward in time." (p. 1509).

The comparative analysis of *Nachträglichkeit* and *retroactivity*, however, requires caution. The image of circular or oscillating temporality more or less covers the two interrelated processes of self-constitution (i.e. the impact of the past on the present and vice versa), but there are also significant differences and divergences. As we have seen above, Freud's method is dominated by the excavation of hidden traumatic memories, whereas the processes of affective awakening and subsequent retroactive self-understanding are mostly initiated by reminder cues or occurr spontaneously. There is also a crucial difference in the temporal schemata: retroactivity can be seen as a kind of hermeneutics that sheds new light on a past event, whereas Freud has little to say about the hermeneutic aspect of afterwardsness, instead he tries to uncover the impact of the unprocessed trauma. Freud also takes pains to adhere to the role of traumatic events and repression, despite the questionable status of these events. In *The Neuro-Psychoses of Defence* (1894) he contends that: "[The] ego was faced with an experience, an idea or a feeling which aroused such a distressing affect that the subject decided to forget about it because he had no confidence in his power to resolve the contradiction between that incompatible idea and his ego by means of thought-activity" (Freud, 1962a [1894], p. 47) In the same paper, the distinction between *traumatic* and *auxiliary* moments appears, which will also be utilized in the *Studies*. However, the status and significance of the two temporal phases are not the same in the two papers. In the *The Neuro-Psychoses of Defence* (1894), repression and symptom formation occur together in one specific traumatic moment, but in the *Studies*, the temporal profile of *Nachträglichkeit* is articulated. The role of the auxiliary moment in the *Defence* is to replenish the weakened traumatic memory-trace with fresh affect and to re-establish the associative link between two similar events. In the *Studies*, however, the traumatic moment is only an "obscure inscription" and the auxiliary moment is the locus of repression and symptom formation (Fletcher, 2013, pp. 61–63). According to Fletcher's analysis, an intricate temporal schema of trauma-genesis can be observed in the *Studies*. The role of the auxiliary moment is not only the repetition or revival of an earlier trauma, but it also has its own impact; that is, an auxiliary moment can later be seen as a first moment.

It is not a misleading metaphor to speak of the concatenation of traumatic moments. Fletcher accentuates that Freud observes the interplay of traumatic moments. Let us start from an auxiliary moment that acts back on an earlier traumatic moment by virtue of the fact that it conveys a new context and understanding to the earlier event. The very act of understanding the former event (e.g., sexual harassment by the father in the case of Katharina) eventuates the deferred production of symptoms (e.g., disgust and repudiation, etc.) in the auxiliary moment. But the auxiliary scene of recognition and understanding can also be traumatic in itself and can bring about deferred symptoms. That is, in the third (traumatic) moment, the deferred symptoms of the auxiliary moment may come to the surface (Fletcher, 2013, p. 63).

The essential component or engine of this temporal dynamics is the presence of unintegrated events. There are "elements of consciousness", as Freud noted in the *Studies*, that are dissociated from the ego. Fletcher (2013) points out the crucial problem that the splitting off the remembered but not understood events does not necessarily lead to hysterical symptoms. Therefore, the interpretative auxiliary moment is significant in symptom formation (p. 62). It might be an intriguing question to ask whether milder dissociation or splitting plays a role in retroactive understanding. In my view, it certainly does; affective awakening exerts its influence not only on explicit recollections, but also on the twilight zone of consciousness where the implicit sense becomes explicit meaning. Let us turn to the process of affective retroactive awakening in the living present, which means the awakening and integration of peripherical and unnoticed perceptions (i.e., in the retentional flow or in distant memories). Similar to the process of Freud's afterwardsness, retroactive awakening may also include the integration of puzzling or disturbing events. That is the deferred way of understanding is at the heart of retroactivity. Despite the superficial similarity between the retroactive awakening of the past and the process of afterwardsness, it must be admitted that repression and the accompanying symptom formation pose a challenge to the inclusion of *Nachträglichkeit* into the general, phenomenological model of retroactivity. For example, Laplanche and Pontalis (1973 [1967]) clearly demarcate the problem of deferred action from the philosophical accounts of conscious reconstruction of the past. The authors argue that the phenomena of deferred action (i.e., *Nachträglichkeit*) can be approached at least from three interwoven perspectives: first, the traumatic event is basically an unassimilated experience that requires further analysis; second, there is the case of retrospective understanding generated by an event in the present, for example, organic maturation; third, the phenomena of *après-coup* is intimately tied to human sexuality (p. 112).

Furthermore, let me reiterate the basic theoretical distinction between Freud and Husserl: there is no room for dual processing in the Husserlian model of consciousness; i.e., the retentional modification gives room only to the system of sedimented empty retentions. Husserl did not distinguish between consciously processed and unconsciously stored experiences; there is no place for two distinctively different encoding processes, but only for gaps and absences in the retentional flow of time-consciousness. Another objection to the direct comparison is the widely held belief that repression for Husserl is restricted to the perceptual organization of the phenomenal field, however, as Smith has demonstrated, the late Husserl acknowledged the problem of repressed complexes. Thus, the comparison between *Nachträglichkeit* and the phenomenology of retroactivity may lead to serious inconsistencies.

According to the current proposal, however, the concepts of the affective unconscious and affective identity can open up a new horizon for the comparison or even the partial synthesis of *Nachträglichkeit* and retroactive self-understanding. In my view, the psychoanalytical considerations of *Nachträglichkeit* complement to the idea of affective identity. The general idea behind the comparison is that the constitution of the affective core self (i.e., affective identity) occurs at the intersection of future-oriented and past-oriented intentional accomplishments of consciousness.

The self is constituted by the retrospective and prospective intentional relations. That is, the ecstatic openness to the unpredictable future and the retroactive reconstruction of the indeterminate are fundamental intentional relations in the constitution of our affective identity.

With regard to affective identity, Kozyreva argued that consciousness strives for *coherence* and that the affective awakening of the past in the horizontal structure of the present is a spontaneous, passive, pre-intentional accomplishment. However, as we have seen in the previous chapters, in the dialogues between phenomenology and psychoanalysis, the issue of repression frequently resurfaces as a phenomenologically intractable problem. Investigating of the role of *Nachträglichkeit* in the constitution of traumatic subjectivity can be one of the productive attempts to mitigate the tension between the two paradigms, and consequently, the Freudian insights have much to offer in explaining the diachronic unity of the self.

Let us return to the temporal schema examined by Fletcher. In the *Studies*, the the incubation of the traumatic moment will later lead to a "deferred traumatic reaction" (Fletcher, 2013, p. 63). This process was characterized by Laplanche as a "time bomb". Let us now take a careful look at the case study of Katharina. Katharina saw her father lying on top of her cousin, which caused a hysterical aura (breathlessness, a sensation of blankness, pressure in the eyes and chest, and hammering and buzzing in the head). Furthermore, Katharina's vomiting was interpreted by Freud as the symptom of conversation, representing the "moral disgust". Fletcher puts heavy emphasis on the Freud's observation that the scene, as an "auxiliary scene," reminded Katharina of her father's earlier sexual advances, which had been rejected without proper understanding:

> In other words, before the process of repression and its consequent symptom formation can occur, an even earlier and more obscure inscription must have taken place, which is only later reactivated and plays its part in the deferred production of the hysterical symptom. It is this earlier moment that Freud now calls the traumatic moment. The auxiliary moment has its effect only after an interval during which it acts back on the first moment interpreting its significance in a new context (Fletcher, 2013, p. 63)

The whole scenario, of course, is embedded in the context of sexual development, i.e., the event in the pre-sexual period of life attains its "traumatic power" through the auxiliary scenes. Freud's temporal schema can be approached from different angles. Fletcher's above-mentioned interpretation based on the picture of the concatenation of traumatic moments: what counts as the *first* and the *auxiliary* moment is relative in different periods of life. In other words, the traumatic force of the first moment can be reactivated by the auxiliary moment, but this second scene can be traumatizing in itself, leading to repression and the reactivation of its belated traumatic impact in the near or distant future. In addition, Fletcher contends that if the traumatic power of the auxiliary scene had been exploited more dramatically, then the whole conceptual distinction between the two, or the postulation of the first scene, might have been dubitable (Fletcher, 2013, p. 63).

At first sight, it seems that *Nachträglichkeit* is inseparable from the threshold of puberty. Fletcher, however, shows that there is a general "temporal dialectic of deferral and retroaction" that is not necessarily closely tied to the issues of asexual or infantile period of development. For example, in the case study of the *Wolf Man* (1955 [1918]) the essential role of puberty is absent (Fletcher, 2013, p. 74). In one of his earlier papers, *The aetiology of hysteria* (1962d [1896]), Freud is uncertain whether deferral or retroaction is the main traumatogenetic factor. In the following quotation he emphasizes the role of recent experiences in symptom formation:

> Let us go back to the part played by the infantile sexual scenes. I am afraid that I may have misled you into over-estimating their power to form symptoms. Let me, therefore, once more stress the fact that every case of hysteria exhibits symptoms which are determined, not by infantile but by later, often by recent, experiences. Other symptoms, it is true, go back to the very earliest experiences and belong, so to speak, to the most ancient nobility (Freud, 1962b [1896], p. 214).

That is, the importance of the later scene cannot be undermined by the infantile scene, which has been characterized as the causal factor. In explaining symptom formation, the infantile and the later scenes are equally important; that is, hysterical symptoms are *overdetermined*. This means that the idea or affect that specified the symptoms can be called up "by combination of several factors" and can be "aroused from various directions simultaneously" (Freud, 1962d [1896], p. 216). However, in this same paper Freud (1962d [1896]) is inclined to accept the "determining power of the infantile scenes," which are concealed by later scenes and can be disclosed by the emergence of mediating associations between a stimulated hysterogenic point and a memory trace (p. 218) Freud remarks that the assumption of unconscious memories is a counterintuitive step for philosophers. For him, unconscious memories are memory traces that retain their "affective power" and "posthumously produce" and maintain hysterical symptoms. These ideas or affections are not transformed into powerless memories, but rather they remain and persist in the unconscious (p. 213, 219). The Husserlian interpretation of this scenario can be based on the process of the "backward radiating traumatic force" that takes place owning to the reinterpretation of early sexual scenes in virtue of puberty. However, Freud only partially accepts the constitutive role of present knowledge in the genesis of trauma, he is more inclined to search for the "time bomb" in the unconscious. The aetiology of infantile sexual abuse pulls him toward the view that the hidden traumatic scenes play a determining role in symptom formation.

In his meticulous analysis, House has convincingly shown that Freud, in his seduction papers (1896), attributed the retrospective traumatogenetic impact of secondary scenes to the deferred action of the first scenes. As a result, the "double temporal dimension of afterwardsness" is reduced to the factor of deferral (House, 2017, p. 258). Phenomenologically speaking, the bidirectional movement of retroactive self-understanding (in which the past breaks into the present and vice versa) was restricted only to the unidirectional determining role of the past. In this respect, the role of the present is not the complete re-schematization of the past: there is no way for the radical step of retroactive cancellation of the earlier meaning-structure, but only for the affective develoment and authentic recognition of the traumatic event. Of, course the true meaning of the event unfolds in the present and stirs up emotions, but the traumatic event is embedded in the past and awaits for conscious recognition[17]. Nevertheless, despite of Freud's uncertain vacillation between the importance of *deferral* and *retroactive* processes, the return of the repressed (i.e. unconscious memories) is still worthy of phenomenological consideration and even of integration.

As shown in Chap. 3, there are several discrepancies between the phenomenological and psychoanalytical ideas of the unconscious, and it is a common element in both classical and contemporary approaches to reinterpret the Freudian insights in phenomenological or hermeneutic terms. Moreover, the theoretical value of *Nachträglichkeit* cannot be restricted to a superficial analogy with the Husserlian notions of "affective awakening" or "backward radiating affective force". In my view, "afterwardsness" may lead us to a new way of understanding the phenomenology of affectivity and affective identity. From the perspective of traumatization, the affective awakening of the *distant past* is of crucial importance. However, this mechanism is not identical with the simple re-schematization or reconsideration of a former meaning-structure.

In order to illustrate the differences between retroactive understanding and *Nachträglichkeit,* I borrow another thought experiment from Peter Goldie. I contend that Goldie's example clearly illustrates the differences between milder emotional traumas and traumas which may lead to dissociation and repression. Goldie's (2012) tells a fictional story in which he goes to see a football match, but – stupidly enough – buys a forged ticket from a conman, and he has to go home full of anger and resentment. The ironic gap opens up between the two perspectives due to emotional evaluation (p. 39). In a later moment he considers his own self-pity rather pathetic, and he cannot fathom how foolish he was to trust in that person. He retrospectively realizes that there were clear signs in the conman's behavior that indicated that something was not right. In Goldie's (2012) own words: "Now that I know that he was a conman, it is in fact no easy feat to remember him as he then seemed to me; the shifty way he kept looking over his shoulder is now clear evidence to me of what wasn't known then – how could I have missed it at the time?" (p. 52) On the one hand, Goldie's example is intended to show the ironic distance between the now and then perspectives, but, on the other hand, it is also a good example of the evaluative aspects of retroactive understanding. In retroactive

[17] Fletcher shows Freud's striking ambivalence in his analysis of the *Wolf Man* (1918): „The complexities of this second moment of Freud's temporal schema, the wolf dream, are thereby reduced to a delayed 'understanding' of a primal scene that carries its own meaning – castration always already inscribed within it, and needing only to be 'recognized' and 'understood' and defended against. Thus the double temporal dimension of afterwardsness is thereby reduced to the single dimension of deferral or postponement, by the privileging of the all-determining moment of the parental scene and its pregiven meaning." (Fletcher, 2013, p. 258) However, there are also hesitations in this case study. Fletcher shows that there are even contradictory statements in one paragraph of the text. I do not intend to analyze the complicated story of the *Wolf Man* in detail, but the crucial point is that Freud speaks of both a deferred revision (i.e., resignification or resymbolization) and the activation of an image in a new context of understanding (Fletcher, 2013, p. 254). Again, there is the vacillation occurs between the *latent meaning* that seek to break throw the boundaries of the ego and the genesis of a *reconstructed meaning* in the living present.

self-understanding, it is difficult to differentiate clearly between affective and cognitive schematization, since the two processes may run unnoticed in parallel. In the Freudian context several, factors can widen the scope of the phenomenology of retroactive self-understanding. First, it is a crucial difference that in the case of Freudian retrospection, the new understanding of the first (traumatic) scene is initially unavailable for consciousness. The first traumatic scenes are usually undigested, dissociated experiences. As Fletcher points out: they are untranslatable, unassimilated foreign bodies. Laplanche describes the status of the traumatic scenes as if they were waiting in a kind of limbo; they are at the corner of the preconscious but do not appear as conscious or even properly repressed states (Laplanche, 1976, p. 41, cf. Fletcher, 2013, p. 76).

Things become more complicated when we bring into consideration the interplay between the two scenes. Laplanche (1976) indicates that Freud ends up failing to explain the original traumatic event. So where does the *traumatic impact* come from? To answer this question, Laplanche turns to the case of Emma. He argues that the first scene triggers nothing, there is no excitation, reaction, or symbolization. He is also quite skeptical about Emma's description of the second scene. It is very likely that it was only a banal scene of two assistants laughing, yet, this second scene triggered the excitation by awakening the first scene. Laplanche (1976) accentuates that Freud turns his attention from physical to *psychical trauma* when he claims that the first memory – triggered by the second event in the case of Emma – acts as an internal alien entity, a foreign body. In short, what is traumatic is not the event itself, but the memory that was constructed by the inscription of the first scene into the second (pp. 41–42). The final blow to the authenticity of the traumatic event comes when Laplanche (1976) argues that the traumatic impact – or affective force in phenomenological terms – is generated by an introjected, fantasized scene (p. 42)[18].

As far as I am concerned, it seems to be that Laplanche's interpretation leaves room for the integration of *Nachträglichkeit* into the unusual intentional state of retroactive self-understanding. The genesis of the trauma could be the end result of the superimposed scenes, and it is quite difficult to locate the upsurge of the affective or traumatic force in one particular event. But how can we deal with the repressed and the return of the repressed? In my view, one viable way to integrate the Freudian insights is to stipulate that the affective force of the trauma can be as mobile and dynamic as the Freudian concept of libido intuitively suggests. Of course, phenomenology has reservations about the naturalistic concept of "libidinal cathexes", but if we assume a certain kind of affective unconscious in which affective forces can "radiate" between past and present much like oscillating impulses between nodes of associations, then we are a few steps closer to reconciling

[18] In this chapter, I do not intend to analyze the role of fantasy in trauma-genesis, although the vicissitudes of seduction theory and the false memory debate demonstrate its inevitable role in trauma-constitution. However, in order to give a concise analysis of retroactivity and *Nachträglichkeit,* I have left the problem of fantasy in the background. For a detailed analysis of the role of fantasy in the genesis of trauma, see Fletcher (2013).

retroactivity with afterwardsness[19]. As was discussed previously, the affective relief structure involves the rivalry of affective forces and gives room for suppression and repression in the phenomenal field. A similar effect can be imagined not horizontally, in the phenomenal field, but rather vertically in the retentional oblivion, that is., in the system of sedimented experiences of the distant past. That is, the occurrences of *Nachträglichkeit* lead us to introduce such an affective unconscious, in which through the affective awakening of repressed affects a spontaneous, implicit meaning bestowing process takes place, which facilitates the interplay and reconciliation of the present and the past. In this respect, the affective and cognitive schematization processes are the two sides of the same coin of retroactive self-understanding. From the viewpoint of self-constitution, it can be argued that an implicit affective and cognitive schematization takes place that tries to determine the indeterminate past with the help of the actual insights of the present. The crux of this argument is to propose a phenomenological unconscious that, on the one hand, allows the dissociation between affect and representation, and on the other hand, adheres to the main mechanism of retroactivity; that is, the traumatic past is constituted in the living present when the past (including its affective force) breaks into the present, and, at the same time, the present influences the appearing memory.

What I wanted to show with the short introduction to the problem of *Nachträglichkeit* is that repression, displacement, defense, and even fantasy can play a crucial role in the constitution of psychic trauma. For Freud, the belated process of trauma genesis is intimately tied to sexuality; however, I have argued that unintegrated life events are not restricted to sexuality, and that the problem of repression may complement the phenomenology of retroactivity. From a phenomenological angle, in the case of the awakening of the past an empty retention emerges from sphere of sedimentation, but, occasionally, conscious recollection is impeded by the resistance of the ego, and repression, displacement, conversion may occur – as Freud has already suggested. The integration of the Freudian repression into the phenomenology of affective retroactive awakening provedes further support for Welsh's active repressed retentions. Not only can memories of the distant past be uncovered by retroactive awakening and reinterpretation, but unintegrated traumatic affections await the right circumstances to express their full affective (i.e., traumatic) potential and meaning. The symptoms of Freud's hysterical patients or the avoidance behavior of PTSD patients may also be the signs of the invisible presence of active repressed retentions.

However, the phenomenological unconscious that was proposed in this work is not identical with the Freudian repressed or id. I believe that Welsh's idea of active repressed retentions can be harmonized with recent studies on the phenomenological implicit unconscious. An intriguing formulation of an implicit unconscious can be found in Summa's works. She contends that the sphere of sedimentation is not a passive storage of information, but rather an "active moment of our self-experience"

[19] Tibor Sutyák, in his book on Freud, suggested that there is a close analogy between libidinal cathexis and the regulation of attention in the intentional movement of consciousness (see Sutyák, 2008).

(*ein wirkendes Moment unseres Selbsterlebens*) from which the affective character of a concrete memory crystallizes and attracts conscious attention. Summa puts emphasis on the affective character of remembering and she proposes a dynamic, present-oriented unconscious with the main function of selectively constructing an autobiography (Summa, 2016, p. 315). From this structure of sedimentation, traumatic recollections can spontaneously surface: foreboding atmospheric qualities of situations, obscure fantasies, or feelings in the lived body may be the premonitions of imminent rememberings (Summa, 2016, pp. 319–320). Welsh's active repressed retentions and Summa's active structure of sedimentation may also, at least in part, provide a phenomenological background for anchoring the dissociated or repressed elements that may affect the ego in a deferred way.

Another convergence can be drawn from the bidirectional movement between past and present, which is dominant in both retroactivity and *Nachträglichkeit*. Ullmann argued that retroactive awakening is essentially a mutual modification between the present and the past. Mutual modification here means that the experience of lived time is not limited to the retentional oblivion of time-consciousness, but the retentional flow is also permeated with rising, changing, and articulating meaning-structures (Ullmann, 2010, p. 287). Therefore, there is room for undigested, active repressed retentions in the past that are waiting for the right context to unfold their affective or traumatic potential.

Retroactivity is by definition a constitutional process of consciousness. And how does it relates to trauma? My proposal is that a special kind of traumatic retroactivity can be introduced as the subtype of affective retroactive awakening. The mutual affective and cognitive relationship between the past and the present, that is, as emphasized earlier, the upward flow of the past into the present and the downward flow of the present into the past, means that a traumatic scene is constituted at the heart of the living present. This formulation implies that trauma is constituted by the synthesis of memory fragments and impulses stemming from the present. The result is a somewhat hybrid experience of the "foreign body" or "inner alterity" at the core of subjectivity, which may bolster the general claim that we are all traumatized subjects in one way or another. Thus, the inclusion of the Freudian insights may shed light on the yet undiscovered layers of retroactive understanding. Ullmann highlighted an important element when he emphasized that in retroactive self-understanding it is not only the *past* that is altered, but also the self; the self of the actual present. I would like to add here that retroactive self-understanding may even be able to reconfigure the affective relief structure of the self by integrating the newly found (i.e., constructed) experiential content constituted by the intentional relation between the past and the present.

In addition, we can speculate that the affective relief structure of the living present is imbued with reconstructed memories and this constructive process is often occurs at the back of consciousness, under explicit awareness. The previous chapters have tried to show that implicit (body) memories and *Nachträglichkeit* can shed new light on the affective core of the self by providing a glimpse into the fragments of unprocessed experiences. The results lead us to the phenomenological notion of traumatic subjectivity, which includes unassimilated experiences at the pre-reflective level of affectivity. As Ullmann contends: “The notion of the traumatic subject means that we are affected in the pre-reflective affective layer of our being long before the constitution of knowledge and convictions; and this affective sublayer continue to carry the gaps and fractures of the affective being.” (Ullmann, 2010, p. 108). If we can somehow reflect on the process of retroactivity in order to observe the underlying dynamics of affectivity and meaning-constitution, then we arrive at the problem of affective identity that lies at the heart of the notion of the (less) minimal self. The notion of affective identity is indebted to Freudian psychoanalysis and incorporates the elusive problem of trauma as well as the process of traumatization.

Ullmann (2010) argued that psychoanalysis has much to say concerning the problem of affective schematization and Fletcher also interpreted Freud’s work as an insight into the affective organization of the subject. More precisely, Fletcher (2013) argued that while Freud dismissed Charcot’s conception of trauma[20] and considered the unconscious as a pathological formation partly constituted by traumatic experiences, he simultaneously sought to uncover the “affective organization of subjectivity as such,” from which the notion of trauma as a foreign body (and later as infiltration in the ego) emerged (p. 54)[21]. In the following chapter, I will focus more closely on the bodily processes of retroactive meaning-constitution, which will, hopefully, further elaborate the problem of retroactive self-understanding.

References

Bernet, R. (2000). The traumatized subject. *Research in Phenomenology, 30*, 160–179.

Birksted-Breen, D. (2003). Time and the *après-coup*. *International Journal of Psychoanalysis, 84*(6), 1501–1515.

[20] With regard to the genesis of trauma, Freud shifts the emphasis from Charcot’s *unconscious hypnoid state* to the role of undischarged, *fixated affect*. Undischarged excitations in the psychical apparatus will be the psychical equivalents of trauma. This kind of theoretical development has serious ramifications for the aetiology of neurosis: “What distinguishes Freud’s account of a *psychical* representation paralysis from Charcot’s conception is his development of the role of affect and his conception of psychical trauma in terms of accumulated, undischarged affect. Affect for Charcot was the state of fright accompanying the physical trauma, which constituted a quasi-hypnotic precondition and medium in which paralyzing autosuggestions could arise. For Freud, affect is an active causal agent conceived as a “quota of affect” regulated by the principle of constancy. […] The psychical trauma is constituted by the undischargeable surplus of affect, an excess that prevents the psychical system from returning to its optimum equilibrium or state of constancy prior to its disruption by the traumatic event.” (Fletcher, 2013, p. 30)

[21] “As Freud describes the complex organization of memory files, of zones and degrees of resistance, of nucleus and periphery, of nodal points and points of convergence, it begins increasingly to look like a whole cartography of the subject, of the mental and affective organization of *subjectivity as such*, conscious, preconscious, and unconscious, is being laid out.” (Fletcher, 2013, p. 54)

Bisteon, G., Vanheule, S., & Craps, S. (2014). Nachträglichkeit: A Freudian perspective on delayed traumatic reactions. *Theory & Psychology, 24*(5), 668–687.

Blaney, P. H. (1986). Affect and memory: A review. *Psychological Bulletin, 99*(2), 229–246.

Bower, G. H. (1981). Mood and memory. *American Psychologist, 36*(2), 129–148.

Brewin, C. R., & Andrews, B. (1998). Recovered memories of trauma: Phenomenology and cognitive mechanism. *Clinical Psychology Review, 18*(8), 949–970.

Browning, D. L. (2018). Laplanche on Après-coup: Translation, time, and trauma. *Journal of the American Psychoanalytic Association, 66*(4), 779–794.

De Sousa, R. (2017). Memory and emotion. In S. Bernecker & K. Michaelian (Eds.), *The Routledge handbook of philosophy of memory* (pp. 154–165). Routledge.

Eickhoff, F.-W. (2006). On Nachträglichkeit: The modernity of an old concept. *International Journal of Psychoanalysis, 87*(6), 1453–1469.

Fischmann, T., Russ, M. O., & Leuzinger-Bohleber, M. (2013). Trauma, dream, and psychic change in psychoanalysis: A dialog between psychoanalysis and the neurosciences. *Frontiers in Human Neuroscience, 7*(877), 1–15.

Fletcher, J. (2013). *Freud and the scene of trauma*. Fordham University Press.

Freud, S. (1955 [1918]). From the history of an infantile neurosis (the *Wolf Man*). In J. Strachey (Ed. & Trans.), The standard edition of the complete psychological works of Sigmund Freud (Vol. 17) (pp. 1–124). The Hogarth Press.

Freud, S. (1961 [1930]). Civilization and its discontents. In J. Strachey (Ed. & Trans.), *The standard edition of the complete psychological works of Sigmund Freud* (Vol. 21, pp. 64–148). The Hogarth Press.

Freud, S. (1962a [1894]). The neuro-psychoses of defence. In J. Strachey (Ed. & Trans.), *The standard edition of the complete psychological works of Sigmund Freud* (Vol. 3, pp. 45–61). The Hogarth Press.

Freud, S. (1962b [1896]). Heredity and the aetiology of the neuroses. In J. Strachey (Ed. & Trans.), *The standard edition of the complete psychological works of Sigmund Freud* (Vol. 3, pp. 143–156). The Hogarth Press.

Freud, S. (1962c [1896]). Further remarks on the neuro-psychoses of defence. In J. Strachey (Ed. & Trans.), *The standard edition of the complete psychological works of Sigmund Freud* (Vol. 3) (pp. 162–185). The Hogarth Press.

Freud, S. (1962d [1896]). The aetiology of hysteria. In J. Strachey (Ed. & Trans.), *The standard edition of the complete psychological works of Sigmund Freud* (Vol. 3, pp. 191–221). The Hogarth Press.

Freud, S. (1966 [1895]). Project for a scientific psychology. In J. Strachey (Ed. & Trans.), *The standard edition of the complete psychological works of Sigmund Freud* (Vol. 1) (pp. 295–398). The Hogarth Press.

Freud, S., & Beruer, J. (1955 [1895]). Studies on hysteria In J. Strachey (Ed. & Trans.), *The standard edition of the complete psychological works of Sigmund Freud* (Vol. 2, pp. 294–398). The Hogarth Press.

Fuchs, T. (2012). The phenomenology of body memory. In S. C. Koch, T. Fuchs, M. Summa, & C. Müller (Eds.), *Body memory, metaphor and movement* (pp. 9–22). John Benjamins Publishing Company.

Fuchs, T. (2015). From self-disorders to ego disorders. *Psychopathology, 48*(5), 324–331.

Fuchs, T. (2019). The interactive phenomenal field and the life space: A sketch of an ecological concept of psychotherapy. *Psychopathology, 52*(2), 67–74.

Golden, B. K., & Bergo, B. (Eds.). (2009). *The trauma controversy: Philosophical and interdisciplinary dialogues*. SUNY Press.

Goldie, P. (2003). One's remembered past: Narrative thinking, emotion, and the external perspective. *Philosophical Papers, 32*(3), 301–319.

Goldie, P. (2012). *The mess inside: Narrative, emotion, and the mind*. OUP.

Hacking, J. (1995). *Rewriting the soul: Multiple personality and the science of memory*. Princeton University Press.

House, J. (2017). The ongoing rediscovery of Après-Coup as a central Freudian concept. *Journal of the American Psychoanalytic Association, 65*(5), 773–789.
Hunt, N. C. (2010). *Memory, war and trauma*. CUP.
Husserl, E. (2001). *Analyses concerning passive and active synthesis: Lectures on transcendental logic*. (A. J. Steinbock, Trans.). Kluwer Academic Publishers.
Jung, C. G. (1961 [1913]). *Freud and psychoanalysis: The theory of psychoanalysis*, (R. F. C. Hull, Trans.) In C.W. (Vol. 4, pp. 83–228). Routledge & Kegen Paul.
Kozyreva, A. (2017). *Phenomenology of affective subjectivity: Analyses on the pre-reflective Unity of subjective experience*. Universitäts Bibliothek.
Kozyreva, A. (2018). Non-representational approaches to the unconscious in the phenomenology of Husserl and Merleau-Ponty. *Phenomenology and the Cognitive Sciences, 17*(1), 199–224.
Lacan, J. (2006 [1953]). The function and field of speech and language in psychoanalysis. In *Écrits: The first complete translation in English* (pp. 197–268). (B. Fink, Trans.). Norton.
Laing, R. D. (1978). *The divided self*. The Penguin Force.
Laplanche, J. (1976). *Life and death in psychoanalysis* (trans. Jeffrey Mehlam). The John Hopkins University Press.
Laplanche, J. (1999). Notes on afterwardsness. In J. Laplanche (Auth.), *Essays on otherness* (pp. 264–269). Routledge.
Laplanche, J. (2017 [1989–1990]). *Problématiques VI: Après-coup*, transl. J. House. The Unconscious in Translation.
Laplanche, J., & Pontalis, J.-B. (1973 [1967]). Deferred action; deferred. In D. Nicholson-Smith (Trans.,), *The language of psycho-analysis* (pp. 111–114). The Hogarth Press and the Institute of Psycho-Analysis.
Larrabee, M. J. (1995). The time of trauma: Husserl's phenomenology and post-traumatic stress disorder. *Human Studies, 18*(4), 351–366.
Levine, P. A. (2015). *Trauma and memory: Brain and body in a search for the living past*. North Atlantic Books.
Loftus, E. F., & Kaufman, L. (1992). Why do traumatic experiences sometimes produce good memory (flashbulbs) and sometimes no memory (repression)? In E. Winograd & U. Neisser (Eds.), *Affect and accuracy in recall: Studies of "flashbulb" memories* (pp. 212–223). Cambridge University Press.
Marosán, B. P. (2020). *Kontextus és fenomén II. Az igazság problémája az időtudat és a keletkezés husserli filozófiájában*. L'Harmattan.
Merleau-Ponty. (2012). *The phenomenology of perception*. (D. A. Landes, Trans.). Routledge.
Mishara, A. (1990). Husserl and Freud: Time, memory and the unconscious. *Husserl Studies, 7*, 29–58.
Neisser, U. (2014 [1967]). *Cognitive psychology* (Classic Edition). Psychology Press.
Noel-Smith, K. (2016). *Freud on time and timelessness*. Palgrave Macmillan.
Ratcliffe, M. (2008). *Feelings of being: Phenomenology, psychiatry and the sense of reality*. OUP.
Ratcliffe, M. (2017a). *Real hallucinations: Psychiatric illness, intentionality, and the interpersonal world*. MIT.
Ratcliffe, M. (2017b). Selfhood, schizophrenia, and the interpersonal regulation of experience. In C. Durt, T. Fuchs, & C. Tewes (Eds.), *Embodiment, Enaction, and culture: Investigation the constitution of the shared world* (pp. 149–172). The MIT Press.
Saks, E. R. (2015). *The center cannot hold: My journey through madness*. Hachette Books.
Schacter, D. L. (1996). *Searching for memory: The brain, the mind, and the past*. Basic Books.
Schacter, D. L. (2001). *The seven sins of memory: How the mind forgets and remembers*. Houghton Mifflin Harcout.
Smith, N. (2010). *Towards a phenomenology of repression – A Husserlian reply to the Freudian challenge*. Stockholm University Press.
Steinbock, A. J. (2004). Affection and attention: On the phenomenology of becoming aware. *Continental Philosophy Review, 37*, 21–43.
Summa, M. (2014). *Spatio-temporal intertwining: Husserl's transcendental aesthetic*. Springer.

Summa, M. (2015). Are emotions recollected in Tranquility? Phenomenological reflections on emotions, memory, and the temporal dynamics of experience. In M. Ubiali & M. Wehrle (Eds.), *Edmund Husserl on: Feeling and value, willing and action* (pp. 163–181). Springer.
Summa, M. (2016). Gefühl, Emotion und Erinnerung beim traumatisierten Selbst. Eine Frage der Kohärenz? In R. Esterbauer, P. Schmidt, A. Paletta, & D. Duncan (Eds.), *Bodytime. Leib und Zeit bei Burnout und in anderen Grenzerfahrungen* (pp. 307–329). Verlag Karl Alber.
Sutyák, T. (2008). *A Freud-elv – A pszichoanalízis mint lélekfilozófia*. Kalligram.
Szabó, A. (2016). *Phenomenology and psychosomatics: A Jungian and Post-Jungian approach* (dissertation). Retrieved July 09, 2019, from https://dea.lib.unideb.hu/dea/handle/2437/229953
Tengelyi, L. (2007a). *Tapasztalat és kifejezés*. Atlantisz.
Tulving, E. (2002). Episodic memory: From mind to brain. *Annual Review of Psychology, 53*(1), 1–25.
Ullmann, T. (2010). *A láthatatlan forma*. L'Harmattan.
Ullmann, T. (2012a). Az idő mint a szubjektivitás alapja és horizontja. In T. Ullmann (Ed.), *Az értelem dimenziói* (pp. 91–109). Könyvpont – L'Harmattan.
Ullmann, T. (2012b). Fantázia és intencionalitás. In T. Ullmann (Ed.), *Az értelem dimenziói* (pp. 125–149). Könyvpont – L'Harmattan.
Ullmann, T. (2015). A narratív, a traumatikus és az affektív szubjektivitás. In I. Bujalos, M. Tóth, & T. Valastyán (Eds.), *Az identitás alakzatai* (pp. 21–37). Kalligram.
Ullmann, T. (2017). Phenomenology of experience and the problem of the unconscious. In M. Gabriel, C. Olay, & S. Ostritsch (Eds.), *Welt und Unendlichkeit: Ein deutsch-ungarischer Dialog in memoriam László Tengelyi* (pp. 141–161). Verlag Karl Alber.
Welsh, T. (2002). The retentional and the repressed: Does Freud's concept of the unconscious threaten Husserlian phenomenology? *Human Studies, 25*, 163–183.
Williams, H. L., Conway, M. A., & Cohen, G. (2008). Autobiographical memory. In G. Cohen & M. Conway (Eds.), *Memory in the real world* (3rd ed., pp. 21–90). Psychology Press.
Yaw, M. (2015). Husserl and PTSD: The traumatic correlate. *Journal of Phenomenological Psychology, 46*, 206–226.
Yiassemides, A. (2014). *Time and timelessness: Temporality in the theory of Carl Jung*. Routledge.
Zahavi, D. (2017). Thin, thinner, thinnest: Defining the minimal self. In C. Durt, T. Fuchs, & C. Tewes (Eds.), *Embodiment, enaction, and culture: Investigation the constitution of the shared world* (pp. 193–201). MIT Press.
Zahavi, D. (2020). Consciousness and (minimal) selfhood: Getting clearer on for-me-ness. In U. Kriegel (Ed.), *The Oxford handbook of the philosophy of consciousness* (pp. 635–653). OUP.

Chapter 7
The Affective Core Self and Affective Identity

Abstract The aim of the chapter is to introduce two complementary types of retroactive constitution of sense, namely the implicit or bodily and the conscious or narrative ways of retroactive self-understanding. Drawing on psychoanalytical and phenomenological insights, the chapter argues that an affective identity is constituted by the underlying processes of implicit-bodily and conscious-narrative self-understanding. The distinction between the implicit and explicit process of retroactive sense-making entails a phenomenologically oriented model of psychointegration, including Freud's insights into the afterwardsness of psychic trauma. Thus, the chapter also reflects on the complex relationship between the traumatic event and traumatic experiences. Finally, the chapter clarifies the concept of the affective core self.

Keywords Trauma · Bodily retroactivity · Retroactive self-understanding · Affective relief · Body memory

The aim of this chapter is to provide a synthesis and to differentiate between two levels of retroactive constitution. This distinction will narrow the gap between narrative self-understanding and the passive-implicit modes of affective retroactive awakening. In order to flesh out the idea of the affective core self, I will also refer to the characteristics of body memory and afterwardsness.

Both psychoanalysis and phenomenology have suggested that the main characteristic of the self is the constant striving for inner coherence. In Freudian psychoanalysis, defense and repression are the two key mechanisms that maintain the integrity of the ego. However, the ego has to surrender to the tyranny of the past or to the return of the repressed. Body memory and the phenomenology of retroactivity represented the problem of the persistent past that exerts a pre-reflective influence on the living present. The organic, indeterminate yet still powerful past can be influential through the horizontal unconscious of implicit body memory and through

L. Horváth, *The Affective Core Self*, Contributions to Phenomenology 130,
https://doi.org/10.1007/978-3-031-56920-3_7

the "see-saw" effect of retroactive self-understanding (including *Nachträglichkeit*).[1] The unusual intentional state of retroactive self-understanding may have the power to *modify* or even *rewrite* the dispositional structure of implicit body memory and, consequently, the affective relief structure. The resulting change in the "invisible nervation" of the affective unconscious could be a hardly noticeable, or a more pronounced and lasting effect. The mutual interplay between the implicit "felt senses"[2] of body memory and the usually explicit narrative self-understanding presupposes the twilight zone of emerging new meanings. The retrospective insights of *change*, that can be articulated in narrative forms or felt through the lived body and lived time, could initiate a ripple effect in the implicit structure of the sense of the self.

Ullmann (2010) argued that retroactivity is one of the most fundamental but hidden activities of consciousness, which can be understood as a countercurrent in the unidirectional stream of inner time-consciousness. Kozyreva (2017) argued that the past is always connected to the living present not only by means of explicit recollection, but also through affective awakenings. The horizontal structure of consciousness implies that consciousness "can hardly stand 'blind spots' and it shows a strong inclination towards coherence." (p. 211) A similar motif can be found in Fuchs' arguments about the horizontal unconscious, and Ullmann also defines his notion of affective identity in terms of blind spots. Ullmann argues that the affective unconscious can be indirectly inferred by observing the limits of narrative self-understanding. The affective unconscious is an "unconsciously schematized substratum of subjectivity" (Ullmann, 2017, p. 159).[3] Moreover, his affective unconscious is not an invisible system or reservoir, but rather a "prefigured implicit structure of sense" that organizes our first-person phenomenal field in different ways, based on social determinants and emotional learning processes (Ullmann, 2017, p. 160) It is also noteworthy that the blind spots of our affective life can be indirectly circumscribed by our typical crises in intersubjective relations (cf. Ullmann, 2015, p. 33).

Fuchs (2012) argued that blind spots of consciousness are inadvertently repeated behavioral patterns or missed opportunities in our life. In agreement with Freud, Fuchs also claims that repression has a significant role in the constitution of these

[1] Laplanche introduces the metaphor of see-saw movement between the two moments that symbolizes the indeterminate source of trauma: "…it may be said that, in a sense, the trauma is situated entirely in the play of "deceit" producing a kind of seesaw effect between the two events." (Laplanche, 1976, p. 45)

[2] For the explanation of "felt sense," see Sect. 4.3 and footnote 109. In this particular context "felt sense" mainly refers to obscure *Leib*-experiences.

[3] "Affectivity is not a kind of emotional »state« which is experienced in an evident and immediate way. Affectivity has a dynamic character. It is not a homogeneous sphere of articulated emotions but a kind of *schematized affectivity*, which is nothing else but the subject's unconscious. An event, a gesture or a remark can function as a sign, in other words, they can launch a series of emotional responses to a concrete situation. These responses fit to a kind of emotional pattern or configuration (a *Gestalt* or a schematic structure) – but this structure is not on the conscious level where we elaborate causal explanations and integrate new events into our narrative life story." (Ullmann, 2017, p 159)

blind spots. That is, the "blind spot in self-awareness" results from "active and emotionally charged repression" (p. 77). At first glance, the expression of a blind spot seems to allude to some tangible or easily definable trait or behavior. From the first-person perspective, however, blind spots are only recognizable in retrospect. An outburst of anger may vaguely allude to a general sense of irritability; a general sense of being-in-the-world or being-with-others. The retrospective understanding of our blind spots sheds light on their *dispositional* character. Similar to the Heideggerian moods, these dispositions may render the person to feel, perceive, or act in a certain or even predeterminate way under certain circumstances (cf. Fuchs, 2013, p. 618).

The phenomena of retroactive self-understanding may shed some light on this underlying affective infrastructure. In the previous chapter, retroactivity was defined as a special kind of intentional relation that enables the resurrection of distant memories on the basis of new information from the living present. The simplest example of retroactive awakening is when we recognize that a certain tune was audible at the background while we were immersed in a conversation with someone. In this case, the affective force, metaphorically speaking, leads backward into the retentional flow and awakens earlier impressions, it makes sense to the earlier unrecognized sounds (i.e., a *Zeitgestalt* begins to take shape in the just-past). Ullmann (2010) argues that *retroactivity in the present* means the awakening of sensations in the just-past retentional flow, but we can speak of the role of retroactivity in recollection where a kind of Gestalt-switching or sense-bestowing is at work; that is, a distant memory can also be recognized by a totally new sense-making process due to events in our actual living present (p. 280). Therefore, retroactivity can be understood as the *awakening* of past impressions and the *cancellation* of the former sense. The phenomenon of retroactive cancellation has already been discussed in Sect. 6.3. A simple example is a situation in which we see a stranger approaching and we believe that she is "Jane" because we have arranged to meet her. However, as the stranger comes closer, we suddenly realize that the person is actually a stranger and not Jane (Larrabee, 1995, p. 360). We retroactively change the passively attributed meaning in the retentional flow, and thus the earlier sense undergoes a sudden *Gestalt*-switch.

In the previous chapters I proposed the idea of traumatic retroactivity, as the type of retroactive awakening, that includes the phenomena of intrusive memory and afterwardsness. The comparative analysis between *Nachträglichkeit* and retroactivity has broadened the scope of retroactive awakenings. I suggest that psychoanalysis, and in particular Freud's early trauma theory, can contribute to the understanding of how a "new past experience" emerges from the retroactive cancellation of the earlier meaning and context of a memory. But it is not only revision that creates a newly articulated or found past in the living present. In several nearly pathological or explicitly pathological cases, the past is in a state of constant unrest and occasionally bursts into the present. Of course, the reminder and retrieval cues, which can be found in the living present, are important factors in the reanimation of the past. As we have seen, Larrabee has introduced the notion of flash-forward experience to accentuate the hybrid character of PTSD flashbacks (see Sect. 6.3). (The flash-forward experience is the reenactment of a past experience in the present.) The intrusive symptoms, the symptoms of hysterics, which were examined and theorized by Freud, and the enactments of transference during analysis may illustrate the unrest and the awakening of the powerful but yet indeterminate past. To put it more drastically, *Nachträglichkeit* symbolized the return or even the intrusion of the non-representational past in the analytic process of (re)construction. The phenomenon of affective retroactive awakening also assumes the emergence of the past through re-evaluation and re-interpretation.[4]

I tried to integrate the problem of repression into the idea of retroactive self-understanding. I suggest that the Freudian *Nachträglichkeit* draws more attention to the non-representational occurrences of the past because it deals explicitly with the problem of repression and resistance and the problems of symptom formation. By contrast, the retroactive awakening of the distant past can be characterized as the revision or reevaluation of autobiographical memories. However, we have also seen, that contemporary scholars and even Husserl thought that affective retroactive awakening includes an "implicit intentionality", sense formations emerge even when they cannot reach the level of reproductive consciousness. Thus, both the Freudian afterwardsness and retroactive awakening can open the gates to the implicit (*sub*-conscious) domain of consciousness. In addition, I argue that the characterization of retroactivity can be refined with two intertwined levels of retroactive self-understanding (i.e., the conscious and the implicit). The crucial question that I want to address is whether it is possible to speak of affective awakening and retroactive self-understanding at the non-representational level of the self? I suggest that implicit body memory will prove useful in answering this question.

The phenomenological and psychoanalytic insights discussed in the previous chapters converge to the view that an implicit (and bodily) retroactive self-understanding is at work beneath the surface of explicit intentional consciousness. As Fletcher has convincingly demonstrated, Freud was unable to abandon the notion of determinism, and consequently he vacillated over different meanings of *Nachträglichkeit,* but in the end he concluded that a concrete traumatic experience must exist in one way or another. Contrary to the idea of the intentional consciousness, which can be divided into the focal and peripheral regions of conscious awareness, Freud's models imply the image of a stratified psyche. In *Civilization and its Discontents* (1961 [1930]), the metaphor of Rome suggests the layered notion of the psyche, and Freud struggles with the difficulty of describing the history of the self in pictorial terms. While he raises the possibility of the complete annihilation of the past, he finally contends that: "We can only hold fast to the fact that it is rather the rule than the exception for the past to be preserved in mental life." (p. 72). The

[4] According to Ullmann's claim, retroactive understanding may generate a new *content*. This statement dose not invalidate the claim that retroactivity is primarily a conscious intentional state and strongly associated with explicit recollection. The main point of the claim is only that the process of re-schematization radically alters the past experience; it creates, so to speak, an *emergent* or *hybrid* intentional state. As Ullmann argues: we do not see two incongruent interpretations of the same memory, but rather two memory occurs with quasi-independent content (cf. Ullmann, 2010, p. 279).

stratified layer of the unconscious lends support to the idea of determinism, which is also present in the mechanism of *deferred action* of the repressed. The lurking motif of determinism was accentuated by Laplanche's "time bomb" metaphor.

Despite the role of the repression in the constitution of the Freudian unconscious, there is a similarity between the phenomenological notion of the affective unconscious and the dynamic unconscious. For start, if we accept Fletcher's claim that Freud tried to uncover the "mental and affective organization of subjectivity" by the elaboration of a traumatic scenography in which memory and fantasy were intimately intertwined, then the phenomena of retroactivity and *Nachträglichkeit* point to the same latency of the self.[5] Things become more complicated when we try to characterize the *content* of the affective unconscious in question. The affective unconscious and thus the affective core self may consist of procedural knowledge, skills, capacities, reactive behavioral and emotional patterns, habits, preferences, internalized social norms, childhood traumas, recent traumas, and preconscious associative processes and so on. Furthermore, the affective unconscious may contain primary existential feelings, affective atmospheres, and moods including even the pathological delusional mood. In sum, it seems easy to introduce a reified, objectified notion of the affective unconscious, which is the sum total of the pre-intentional and pre-conscious features mentioned above. When we attempt to list the contents of the non-representational, passive-implicit realm of consciousness, a reified notion of the unconscious may emerge.

In order avoid the trap of objectification, I side with Kozyreva (2017), who argus that the true role of phenomenological investigation is not to uncover the realm of the implicit unconscious, but rather to examine emerging intentional states in the in-between region of implicit and explicit cognition. In this respect, emotional and traumatic memories are of particular importance (p. 183). The task of the genetic phenomenological approach, in this context, is to examine the affective organization of the living present and the underlying implicit constitutional activity of retroactive self-understanding. I also sided with Ullmann who sees retroactivity as an independent intentional relation that gives us glimpses of the latent yet all-encompassing process of change. I will attempt to complete the recent discussions of retroactivity with the idea of a two-tiered retroactive self-constitution that includes both the conscious (narrative) revision of the past and also the pre-reflective sense-making process of implicit body memory. The intertwined aspects of explicit-narrative and implicit-bodily meaning-bestowing processes may also provide a new phenomenological framework for taking into account the temporal schema of *Nachträglichkeit* (i.e., the two aspects of *reactivation* and *revision*).

The most typical form of retroactive affective awakening is the reinterpretation of a particular event or series of events on the basis of new knowledge from the actual present. I have already illustrated this with Goldie's illuminating thought

[5] "As Freud describes the complex organization of memory files, of zones and degrees of resistance, of nucleus and periphery, of nodal points and points of convergence, it begins increasingly to look like a whole cartography of the subject, of the mental and affective organization of *subjectivity as such*, conscious, preconscious, and unconscious, is being laid out." (Fletcher, 2013, p. 54)

experiments. For further illustration, let us consider the common scenarios of breaking off a close friendship or leaving a partner. The typical response in these cases is to re-schematize past experiences. For example, when Freud and Jung broke off their friendship, Jung later claimed that Janet had been a greater influence on him than Freud. This is partly true, since Jung's complex psychology is inconceivable without the influence of Janet, but it was also obvious that Jung was trying to belittle the role of Freud in his intellectual development. Freud, his early role model, turned into a rival after their split around 1913.[6] Or take the example of a divorce. In this case, the memories and emotions of several happy years can quickly fade away because of the suddenly felt resentment, anger, and other affective impulses. In the injured person's mind, vivid memories of holidays and other blissful memories are suddenly transformed into ephemeral illusions or, in the worst case, into anxiety-provoking memories. Repetitive patterns of self-blame, anxiety, despair, anger, and remorse may spring forth from the person's affective unconscious, and these feelings tend to paint over the pleasant memories of several years, slowly generating an all-enveloping melancholic existential feeling. In these everyday scenarios, the retroactive cancellation and overwriting of earlier meaning accompanies narrative self-understanding. The phrase of "affective retroactive awakening" means that narrative self-understanding is not only a thought process, but can also be initiated in a bottom-up way by initially obscure affective tendencies. It seems plausible to suppose that narrative self-understanding can be initiated by the activation of subliminal affective impulses originating from the person's affective unconscious. To illustrate this passive-associative process, the metaphor of the "affective force" has been adopted from Husserl's *Analyses* of passive synthesis.

The idea of the affective unconscious can be extended to include the slowly developing affective atmospheres such as the delusional mood. The delusional mood can be interpreted as a holistic affective awakening that schematizes and colors the subject's entire phenomenal field. As discussed earlier, bizarre delusions can result from the search for meaning in order to get rid of the strange, uncanny atmosphere of being-in-the-world, which is filled with uncertainty and tension.[7] The incomprehensibility of the delusional mood suggest that some kind of implicit, non-representational alteration occurs in the subject that transforms the perception of reality (cf. Henriksen, 2013, p. 109). Without intersubjective and narrative understanding, delusional mood can lead to other experiences of estrangement, including hyperreflexivity, hallucinations, and other symptoms of psychosis.

It is crucial to see that even the undifferentiated affective atmospheres or the affective impulses of the lived body are not endowed with inherent meaning, rather their hidden "meaning cores" are generated by conscious-narrative understanding,

[6] The relationship between Freud and Jung deteriorated and became irreparable in 1912. In their quarrel, both Freud and Jung imputed psychological disorders to each other, even many years after the split. In December 1912, Jung reflected on the misuse of psychoanalytic insights. In the same year, Freud also wrote about the "abuse" of psychoanalysis in polemics (see Doran, 2017, p. 53)

[7] Sections 5.2 and 5.3 discuss the topic of delusional mood in more depth. Here I use the term only to illustrate the emergence of inchoate affective impulses and atmospheres.

which at the same time could alter their apparently implicit structure of sense.[8] I have not attempted to provide the comprehensive analysis of delusional mood here, only to suggest that the non-representational affective atmospheres can be understood as manifestations of the affective unconscious. It has been mentioned that Kozyreva, following Husserl, argued that every conscious remembering is initiated by latent, unrecognized affective awakenings (Kozyreva, 2017, p. 189). In other words, it is reasonable to claim, at least from a phenomenological point of view, that retroactive understanding is fueled by the pre-egoic (i.e., self-affective) dynamics of the affective unconscious. Delusional mood may be one instance where the affective unconscious, as it were, is completely detached from consciousness, and the "deferred action" of the repressed material may be another instance in which a radical gap is constituted between the implicit and explicit processing.

The question, however, is how does the affective unconscious give rise to affective awakenings? If we accept the thesis, implied by Husserl and further elaborated by Kozyreva, that affective awakening is an all-encompassing pre-intentional accomplishment that occurs in various forms, the question arises: what is the driving force of this passive-associative phenomenon? How and why do self-affective awakenings emerge in the affective unconscious? In the previous chapters, Welsh's notion of "active repressed retentions" referred to the dynamic, sedimented structure of sense that includes spontaneous affective fluctuations.

It has been argued that spontaneous traumatic affections may also rise and fall at the periphery of consciousness. As Summa (2016) has summarized: "Generally, traumatic rememberings at first appear in implicit and spontaneous manner in the form of an emotionally strongly charged bodily feeling or in a still foreboding representation." (p. 319). The crux of her argument is that prior to the explicit, narrative elaboration of a traumatic event, traumatic rememberings occur involuntarily and implicitly. These implicit traumatic rememberings are often inaccurate (*ungenau*) or appear as false memories (p. 319.). Nevertheless, there is a pre-thematic, non-representational self-organizing process that tries to give answers to unassimilated but shocking, or at least surprising, events. But what can be the drive of this self-organizing process? From a phenomenological point of view, the answer may lie on the side of the self. Summa argues that because of the traumatic events, the self is in constant state of tension. This tension is the result of two conflicting motives: on the one hand, the subject is trying to find an answer to a totally foreign experience, and on the other hand, the subject is unable to find a reassuring answer to the event that alien to consciousness. The result is fixation and the distortion of the remembered traumatic event occurs. By referring to Freud, Summa argues that the alienation from the past and the future is bound to happen (cf. Summa, 2016, p. 318). What I wanted to emphasize with these considerations is the fact that the spontaneous affective awakenings of the (traumatic) past can lead to fixation or progression. The

[8] I refer here to Jopling's (2000) description of the crystallization of sense, discussed in Sect. 2.3. Furthermore, the problem of the *meaning core*, as we have seen in relation to body memory, is also a recurrent theme in working with implicit memory in therapeutic context.

crucial point here is to find the right context for integration that facilitates the responsiveness of the self.

The following section will argue that affective retroactive awakening and retroactive self-understanding plays a crucial role in the constitution of a (less) minimal self. I suggest that the processes of the retroactive awakening, including the acceptance or cancellation of the emerging sense, can significantly alter the phenomenal character of the affective core self, which is a (less) minimal self, including its bodily foundation. The affective infrastructure of the self (i.e., the affective relief structure) is not only the network or, with Ullmann's phrase, the "invisible nervation" of dispositions, habits, traits, and reactive patterns in general, but it also has the propensity to change and reorganize through the processes of retroactive awakening and retroactive understanding. How does this dynamic structure of sense behave in traumatic situations? Recent studies of body memory and implicit memory suggest a general phenomenological idea. I suggest that, in cases of physical or psychological shock, there is an implicit or, phenomenologically speaking, a self-affective response that can be further elaborated through conscious-narrative understanding, which may or may not include the retroactive cancellation and revision of a previous structure of sense.

Let us suppose that in case of retroactive self-understanding, a sudden explosion, a far-ranging shockwave in time-consciousness results from a shocking experience. The shocking event engenders a widespread process of affective schematization that cannot localized in the living present, but, with Husserl's phrase, radiates back and forth in time-consciousness, eventuating the processes of retroactive and/or proactive constitution of sense: the past appears in a new light and the future is deprived of certain possibilities.[9] If we accept the argument that the integration of traumatic events can create a ripple effect in the affective unconscious and reorganize its elements, then the phenomenal character of the (less) minimal self can also change. I argue that this change does not affect the formal notion of the (more) minimal self, but it does affect the structure of the (less) minimal self, including the self-world relation (i.e., the phenomenal character and the affective atmosphere of the self's being-in-the-world).

Ratcliffe's insights about bodily grounded existential feelings are particularly relevant here: the ripple effect in the affective unconscious means that the basic existential feelings of being-in-the-world (e.g., felt familiarity or alienation) and being-with-others can be significantly altered. The phenomenal character of the self may undergo subtle or more pronounced changes following the integration of

[9] With the metaphor of "shockwave in time-consciousness" I only wanted to demonstrate the sudden impact of retroactive and/or pro-active cancellation of meaning. It is noteworthy that several psychopathological outcomes can be the result of traumatic events. Traumatized people are unable to imagine an open future and become stuck in the repetition of painful experiences. Both proactive and retroactive schematization fail to function properly. Ratcliffe, in his analysis of trauma, mentions the states of "foreshortened future" and "narrative foreclosure" in relation to traumatized victims. The general pattern in these cases is that the habitual certainty, including the "habitual style of anticipation", is lost (cf. Ratcliffe, 2017, 116–121).

anxiety-provoking implicit memories. The ripple effect, the reorganization of the affective unconscious, does not obliterate the phenomenological unity of the self. The diachronic unity of the self is preserved, but something has changed in the latency of the affective unconscious: the world and the subject's relation to others are altered, the significant memories of the past and the possible projects of the future are reorganized in a new way. In other words, the invisible nervation of the affective unconscious is reorganized in such a way that the new structure of sense, on the one hand, reformulates the distant memories and on the other hand, causes subtle changes in the holistic phenomenal character of the self.

The problem of retroactivity lies at the heart of self-understanding and points to the unresolved question about the relationship between the (minimal) *self* and the *person*. To clarify the issue, I will follow Fuchs' the distinction. For Fuchs, the Ego (i.e., the person) refers to reflective self-consciousness and autobiographical memory, whereas the self refers to pre-reflective self-awareness and the background feeling of the body.[10] I do not want to downplay the role of narrative self-understanding in the retroactive constitution of the past at the level of the person. For designating the retroactive constitution of the past during self-understanding, I propose the term conscious or narrative retroactivity. We may refer to the process of the retrospective reevaluation of distant explicit memories as *conscious or narrative retroactivity* that occurs at the personal level. But there can also be a passive sense-bestowing process that can be labelled as *implicit or bodily retroactivity*. By implicit-bodily retroactivity I have in mind the processes of affective awakening and the ensuing explication of the meaning cores of body memories.[11] At the subpersonal level, the ebb and flow of disturbing meaning cores can occur prior to conscious evaluation.[12]

The separation of the two mechanisms is mainly for theoretical reasons. Based on the previous considerations, I propose that in the case of trauma-processing, the ensuing oscillatory movement in time is often accompanied by implicit or bodily retroactivity. Larrabee's flash-forward experiences and other kinds of enactments in the living present can also be subsumed under the process of implicit-bodily retroactivity. Similarly, Kozyreva has made a close connection between body memory and the "implicit or non-objectifying intentionality of affective awakening" (Kozyreva, 2018, p. 221). Based on the phenomenology of body memory, Kozyreva speaks of "non-representational past relation". Furthermore, following Husserl's theory of affectivity, she shows that *affective awakening* and *explicit recollection* are

[10] As we have seen earlier, Fuchs established a close connection between the self and bodily feelings. And a Husserlian legacy came to the fore when he defined the feeling body and its interoceptive, proprioceptive, and kinaesthetic awareness as an instance of auto-affection (Fuchs, 2015, p. 325).

[11] According to Fuchs, explication means

[12] For instance, the psychosomatic aftereffects of railway accidents can be characterized as belated effects of implicit retroactivity. This type of "nervous shock" occurred without any actual anatomical damage and plagued the neurologists in the late nineteenth century (Fletcher, 2013, p. 19). The implicit retroactive awakening of the accident presupposes a hidden, undigested *meaning core*, awaiting for, abreaction, recognition, and suitable narrative processing.

closely related, but not identical processes. The relationship between them can be seen as constitutive: "Clearly, not all awakenings reach the level of actual memories, but all rememberings start as affective awakenings, and these latter can be seen as tendencies towards reproductive intuitions." (Kozyreva, 2017, p. 189) The above distinction between implicit and conscious retroactivity reflects Husserl's and Kozyreva's distinction between affective awakening and explicit recollection; that is, in addition to the implicit retroactivity of the lived body, I can also imagine a conscious, narrative retroactivity on the personal level that gives new meaning to distant memories from the actual, new perspective.

It seems difficult to draw a clear line between explicit recollection and retroactive self-understanding, since the retrieval of episodic memories is also a reconstructive and evaluative process. Interests in the present influence the process of recollection. Goldie provided illuminating examples of autobiographical memories in which cognitive and affective evaluation of past experiences occurs through the process of narrative thinking. The main difference between explicit recollection and conscious or narrative retroactivity may lie in the explicit evaluative processes. That is, conscious-narrative retroactive awakening involves not only the reconstruction of an autobiographical memory, but a definite re-interpretation and re-evaluation is also taking place. Not every explicit recollection is a reinterpretation, but every retroactive awakening is also a determined reevaluation and reinterpretation. Nevertheless, both explicit recollection and conscious retroactivity are based on constructed episodic memories. Memory research has highlighted the inherent, unconscious biases in retrieved episodic and autobiographical memories due to failures in source monitoring and expectations stemming from the present. Conscious retroactivity requires (re)constructed (episodic or semantic) memories upon which the affective evaluation exerts its influence. Schacter (1996) argued that despite the presence of constructive processes of memory retrieval, the core components of autobiographical memory (especially the frequently retold life events) are more or less intact (p. 299). However, the main purpose of conscious retroactivity, however, is not to verify the recollected memories, but to reinterpret their meaning according to the new circumstances and context.

Following Kozyreva and Ullmann, I argue that the role of retroactivity is not only to revive explicit recollections of the past with fresh affective force, but also to play with alternative perspectives in order to gain a new understanding of the same event. In this respect conscious retroactive self-understanding may involve both an affective and a cognitive function. The former may initiate remembering, the latter may be a intentional attempt to gain self-knowledge; in this respect, conscious retroactivity is closely related to idea of narrative self-understanding, which also often shifts or mobilizes different perspectives to interpret the same autobiographical content. In contrast to conscious or narrative retroactivity, implicit or bodily retroactivity will bring us closer to the non-representational realm of the phenomenological unconscious.

Fuchs arrived at the conclusion that bodily affective impulses inherently contain a hidden meaning core. This work has already suggested that implicit body memory can be integrated into the affective relief structure of the self. The non-representational impulses and potential meaning cores of the lived body are embedded in the relief structure. (This argument was developed in Chap. 4.) The relief structure is the hidden dimension of meaning formation; basic drives, affections, sensations strive for conscious recognition, and they are transformed and elaborated by the representative capacities of reproductive intentionality (i.e., the acts of remembering and imagination). Ullmann proposed his own notion of affective relief which is a dimension of a not yet fully articulated sense with vivacity or feeling tone (Ullmann, 2010, p. 249).[13] I would like to add that implicit or bodily retroactivity can be the self-organizing activity of these pre-intentional and non-representational elements. I define implicit retroactivity as a pre-reflective and pre-intentional activity that is closely related to the various manifestations of implicit body memory, such as pain memory, intercorporeal, and situational memory.[14] Psychosomatic symptoms, disturbing existential feelings, moods and atmospheres could also be the signs of an impending implicit-bodily retroactive awakening. I also suggest that the affective core self can be understood as the condensation of the affective relief structure. Metaphorically speaking, the affective relief symbolizes the spatially extended "map" or "pattern" of the dormant yet dominant pre-reflective and pre-intentional elements of the self. These motives, traits, habits, and dispositions come to the fore in certain circumstances and remain in the background in the absence of actuating cues in the current situation. The notion of the affective core self refers to the temporal continuity and bodily grounding of the affective relief structure. In what follows, I will try to find scenarios that in which the circular activity between implicit-bodily and conscious-narrative retroactivity occurs. For start, let us turn back to the problem of traumatized subject.

Ullmann (2015) argued that trauma does not only means an unrepresentable, repressed event in the past, but also the disintegration of the subject's worldview: "The trauma violates the well-known and secure world of the subject who cannot integrate it into his or her world without destroying that world." (p. 154) Regarding traumatic experiences, Ratcliffe has pointed out that the consequences of trauma are the loss of trust and the disturbance of the modalities of intentionality (Ratcliffe, 2017, p. 118). Gusich (2012) proposed a similar idea when she emphasized the role of the "traumatic force" and the affective strength of the situation (Bernet, 2000, p. 169).[15] In a severe traumatic situation, a self-protective denial of the event takes

[13] The problem arises that this dynamic notion of the phenomenological unconscious is burdened by the drawbacks of constructive phenomenology. However, a survey of the field of consciousness studies can provide possible alternatives for expanding the horizon of consciousness to incorporate the affective unconscious. For example, similar to focusing therapy, mentioned by Fuchs, or psychedelic therapies may be of some use in this regard (****). But the main goal of this chapter is to show the presence of the affective unconscious in the ordinary circumstances of being-in-the-world.

[14] For the types of body memory, see Fuchs, 2012.

[15] There is a certain similarity between the notion of the *affective strength* of the trauma and Freud's idea according to which the psychic apparatus is a closed homeostatic system that tries to maintain its equilibrium. Freud's concept of trauma is constantly evolving, and his preoccupation with psychic trauma does not mean that he neglected physical trauma. Trauma, for Freud, is the breaching or breaking of the boundaries of the psychic apparatus; that is, trauma has both quantitative and economic dimensions (see Fletcher, 2013, p. 30).

place in order to reduce the affective strength of the situation. The traumatic event cannot be integrated into the subject's horizon of sense, and temporal disorientation (i.e., the fixation of awareness to the traumatic event) also ensues (Gusich, 2012, p. 510).

In order to represent the activity of retroactive self-understanding, I have to restrict the meaning of trauma to psychic, and, more specifically, to emotional trauma that caused by unacceptable, shocking life-events.[16] According to the psychoanalytic framework, emotional trauma and the problem of affect-transformation are tied together. For example, Stolorow (2008) defines *affect* as "subjective emotional experience" that is regulated from birth. Thus, affect or emotion is not located in some kind of Cartesian interiority, but has its own developmental trajectory and is regulated or misregulated in the child-caregiver relationship. Emotional trauma, in this sense, overwhelmes the affect-integrating capacity of the psyche. As Stolorow puts it: "Trauma is constituted in an intersubjective context in which severe emotional pain cannot find a relational home in which it can be held. In such context, painful affect states become unendurable – that is, traumatic." (p. 114). Stolorow further claims that therapeutic vignettes are clearly demonstrate the process of affect-transformation. Patients often experience the impact of a traumatic experience in bodily ways; these pre-symbolic, exclusively somatic states (e.g., shaking and blushing, etc.) need to be translated into a narrative form. For example, the patient begins to talk about the terror she experienced as a result of childhood abuse (Stolorow, 2008, p. 115). The observation that the transformation of affect takes place in the therapeutic situation is particularly important for implicit-bodily retroactivity. Stolorow argues that the "desomatization and verbalization" of affect is not actuated endogenously in the interiority of the subject, but rather by the mediation of the therapeutic situation (Stolorow, 2008, p. 116). Somatization, or to put it phenomenologically, *Leib*-experiences may operate implicitly from the first-person perspective, but they could be transformed into meaningful content in the appropriate, intersubjective context. In addition, Stolorow (2008) draws a close analogy between the affective state of the emotional trauma and Heidegger's anxiety: the traumatized person is pushed toward the authentic state of being-toward-death (p. 119). In the following pages I intend to reflect on the temporal dynamics of the integration of awakened affects.

First, let us examine the "otherness" of the traumatic situation. In the case of unacceptable traumatic events, psychosomatic reactions and conscious narration can occur simultaneously. When something unbearable happens (e.g., a near-fatal accident or any kind of psychophysical shock), one may say to oneself that "It can't be possible!" or "I can't believe that!", etc. Gusich (2012) argues that an important feature of trauma is "strange disbelief," that is, we cannot believe and understand what just happened in the traumatic situation: "We encounter the traumatic situation and may even participate in it, but we nonetheless withhold our assent from what

[16] Gusich (2012) argues that the affective strength of the traumatic situation is accompanied by a significant emotional appraisal. Disbelief in the traumatic event serves the purpose of eliminating the affective strength of the event (p. 510).

happening." (p. 506). Gusich's (2012) argument concerns situations in which we withhold our assent to negative and dire cognitive judgments. The process of withholding is intentional and functions as a self-protective process. Another important feature of trauma, closely related to the first, is the inability to categorically comprehend the traumatic event. The "abnormal" or "improbable" element of the traumatic situation hinders integration (p. 506).

The concept of implicit retroactivity focuses on the deferred processing of the unbearable novelty of a traumatic event. Based on the complex nature of body memory, I suggest that a special kind of processing can take place in the lived body (*Leib*) through affective experiences and psychosomatic symptoms. In the case of anxiety and shock-provoking situations, the predictive capacities, i.e., the protentional dynamics, of consciousness collapse and the undigested traumatic event is accompanied by anxiety, fear, anger, or other affective response. We could say that the unbearable novelty is like a dagger that thrusts into the affective core self and engenders a radical alteration on the fundamental level of being-in-the-world. Summa (2016) refers to Waldenfels (2002) when she proposes the distinction between the *traumatic event* and *traumatic experience*. She claims that the traumatic event is always ahead of the traumatic experience as "frightening astonishment" (*schreckende Überraschung*). The traumatic experience is basically the after-effect (*Nachwirkung*) of the traumatic event, i.e., the shocking element of surprise. In short, the occurrence of trauma can be recognized by reflecting on a sudden time lag (*Zeitverschiebung*) (Summa, 2016, p. 319). It seems reasonable to say that the belated processing (i.e., afterwardsness) is an inherent feature of trauma-constitution, even in those cases where repression and symptom formation have not yet activated. From phenomenological perspective, the sudden traumatic impact can lead to subversive changes in the person, including the alteration of his or her affective core self.

The task of Chap. 6 was to incorporate Freud's *Nachträglichkei* into the broader phenomenological concept of retroactive trauma. The afterwardsness or the deferred action of the trauma is not necessarily just a "radically hermeneutical concept." as, for example Rabelhofer (2019) suggests the analysis of *Nachträglichkeit*. Bisteon and his colleagues cite recent studies that have shown the belated onset of PTSD symptoms. Inspired by these studies, the authors define *Nachträglichkeit* as an effect in which a potentially traumatic experience can be delayed (even by several years) and which requires a second moment (or trigger) to occur. It is important to add that the first moment leaves behind a not yet processed or understood "mnemic trace" (Bisteon et al., 2014, pp. 674–676). That is, the retroactive or belated traumatization does not mean that the subject is traumatized only by understanding something that was previously a blind spot in his or her consciousness. The phenomenological concept of retroactive trauma adheres to Lacan's and Bernet's considerations: the delayed onset of trauma requires a "shock without signification" for the subject (Bernet, 2000, p. 163).

I propose that there is a correlation between the subversive nature of an imminent traumatic event and the self-affective reaction (i.e., implicit-bodily retroactivity) of the core self. My central contention is that the time lag in the constitution of traumatic experience can be approached and interpreted through both implicit and conscious retroactive awakenings. It is possible that certain disturbing events will evoke implicit-bodily retroactivity and then the delayed conscious-narrative processing. For example, in the case of a street robbery, I would immediately feel intense fear and humiliation in a self-affective way after the event but I would only be able to consciously understand the situation later. In this case the affection of intense fear emerged through implicit retroactive awakening with its accompanying reactions in the lived body. Or let us take another example to illustrate the onset of implicit retroactivity and the delay of conscious retroactivity. Imagine you're walking along the sidewalk in the city center, next to the library, and you're busy with your phone. Suddenly we hear the sound of brakes screeching, look up and see a car approaching us on the pavement at high speed. Suddenly our sense of time slows down and we manage to jump against the wall to avoid being hit by the car. After barely escaping a hit-and-run accident, the car hits a light pole behind us, which luckily for us, tips into the street rather than the pavement. We are surprised by the fact that we escaped the potential accident, but soon afterwards we are overcome with anxiety and fear of death, and it becomes clear to us, belatedly, that we could have been killed in this situation. It is important to note that these examples have nothing to do with the above-mentioned delayed onset of PTSD symptoms examined by Bisteon and his colleagues. The short stories only illustrate the potential presence of traumatic retroactivity.

It may be objected that the reference to psychosomatic symptoms in a phenomenological framework displays an unreflective naturalism. However, the objection can be countered with the help of the lived body (*Leib*) and body memory. From a phenomenological point of view, psychosomatic symptoms can be interpreted as the internal, first-personal experiences of the subject's lived body. As I have indicated in Chap. 4, body memory can be interpreted as a set of implicit processes, including the subjectively felt reactions of the lived body in a situation. That is, body memory exhibits the peculiar characteristics of the *Leib*, which designates our first-person access to and the medium of our own bodily being.[17] In this context, psychosomatic symptoms simply refer to the self-affective reactions of the lived body, which may occur consciously or unconsciously and reflect the interpersonal situations and environmental atmospheres in a pre-reflective way.

To emphasize the pre-reflective activity of the lived body, it may be useful to speak of *Leib-reactions, Leib-experiences,* and even *Leib-analysis* in therapeutic context.[18] Gallagher, for example, has shown the immense importance of

[17] According to Fuchs (2013), the lived body is a vital process and, simultaneously, a felt body that constitutes the space of resonance for every mood and emotion.

[18] Szabó (2016) proposes that "Leibanalysis" lies on the border between the instinctual and representational poles of experiential reality and harnesses the malleable aspect of the lived body. In his synthetizing attempt, he argues that the Leib is the *interface* or *node* thorough which symbolization occurs and through which psychotherapeutic interventions can be actuated. In this context, Jungian psychology can be harmonized with the phenomenology of the lived body. The *Leib* lies at the intersection of the psychic, the bodily, and the sociocultural aspects of the individuum (Szabó, 2016, pp. 114–115).

somaesthetic experiences (e.g., pain, hunger, warmth, etc.) in the Husserlian oeuvre; these are essentially sensual-hyletic data that can come to the focal point of awareness (Gallagher, 1986, p. 142; cf. Szabó, 2016, p. 42) Larrabee, following Husserl, argues that some kind of passive-associative process is at work during psychosomatic reactions; that is, even non-somatic experiences may give rise to somatic effects: "For example, the pressure of making a decision by a certain deadline, a seemingly mental process, becomes the pressure of the headache; or a decision to carry a number of inappropriate responsibilities is somatised in a kinked back." (Larrabee, 1995, p. 358). The citation suggests that in case of existential pressures, somatization (i.e., *Leib-experience*) may occur in a passive-associative manner. The argument tries to demonstrate the "ripple effects" of Leib-reactions and Leib-experiences that may anticipate the genesis of new meaning and understanding.

We can imagine scenarios in which the time lag between a traumatic event and its *Leib*-experience is longer than it was in the street robbery thought experiment. Let us return to the survival of the hit-and-run accident. We managed to evade the incoming car, but it is certain that something totally unexcepted happened to us: a car came in front of us on the sidewalk. Luckily we have survived the encounter without a scratch. The event was so unexpected that we did not live through any overwhelming *Leib*-experiences, but our subjective experience of time slowed down.[19] We begin to act as if nothing has happened, and later we tell the story of the accident to someone. But the act of telling the story suddenly filles us with fear of death, anxiety, or a feeling of discomfort in our lived body. In this scenario, our affective core self remains stable, but later we may feel an unbearable shock in our lived body. In this example, unusually, the affective dynamics of implicit retroactive awakening accompanies the narrative recollection of the event; that is, the retroactive understanding of the event brought about the bodily-affective symptoms (i.e., *Leib*-experiences). The scenario intends to demonstrate the slippages and gaps between conscious retroactive self-understanding and implicit-bodily retroactivity that is rooted in the passive-associative reactions of the lived body.

In summary, the general idea is that there is a *gap*, but also a *synchronizing* tendency, between implicit-bodily and conscious-narrative retroactivity; these two processes are interrelated in the meaning-bestowing process. Furthermore, in cases of massive trauma, retroactive self-understanding can lead to the re-schematization of

[19] Contemporary trauma theories have proposed the concept of "peri-traumatic dissociation," which refers to the phenomenon that occurs at the moment of the traumatic event. Similar to repression, dissociation is also a defense mechanism that seals off the full impact of the traumatic event. Dissociation often entails a detached, observer perspective in which the pain and distress associated with the situation are absent. The result can be the emergence of vivid, intense, and repetitive memories, which may indicate the absence of the normal conscious encoding process during the event (cf. Brewin & Andrews, 1998, p. 951). What I wanted to express with implicit-bodily retroactivity is the first-person phenomenology of a milder form of dissociation. That is, we have the conscious memory of the event, but its full traumatic force belatedly emerges. In the example of the accident, we may even have an observer perspective, but the full affective impact of the accident overwhelms us in a deferred way. Suddenly, we realize that we managed to survive the deadly danger without serious injury, and this feeling may belatedly recapture the frightening, anxiety-provoking nature of the event.

the entire affective core self. The nature of the retroactive constitution of sense sheds light on the internal (affective) architecture of the core self and the need for the process of integration through passive (bodily) and conscious (narrative) retroactivity. The notion of implicit retroactivity, which is basically a subversive affective awakening, suggests that the pre-reflective and self-affective (i.e., bodily) tendencies precede narrative meaning-bestowing.

The final question to consider is how the two tiers of retroactivity relate to the presumed affective core self? The main idea behind the argument was that traumatic experiences might be crucial factors in initiating change in the affective core self. Here, the process of change is associated with the genesis of new meaning and the rearrangement of dispositions, preferences, cares, and concerns. In light of the comparative analyses in Chap. 2, one could raise the objection that the reconfiguration of the affective core self is only the task of narrative self-understanding, including the revision of autobiography. By analyzing possible ways of retroactive self-understanding, I have attempted to address the passive, latent level of self-constitution that may contribute to the genesis of the affective core self. The problem of traumatic subjectivity and the brief analysis of phenomenological psychopathology have hopefully shown that the stability of the affective core self is fragile and at the mercy of changing circumstances. Given the embodied nature of the affective core self, it seems evident that interpersonal relationsships play a crucial role in its development. Finally, in order to avoid misunderstandings, let me reiterate that the affective core self, as the condensation of the affective relief structure, is strongly connected to Ratcliffe's intersubjectively constituted and Zahavi's (less) minimal self.

The processes of implicit and conscious retroactivity may have important implications for the distinction between the minimal and narrative self. Zahavi contended that the minimal self is the condition of possibility for the development of the narrative self; that is, despite their interrelatedness on the personal level, the relationship between the two is asymmetrical. Recently, this fundamental opposition has been criticized by Bortolan (2020). First, based on recent and traditional phenomenological studies of affection, she introduced the term "affective minimal self" in order to define the minimal self as a bodily self of which we have a non-observational and non-objectifying awareness (p. 75). In this respect, Bortolan joins Ratcliffe (2008), Colombetti (2011), and others in emphasizing the role of the subjectively lived body in the process of meaning-bestowing. However, Bortolan contends that the minimal self contains characteristic features that can also be attributed to the narrative self. Therefore, the gap between the narrative and the bodily minimal self closes when we pay attention to the evaluative properties of feelings. In her own words: "…the self which, through intentional feelings, is experiences as being affected by certain evaluative properties is not only a bodily subject, but rather an individual who possesses a specific set of values, cares, and concerns, shaped thorough a personal history" (pp. 77–78).

She argues that intentional and non-intentional feelings, like Heideggerian moods or existential feelings, are connected to the narrative self. To put it simply, the subjective pole of affective evaluations is at the core of the person. And this pre-reflective sense of self is not a static first-person perspective, but rather a flexible locus that can be affected and modified by changing narratives. As Bortolan eloquently put it: "Changes in the form or contents of one's narratives, for instance, could influence some key features of one's pre-reflective sense of self. For example, disruptions in the unity and continuity of one's self-narratives may give rise to a fragmentation of one's self-experiences, while a marked shift towards negative self-descriptions could hinder one's felt sense of bodily and agentative potentialities." (Bortolan, 2020, p. 82) The implication is that bodily feelings and narratives are intimately intertwined. Following this line of argument, we can say that the affective core self has its own autonomy and peculiar manifestations through body memory, but it is only artificially separable from narrative self-understanding and the narrative self. In other words, theoretically, the narrative self and the affective core self can be represented in a circular fashion in which mutual modification and influence can occur.

Ratcliffe (2016) also holds the view that bodily feelings and narratives mutually influence each other. Existential feelings, which have both bodily and world-directed components, can put pressure on consciousness. For example, in the case of Jasper's delusional mood or in severe depression, the feelings of dread or emptiness can easily infiltrate first-person narratives. In the delusional atmosphere "The person does not understand what is happening to her and experiences a terrifying feeling of indeterminacy, which fuels a quest for certainty in the form of the delusional narrative that eventually crystallizes. The narrative is shaped by the experience but also changes the experience, giving it a more determinate meaning." (p. 182). The impact of existential feelings can be so powerful that it can hinder the possibility of producing autobiographical narratives. The crux of Ratcliffe's (2016) argument is that existential feelings can influence not only the content, but also the narrative form of self-understanding, and the articulating narrative, as a feedback mechanism, exerts an influence on the emerging existential feeling. That is, existential feelings cannot be separated from their conceptualized and expressed character (p. 179).

The dual intentionality of implicit-bodily and conscious-narrative self-understanding fits well with Ratliffe's considerations. Retroactive self-understanding in trauma-processing may shed some light on the mutual relationship of the minimal and the narrative selves. Or, to put it another way, the implicit and explicit mechanisms of meaning-bestowing processes are inseparable features of selfhood. Of course, the deferred action of trauma and repression has posed a challenge to phenomenological interpretation, but in light of the temporal structure of *Nachträglichkeit* and the enactments of implicit body memory, it seemed plausible to introduce a phenomenological psychointegration theory based on the affective unconscious and the two tiered notion of retroactivity.

References

Bernet, R. (2000). The traumatized subject. *Research in Phenomenology, 30*, 160–179.

Bisteon, G., Vanheule, S., & Craps, S. (2014). Nachträglichkeit: A Freudian perspective on delayed traumatic reactions. *Theory & Psychology, 24*(5), 668–687.

Bortolan, A. (2020). Affectivity and the distinction between minimal and narrative self. *Continental Philosophy Review, 53*, 67–84.
Brewin, C. R., & Andrews, B. (1998). Recovered memories of trauma: Phenomenology and cognitive mechanism. *Clinical Psychology Review, 18*(8), 949–970.
Colombetti, G. (2011). Varieties of pre-reflective self-awareness: Foreground and background bodily feelings in emotion experience. *Inquiry, 54*(3), 293–313.
Doran, C. (2017). Rage and anxiety in the split between Freud and Jung. *Humanities, 6*(3), 53. https://doi.org/10.3390/h6030053
Fletcher, J. (2013). *Freud and the scene of trauma*. Fordham University Press.
Freud, S. (1961 [1930]). Civilization and its discontents. In J. Strachey (Ed. & Trans.), *The standard edition of the complete psychological works of Sigmund Freud* (Vol. 21, pp. 64–148). The Hogarth Press.
Fuchs, T. (2012). Body memory and the unconscious. In D. Lohmar & J. Brudzińska (Eds.), *Founding psychoanalysis phenomenologically* (pp. 69–82). Springer.
Fuchs, T. (2013). The phenomenology of affectivity. In K. W. M. Fulford, M. Davies, R. G. T. Gipps, G. Graham, J. Z. Sadler, G. Stanghellini, & T. Thornton (Eds.), *The Oxford handbook of philosophy and psychiatry* (pp. 612–631). Oxford University Press.
Fuchs, T. (2015). From self-disorders to ego disorders. *Psychopathology, 48*(5), 324–331.
Gallagher, S. (1986). Hyletic experiences and the lived body. *Husserl Studies, 3*(2), 131–166.
Gusich, G. (2012). A phenomenology of emotional trauma: Around and about the things themselves. *Human Studies, 35*(4), 505–518.
Henriksen, M. G. (2013). On incomprehensibility in schizophrenia. *Phenomenology and the Cognitive Sciences, 12*, 105–129.
Jopling, D. A. (2000). *Self-knowledge and the self.* Routledge.
Kozyreva, A. (2017). *Phenomenology of affective subjectivity: Analyses on the pre-reflective Unity of subjective experience*. Universitäts Bibliothek.
Kozyreva, A. (2018). Non-representational approaches to the unconscious in the phenomenology of Husserl and Merleau-Ponty. *Phenomenology and the Cognitive Sciences, 17*(1), 199–224.
Laplanche, J. (1976). *Life and death in psychoanalysis*. (J. Mehlam, Trans.). The Johns Hopkins University Press.
Larrabee, M. J. (1995). The time of trauma: Husserl's phenomenology and post-traumatic stress disorder. *Human Studies, 18*(4), 351–366.
Rabelhofer, B. (2019). Inhabiting a time before time. Freud's concept of trauma as a psychoanalytical figure of thought. *Studia Theodisca, 26*, 57–71.
Ratcliffe, M. (2008). *Feelings of being: Phenomenology, psychiatry and the sense of reality*. OUP.
Ratcliffe, M. (2016). Existential feeling and narrative. In T. Breyer & O. Müller (Eds.), *Funktionen des Lebendigen* (pp. 169–192). De Gruyter.
Ratcliffe, M. (2017). *Real hallucinations: Psychiatric illness, intentionality, and the interpersonal world*. MIT.
Schacter, D. L. (1996). *Searching for memory: The brain, the mind, and the past*. Basic Books.
Stolorow, R. D. (2008). The contextuality and existentiality of emotional trauma. *Psychoanalytic Dialogues, 18*(1), 113–123.
Summa, M. (2016). Gefühl, Emotion und Erinnerung beim traumatisierten Selbst. Eine Frage der Kohärenz? In R. Esterbauer, P. Schmidt, A. Paletta, & D. Duncan (Eds.), *Bodytime. Leib und Zeit bei Burnout und in anderen Grenzerfahrungen* (pp. 307–329). Verlag Karl Alber.
Szabó, A. (2016). *Phenomenology and psychosomatics: A Jungian and Post-Jungian approach* (dissertation). Retrieved July 09, 2019, from https://dea.lib.unideb.hu/dea/handle/2437/229953
Ullmann, T. (2010). *A láthatatlan forma*. L'Harmattan.
Ullmann, T. (2015). A narratív, a traumatikus és az affektív szubjektivitás. In I. Bujalos, M. Tóth, & T. Valastyán (Eds.), *Az identitás alakzatai* (pp. 21–37). Kalligram.
Ullmann, T. (2017). Phenomenology of experience and the problem of the unconscious. In M. Gabriel, C. Olay, & S. Ostritsch (Eds.), *Welt und Unendlichkeit: Ein deutsch-ungarischer Dialog in memoriam László Tengelyi* (pp. 141–161). Verlag Karl Alber.
Waldenfels, B. (2002). *Bruchlinien der Erfahrung*. Suhrkamp.

Chapter 8
Summary and Conclusions

This study provided a comparative analysis of the minimal self and the unconscious. At first sight, such a comparison is a nonstarter, since the *formal* definition of the minimal self alludes to the self-givenness of experiences and the phenomenal character of what-it-is-like-for-me-ness. That is, the minimal self is restricted to self-illuminating intentional states and attempts to avoid the Scylla of the nonconscious mental states and the Charybdis of the unconscious. However, as Zahavi himself has shown in his replies to critics, the minimal self can be understood in at least in two ways: first, the *formal* definition refers to the basic distinction between our own and others' mental states; second, the minimal self is also an *experiential core self* and the target property of psychopathological investigations. I have therefore tried to show that it is incumbent upon any theory of the minimal self to distinguish between the strictly phenomenological and psychopathological utilizations of the term.

As we have seen, the experiential core self, that has permeated the study of schizophrenia, is closely connected to the affective infrastructure of the self and the experience of lived time. The formal notion of the minimal self was also, by definition, grounded in internal time-consciousness,. The Husserlian insights into the inseparable relationship between time and consciousness, and the role of passivity in self-constitution, were also crucial. The affective foundation of the experiential self was clearly demonstrated by the phenomenon of *self-affection* and by the closely related features of *hyperreflexivity*. Zahavi argued for the pervasive and constant dimension of *for-me-ness* or *mineness* of intentional states despite the psychopathology of hyperreflexive awareness and disownership symptoms. The quality of mineness or for-me-ness is common to every intentional act, and this kind of phenomenal character seems to be constantly present at the level of pre-reflective self-awareness. In short, pre-reflective self-awareness and the presupposed for-me-ness of experiences are inseparable. Zahavi took enormous effort to demonstrate the significance of the Husserlian and other phenomenological studies for contemporary philosophy of mind, and it is not an exaggeration to say that he built a bridge between these movements.

L. Horváth, *The Affective Core Self*, Contributions to Phenomenology 130,
https://doi.org/10.1007/978-3-031-56920-3_8

As we have seen, Zahavi traced the problem of pre-reflective self-awareness, and the minimal self, back to Husserl; but Sartre and Merleau-Ponty were also the main proponents of the concept. Zahavi argued that we are not mindless video cameras with nonconscious perspectival representation of the environment, but rather possess a first-person perspective endowed with a subtle phenomenal character. In other words, the phenomenological investigation of "first-personal experiential life" could contribute to Nagel's characterization of phenomenal consciousness. Zahavi adds that the self-referential nature of pre-reflective self-awareness could complement or expand the problem of "what it is like to be" a human being. It is indisputable that Zahavi's undertaking is an unprecedented synthesis that includes phenomenology, narrative self-theories, developmental psychology, and psychopathology.

What I wanted to propose in this study is that the multifactorial and multidimensional analysis of the self could allow more room for the analysis of the role of the unconscious in the process of self-constitution. The current work extended the notion of the (less) minimal self to include the phenomena of passivity, repression, and especially retroactivity. The "affective core self" was introduced to highlight the detour from Zahavi's formal minimal or experiential core self. In a topographical sense, the affective core self was positioned between the narrative self and the formal minimal self, and it was characterized as the ground of change and as the source of new meanings that can be constituted during and after subversive or even traumatic experiences. Thus, there is a connection between the affective core self and the problem of the traumatic subjectivity. The affective core self also tries to remain faithful to the concept of the minimal self, since it is not a reified notion of the self, but rather refers to an ongoing process of self-understanding occasioned by subtle changes in the life history or unexpected traumatic experiences.

Following in the footsteps of Fuchs, Ullmann, and Kozyreva, I have argued that the affective core self is an implicit network or the "invisible nervation", as Ullmann has formulated, of affective responses that characterize and schematize the overall phenomenal dimension of the self. Zahavi's formal minimal self denotes a constantly present *phenomenal feeling component* in our experiential life that remains stable under all circumstances and constitutes the pre-verbal and pre-narrative layer of selfhood. While it is intuitively appealing to assume the fundamental phenomenal character of pre-reflective self-awareness, I have tried to reckon with another layer of the self that is the subject of gradual or dramatic changes. To this end, I analyzed the idea of the affective relief structure and regarded it as a viable alternative to the phenomenological unconscious. The affective relief structure suggested a spatio-temporal but also a dynamic picture of the unconscious and created an opportunity to partially integrate the problem of the Freudian repression. I argued that the affective core self could be regarded as the concentrated form of the affective relief structure in which passive, implicit, pre-egoic processes take place. Thus, the affective core self can be defined as the condensation of affective responses and embodied dispositions that have the propensity for change. Psychodynamically speaking, the affective core self denotes the ground of subtle or radical changes in the structure of personality. However, a serious objection can be raised against the assumption of the affective core self. How can we use the notion of the "core self" legitimately if we identify the affective unconscious with the affective relief structure, which is the invisible network or pattern of procedural skills, dispositions, habits, mindsets, and traits? Metaphorically speaking, the affective core self, as the condensation of the affective relief structure, is only the sum total of passive, implicit processes that play a crucial role in the constitution of personal identity. The bodily nature of the self in another important implication of this core self. The harmonizing attempt that tried to reconcile the notions of body memory and affective unconscious aimed to emphasize the bodily basis of the self. The affective relief structure can be imagined as the topography of the passive-implicit layer of the self. Complementing this topographic view, the affective core self emphasizes the role of the body. The body, i.e., especially body memory, expresses the implicit-passive reactions of the affective unconscious, or conversely, the affective unconscious possesses and envelopes the body.

The currents study also examined the temporal dynamics of the affective core self. To achieve this, I agreed with Ullmann's argument according to which retroactivity plays a crucial role in self-constitution and it reveals the protean nature of the seemingly stable self. I expanded the problem of retroactive self-constitution by exploring body memory and the Freudian concept of afterwardness (i.e., *Nachträglichkeit*). Both phenomena can be approached from the perspectives of meaning-bestowing and self-understanding. The analysis of retroactivity and body memory provided the opportunity to introduce implicit (or bodily) meaning-bestowing processes. The idea of implicit-bodily retroactive constitution of sense is the phenomenological reformulation of affect-transformation that occurs in normal and therapeutic circumstances. By introducing bodily retroactivity, this approach reveals the affective-associative connections between past and present and the implicit sense becomes explicit through exploring felt senses and explicating meaning cores. Additionally, in the context of traumatic subjectivity, I tried to illustrate the conjoint aspects of implicit (bodily) and conscious (narrative) retroactive self-understanding.

In the end, a clear demarcation was made between the formal minimal self and the affective core self. The affective core self was inspired by phenomenological insights, but also connected to psychoanalysis and the problem of the unconscious. The current work, approaches the unconscious from multiple perspectives. The Husserlian concept of affective relief and body memory characterized the horizontal extension of the unconscious as a background or fringe consciousness. The characteristics of implicit body memory highlight the pivotal role of traumatic subjectivity that lies at the heart of the affective core self. The affective core self is not only a network of dispositions or habits acquired through development and socialization, but also the implicit realm of past residues that have an implicit impact on the self and its first-person phenomenal field. Pre-reflective self-awareness may constitute the formal feature of identity. The affective core self was introduced as the second layer of the minimal self, in which a further, more personal individuation takes place. The role of time-consciousness is equally important in both cases of the formal minimal self and the affective core self. The former emphasizes the role of the retentional-protentional structure in the diachronic continuity of the self, while the latter places phenomenological weight on the oscillatory movement of consciousness in time.

In my view, the merit of the affective core self lies in the attempt to harmonize the narrative conceptions of the self and the minimal self. According to Zahavi, the theories of narrative identity do not pay enough attention to the phenomenological nature of the self. Narrative theorists, in turn, find Zahavi's minimal self too narrow and limited, failing to adequately account for heterogeneous modes of self-preservation and self-enrichment (not to mention ethical and hermeneutical concerns). I argue that the affective core self may delineate an intermediate level of self-constitution in which certain aspects of narrative self-understanding (especially the psychoanalytic ways of understanding) and implicit processes of trauma processing may converge. Of course, the notion of the affective core self will not reinvent the wheel; rather, it is a synthetic notion that can be approached from various contemporary discourses of the self. From the narrativist point of view it may designate the complicated problems of self-understanding and self-development, and from the point of view of the formal minimal self, it may sound as a psychoanalytic rather than a phenomenological idea. All things considered, in the current interdisciplinary dialogues, the notion of an extended minimal self, i.e., the affective core self, may serve as the foundation for a – phenomenologically inspired – theory of psychointegration.

Bibliography

Albahari, M. (2006). *Analytical Buddhism: The two-tiered illusion of self.* Palgrave Macmillen.
Albahari, M. (2013). Nirvana and ownerless consciousness. In M. Siderits, E. Thompson, & D. Zahavi (Eds.), *Self, no self?: Perspectives from analytical, phenomenological, & Indian traditions* (pp. 79–113). OUP.
Askay, R., & Farquhar, J. (2006). An intelligible yet enigmatic mutual silence: Freud and Husserl. In R. Askay & J. Farquhar (Eds.), *Apprehending the inaccessible: Freudian psychoanalysis and existential phenomenology* (pp. 165–189). Northwestern University Press.
Barresi, J., & Martin, R. (2011). History as prologue: Western theories of the self. In S. Gallagher (Ed.), *The Oxford handbook of the self* (pp. 33–56). OUP.
Bayne, T. (2010). *The Unity of consciousness*. OUP.
Bayne, T., & Chalmers, D. (2003). What is the unity of consciousness? In A. Cleeremans (Ed.), *The Unity of consciousness: Binding, integration, and dissociation* (pp. 23–58). OUP.
Behnke, E. A. (2008). Husserl's protean concept of affectivity: From the texts to the phenomena themselves. *Philosophy Today, 52*, 46–53.
Bermúdez, J. L. (2017). Memory and self-consciousness. In S. Bernecker & K. Michaelian (Eds.), *The Routledge handbook of philosophy of memory* (pp. 180–191). Routledge.
Bernet, R. (2000). The traumatized subject. *Research in Phenomenology, 30*, 160–179.
Birksted-Breen, D. (2003). Time and the *après-coup*. *International Journal of Psychoanalysis, 84*(6), 1501–1515.
Bisteon, G., Vanheule, S., & Craps, S. (2014). Nachträglichkeit: A Freudian perspective on delayed traumatic reactions. *Theory & Psychology, 24*(5), 668–687.
Blackmore, S. (2018). *Consciousness: An introduction*. OUP.
Blaney, P. H. (1986). Affect and memory: A review. *Psychological Bulletin, 99*(2), 229–246.
Blankenburg, W. (2001). First steps toward a psychopathology of "common sense". (Trans. Aaroon L. Mishara). *Philosophy, Psychiatry, and Psychology, 8*(4), 303–315.
Block, N. (1995). On confusion about the function of consciousness. *Behavioral and Brain Sciences, 18*(2), 227–247.
Block, N. (2007). Consciousness, accessibility, and the mesh between psychology and neuroscience. *Behavioral and Brain Sciences, 30*(5–6), 481–499.
Borda, J. P., & Sass, L. A. (2015). Phenomenology and neurobiology of self disorder in schizophrenia: Primary factors. *Schizophrenia Research, 169*(1–3), 464–473.
Bortolan, A. (2020). Affectivity and the distinction between minimal and narrative self. *Continental Philosophy Review, 53*, 67–84.
Bortolotti, L., & Broome, M. R. (2012). Affective dimensions of the phenomenon of double bookkeeping in delusions. *Emotion Review, 4*(2), 187–191.

L. Horváth, *The Affective Core Self*, Contributions to Phenomenology 130,
https://doi.org/10.1007/978-3-031-56920-3

Bower, G. H. (1981). Mood and memory. *American Psychologist, 36*(2), 129–148.

Bower, M. (2015). Genetic phenomenology, cognitive development, and the embodied/extended mind. *Journal of Consciousness Studies, 22*(9–10), 83–108.

Braude, S. E. (1995). *Multiple personality and the philosophy of mind*. Rowman & Littlefield Publishers.

Brewin, C. R., & Andrews, B. (1998). Recovered memories of trauma: Phenomenology and cognitive mechanism. *Clinical Psychology Review, 18*(8), 949–970.

Brown, K., & Bergo, B. (Eds.). (2009). *The trauma controversy: Philosophical and interdisciplinary dialogues*. SUNY Press.

Browning, D. L. (2018). Laplanche on Après-Coup: Translation, time, and trauma. *Journal of the American Psychoanalytic Association, 66*(4), 779–794.

Brudzińska, J. (2012). Depth phenomenology of the emotive dynamics and the psychoanalytic experience. In D. Lohmar & J. Brudzińska (Eds.), *Founding psychoanalysis Phenomenologically* (pp. 23–52). Springer.

Carruthers, G., & Musholt, K. (2018). Ipseity at the intersection of phenomenology, psychiatry and philosophy of mind: Are we talking about the same thing? *Review of Philosophy and Psychology, 9*, 689–701.

Casey, E. S. (2000). *Remembering: A phenomenological study* (2nd ed.). Indiana University Press.

Chalmers, D. (1995). Facing up to the problem of consciousness. *Journal of Consciousness Studies, 2*(3), 200–219. http://consc.net/papers/facing.pdf

Chalmers, D. (1996). *The conscious mind: In search of a fundamental theory*. OUP.

Chapman, J. (1966). The early symptoms of schizophrenia. *The British Journal of Psychiatry, 112*(484), 225–251.

Churchland, P. S. (1989). *Neurophilosophy: Toward a unified science of the mind-brain*. MIT Press.

Colombetti, G. (2011). Varieties of pre-reflective self-awareness: Foreground and background bodily feelings in emotion experience. *Inquiry, 54*(3), 293–313.

Colvin, M. K., & Gazzaniga, M. (2007). Split-brain cases. In M. Velmans & S. Schneider (Eds.), *The Blackwell companion to consciousness* (pp. 182–208). Wiley-Blackwell.

Conrad, K. (2013 [1959]). *Die beginnende Schizophrenie. Versuch einer Gestaltanalyse des Wahns*. Psychiatrie-Verlag.

Conway, M. A. (1992). A structural model of autobiographical memory. In M. A. Conway, D. C. Rubin, H. Spinnler, & W. A. Wagenaar (Eds.), *Theoretical perspectives on autobiographical memory* (pp. 167–194). Kluwer Academic Publishers.

Conway, M., & Pleydell-Pearce, C. W. (2000). The construction of autobiographical memories in the self-memory system. *Psychological Review, 107*(2), 261–288.

Conway, M. A., & Rubin, D. C. (1993). The structure of autobiographical memory. In A. F. Collins, S. E. Gathercole, M. A. Conway, & P. E. Morris (Eds.), *Theories of memory* (pp. 103–137). Lawrence Erlbaum Associates.

Dainton, B. (2004). The self and the phenomenal. *Ratio, 17*(4), 365–389.

Dainton, B. (2008). *The phenomenal self*. OUP.

Damasio, A. (1999). *The feeling of what happens: Body, emotion and the making of consciousness*. Vintage Book.

Davidson, R., & Harrington, A. (2002). *Visions of compassion: Western scientists and Tibetan Buddhists examine human nature*. OUP.

De Sousa, R. (2017). Memory and emotion. In S. Bernecker & K. Michaelian (Eds.), *The Routledge handbook of philosophy of memory* (pp. 154–165). Routledge.

De Warren, N. (2010). The inner night: Towards a phenomenology of (dreamless) sleep. In D. Lohmar & I. Yamaguchi (Eds.), *On time-new contributions to the Husserlian phenomenology of time* (pp. 273–294). Springer.

Dennett, D. C. (1991). *Consciousness explained*. Penguin Books.

Dennett, D. (1992). The self as a center of narrative gravity. In F. S. Kessel, P. M. Cole, & D. L. Johnson (Eds.), *Self and consciousness: Multiple perspectives* (pp. 103–115). Lawrence Erlbaum Associates.

Doran, C. (2017). Rage and anxiety in the Split between Freud and Jung. *Humanities, 6*(3), 53. https://doi.org/10.3390/h6030053

Dreyfus, G. (2013). Self and subjectivity: A middle way approach. In M. Siderits, E. Thompson, & D. Zahavi (Eds.), *Self, no self?: Perspectives from analytical, phenomenological, and Indian traditions* (pp. 114–156). OUP.

Eickhoff, F.-W. (2006). On Nachträglichkeit: The modernity of an old concept. *International Journal of Psychoanalysis, 87*(6), 1453–1469.

Eshleman, M. C. (2007). The Cartesian unconscious. *History of Philosophy Quarterly, 24*(2), 169–187.

Fasching, W. (2008). Consciousness, self-consciousness and meditation. *Phenomenology and the Cognitive Sciences, 7*(4), 463–483.

Feyaerts, J., & Sass, L. (2023). Self-disorder in schizophrenia: A revised view (1. Comprehensive review: Dualities of self- and world-experience). *Schizophrenia Bulletin, sbad169.* https://doi.org/10.1093/schbul/sbad169

Firth, C. D. (1979). Consciousness, information processing, and schizophrenia. *British Journal of Psychiatry, 134*(3), 225–235.

Firth, C. D., & Rees, G. (2007). A brief history of the scientific approach to the study of consciousness. In M. Velmans & S. Schneider (Eds.), *The Blackwell companion to consciousness* (pp. 9–22). Wiley-Blackwell.

Fischmann, T., Russ, M. O., & Leuzinger-Bohleber, M. (2013). Trauma, dream, and psychic change in psychoanalysis: A dialog between psychoanalysis and the neurosciences. *Frontiers in Human Neuroscience, 7*(877), 1–15.

Fivush, R., & McDermott-Sales, J. (2006). Coping, attachment, and mother-child narratives of stressful events. *Mermill-Palmer Quarterly, 52*(1), 125–150.

Flanagan, O. (1991). *The science of the mind* (2nd ed.). The MIT Press.

Flanagan, O. (1992). *Consciousness reconsidered.* The MIT Press.

Fletcher, J. (2013). *Freud and the scene of trauma.* Fordham University Press.

Freud, S. (1955 [1918]). From the history of an infantile neurosis (the *Wolf Man*). In J. Strachey (Ed. & Trans.), *The standard edition of the complete psychological works of Sigmund Freud (Vol. 17)* (pp. 1–124). The Hogarth Press.

Freud, S. (1957a [1915]). Repression. In *Collected papers* (Vol. 14, pp. 84–96). (J. Riviere, Trans.). Basic Books.

Freud, S. (1957b [1915]). The unconscious. In *Collected papers* (Vol. 14, pp. 98–136). (J. Riviere, Trans.). Basic Books.

Freud, S. (1958 [1912]). A note on the unconscious in psycho-analysis. In *Collected papers* (Vol. 12, pp. 260–267). (J. Riviere, Trans.). Basic Books.

Freud, S. (1961 [1930]). Civilization and its discontents. In J. Strachey (Ed. & Trans.), *The standard edition of the complete psychological works of Sigmund Freud (Vol. 21)* (pp. 64–148). The Hogarth Press.

Freud, S. (1962a [1894]). The neuro-psychoses of defence. In *The standard edition of the complete psychological works of Sigmund Freud* (Vol. 3, pp. 45–61). J. Strachey (Ed. & Trans.). The Hogarth Press.

Freud, S. (1962b [1896]). Heredity and the aetiology of the neuroses. In *The standard edition of the complete psychological works of Sigmund Freud* (Vol. 3, pp. 143–156). J. Strachey (Ed. & Trans.). The Hogarth Press.

Freud, S. (1962c [1896]). Further remarks on the neuro-psychoses of defence. In J. Strachey (Ed. & Trans.), *The standard edition of the complete psychological works of Sigmund Freud (Vol. 3)* (pp. 162–185). The Hogarth Press.

Freud, S. (1962d [1896]). The aetiology of hysteria. In J. Strachey (Ed. & Trans.), *The standard edition of the complete psychological works of Sigmund Freud (Vol. 3)* (pp. 191–221). The Hogarth Press.

Freud, S. (1965 [1933]). The dissection of the psychical personality. In J. Strachey (Ed. & Trans.), *New introductory lectures on psycho-analysis* (pp. 71–100). W. W. Norton & Company.

Freud, S. (1966 [1895]). Project for a scientific psychology. In J. Strachey (Ed. & Trans.), *The standard edition of the complete psychological works of Sigmund Freud (Vol. 1)* (pp. 295–398). The Hogarth Press.
Freud, S. (1990a [1920]). *Beyond the pleasure principle*. (J. Strachey Trans.). W. W. Norton & Company.
Freud, S. (1990b [1923]). *The ego and the Id*. (J. Riviere, Trans.). W. W. Norton & Company.
Freud, S., & Beruer, J. (1955 [1895]). Studies on hysteria. In *The standard edition of the complete psychological works of Sigmund Freud* (Vol. 2, pp. 294–398). J. Strachey (Ed. & Trans.) The Hogarth Press.
Fuchs, T. (2005a). Corporealized and disembodied minds: A phenomenological view of the body in melancholia and schizophrenia. *Philosophy, Psychiatry, and Psychology, 12*(2), 95–107.
Fuchs, T. (2005b). Delusional mood and delusional perception – A phenomenological analysis. *Psychopathology, 38*(3), 133–139.
Fuchs, T. (2012a). The phenomenology of body memory. In S. C. Koch, T. Fuchs, M. Summa, & C. Müller (Eds.), *Body memory, metaphor and movement* (pp. 9–22). John Benjamins Publishing Company.
Fuchs, T. (2012b). Body memory and the unconscious. In D. Lohmar & J. Brudzińska (Eds.), *Founding psychoanalysis Phenomenologically* (pp. 69–82). Springer.
Fuchs, T. (2013a). The phenomenology of affectivity. In K. W. M. Fulford, M. Davies, R. G. T. Gipps, G. Graham, J. Z. Sadler, G. Stanghellini, & T. Thornton (Eds.), *The Oxford handbook of philosophy and psychiatry* (pp. 612–631). Oxford University Press.
Fuchs, T. (2013b). Zwischen Leib und Körper. In M. Hähnel & M. Kanup (Eds.), *Leib und Leben: Perspektiven für eine neue Kultur der Körperlichkeit* (pp. 82–93). WBG.
Fuchs, T. (2015). From self-disorders to ego disorders. *Psychopathology, 48*(5), 324–331.
Fuchs, T. (2017). Self across time: The diachronic unity of bodily existence. *Phenomenology and the Cognitive Sciences, 16*(2), 291–315.
Fuchs, T. (2019). The interactive phenomenal field and the life space: A sketch of an ecological concept of psychotherapy. *Psychopathology, 52*(2), 67–74.
Fuchs, T., & Koch, S. (2014). Embodied affectivity: On moving and being moved. *Frontiers in Psychology, 5*(508), 1–12.
Fuchs, T., & Schlimme, J. E. (2009). Embodiment and psychopathology: A phenomenological perspective. *Current Opinion in Psychiatry, 22*(6), 570–575.
Gallagher, S. (1986). Hyletic experiences and the lived body. *Husserl Studies, 3*(2), 131–166.
Gallagher, S. (2000). Philosophical conceptions of the self: Implications for cognitive science. *Trends in Cognitive Sciences, 4*(1), 14–21.
Gallagher, S. (2005). *How the body shapes the mind*. OUP.
Gallagher, S. (2009). Delusional realities. In M. R. Broome & L. Bortolotti (Eds.), *Psychiatry as cognitive neuroscience: Philosophical perspectives* (pp. 245–266). OUP.
Gallagher, S., & Zahavi, D. (2012). *The phenomenological mind*. Routledge.
Gardner, S. (1991). The unconscious. In J. Neu (Ed.), *The Cambridge companion to Freud* (pp. 136–160). CUP.
Gendlin, E. T. (1981). *Focusing* (2nd ed.). Bantam Books.
Giampieri-Deutsch, P. (2012a). Psychoanalysis: Philosophy and/or science of subjectivity? Prospects for a dialogue between phenomenology, philosophy of mind, and psychoanalysis. In D. Lohmar & D. Brudzińska (Eds.), *Founding psychoanalysis phenomenologically* (pp. 83–103). Springer.
Giampieri-Deutsch, P. (2012b). Perception, conscious and unconscious processes. In F. Barth, P. Giampieri-Deutsch, & H.-D. Klein (Eds.), *Sensory perception. Mind and matter* (pp. 245–264). Springer Verlag.
Gibson, J. J. (1983). *The senses considered as perceptual systems*. Praeger.
Goldie, P. (2003). One's remembered past: Narrative thinking, emotion, and the external perspective. *Philosophical Papers, 32*(3), 301–319.
Goldie, P. (2012). *The mess inside: Narrative, emotion, and the mind*. OUP.

Guillot, M. (2017). I me mine: On a confusion concerning the subjective character of experience. *Review of Philosophy and Psychology, 8*(1), 23–53.

Gusich, G. (2012). A phenomenology of emotional trauma: Around and about the things themselves. *Human Studies, 35*(4), 505–518.

Habermas, J. (1971 [1968]). *Knowledge and human interest*. Boston Press.

Hacking, J. (1995). *Rewriting the soul: Multiple personality and the science of memory*. Princeton University Press.

Hanna, R., & Maiese, M. (2009). *Embodied minds in action*. OUP.

Hauke, C. (2000). *Jung and the postmodern: The interpretation of realities*. Routledge.

Hayne, H., Gross, J., McNamee, S., Fitzgibbon, O., & Tustin, K. (2011). Episodic memory and episodic foresight in 3- and 5-year-old children. *Cognitive Development, 26*(4), 343–355.

Henriksen, M. G. (2013). On incomprehensibility in schizophrenia. *Phenomenology and the Cognitive Sciences, 12*, 105–129.

Henry, M. (1973). *The essence of manifestation*. (G. Etzkorn, Trans.). Kluwer Academic Publisher.

Holenstein, E. (1972). *Husserls Phänomenologie der Assoziation. Zu Struktur und Funktion eines Grundprinzips der passive Genesis bei Edmund Husserl*. The Hague.

House, J. (2017). The ongoing rediscovery of Après-coup as a central Freudian concept. *Journal of the American Psychoanalytic Association, 65*(5), 773–789.

Hunt, N. C. (2010). *Memory, war and trauma*. CUP.

Husserl, E. (1970). *The crisis of European sciences and transcendental phenomenology*. (D. Carr, Trans.). Northwestern University Press.

Husserl, E. (1977 [1925]). *Phenomenological psychology. Lectures, summer semester 1925*. (J. Scanlon, Trans.). Martinus Nijhoff.

Husserl, E. (1989). *Ideas pertaining to a pure phenomenology and to a phenomenological philosophy, second book: Studies in the phenomenology of constitution*. (R. Rojcewicz, & A. Schuwer, Trans.). Kluwer Academic Publishers.

Husserl, E. (1991). *On the phenomenology of the consciousness of internal time (1893–1917)*. (J. B. Brough, Trans.). Kluwer Academic Publishers.

Husserl, E. (1998). Static and genetic phenomenological method (A. J. Steinbock, Trans.). *Continental Philosophy Review, 31*(2), 135–142.

Husserl, E. (2001). *Analyses concerning passive and active synthesis: Lectures on transcendental logic*. (A. J. Steinbock, Trans.). Kluwer Academic Publishers.

Husserl, E. (2014). Grenzprobleme der Phänomenologie. Analysen des Unbewusstseins und der Instinkte. In *Metaphysik. Späte Ethik (Texte aus dem Nachlass 1908–1937)*. Springer.

Hutto, D. D. (2017). Memory and narrativity. In S. Bernecker & K. Michaelian (Eds.), *The Routledge handbook of philosophy of memory* (pp. 192–204). Routledge.

Jackson, F. (1983). Epiphenomenal qualia. *Philosophical Quarterly, 32*(127), 127–136.

James, W. (1983 [1890]). *The principles of psychology*. HUP.

Jaspers, K. (1973). *Allgemeine Psychopathologie (Neunte, unveränderte Auflage)*. Springer.

Jaynes, J. (1986). Consciousness and the voices of the mind. *Canadian Psychology/Psychologie Canadienne, 27*(2), 128–148.

Jaynes, J. (2000 [1976]). *The origin of consciousness in the breakdown of the bicameral mind*. Mariner Books.

Jopling, D. A. (2000). *Self-knowledge and the self*. Routledge.

Jung, C. G. (1961 [1913]). *Freud and psychoanalysis: The theory of psychoanalysis* (Vol. 4, pp. 83–228). (R. F. C. Hull, Trans.) Routledge & Kegen Paul.

Kihlstrom, J. F. (1987). The cognitive unconscious. *Science, 237*(4821), 1445–1452.

Kihlstrom, J. F., Dorfman, J., & Park, L. (2007). Implicit and explicit memory and learning. In M. Velmans & S. Schneider (Eds.), *The Blackwell companion to consciousness* (pp. 525–539). Wiley-Blackwell.

Kihlstrom, J. F., Shelagh, M., Tobias, B., & Tobis, I. (2000). The emotional unconscious. In E. Eich, J. Kihlstrom, G. Bower, J. Forgas, & P. Niedenthal (Eds.), *Cognition and emotion* (pp. 30–86). OUP.

Kozyreva, A. (2017). *Phenomenology of affective subjectivity: Analyses on the pre-reflective Unity of subjective experience*. Universitäts Bibliothek.

Kozyreva, A. (2018). Non-representational approaches to the unconscious in the phenomenology of Husserl and Merleau-Ponty. *Phenomenology and the Cognitive Sciences, 17*(1), 199–224.

Kriegel, U. (2003). Consciousness as intransitive self-consciousness: Two views and an argument. *Canadian Journal of Philosophy, 33*(1), 103–132.

Krippner, S. (1972). Altered states of consciousness. In J. White (Ed.), *The highest state of consciousness* (pp. 1–5). Doubleday.

Krueger, J. W. (2013). The who and the how of experience. In M. Siderits, E. Thompson, & D. Zahavi (Eds.), *Self, no self?: Perspectives from analytical, phenomenological, and Indian traditions* (pp. 27–55). OUP.

Lacan, J. (2006 [1953]). The function and field of speech and language in psychoanalysis. In *Écrits: The first complete translation in English* (pp. 197–268). (B. Fink, Trans.). Norton.

Laing, R. D. (1978). *The divided self.* The Penguin Force.

Laplanche, J. (1976). *Life and death in psychoanalysis.* (J. Mehlam, Trans.). The Johns Hopkins University Press.

Laplanche, J. (1999a). Notes on afterwardsness. In J. Laplanche (Auth.) (Ed.), *Essays on otherness* (pp. 264–269). Routledge.

Laplanche, J. (1999b). Time and the other. In J. Laplanche (Auth.) (Ed.), *Essays on otherness* (pp. 238–263). (L. Thurston, Trans.). Routledge.

Laplanche, J. (2017 [1989–1990]). *Problématiques VI: Après-coup* (J. House, Trans.). The Unconscious in Translation.

Laplanche, J., & Pontalis, J.-B. (1973 [1967]). Deferred action; deferred. In *The language of psycho-analysis* (pp. 111–114). (D. Nicholson-Smith, Trans.). The Hogarth Press and the Institute of Psycho-Analysis.

Larrabee, M. J. (1995). The time of trauma: Husserl's phenomenology and post-traumatic stress disorder. *Human Studies, 18*(4), 351–366.

Levine, J. (1983). Materialism and qulia: The explanatory gap. *Pacific Philosophical Quarterly, 64*, 354–361.

Levine, P. A. (2015). *Trauma and memory: Brain and body in a search for the living past.* North Atlantic Books.

Loftus, E. F., & Kaufman, L. (1992). Why do traumatic experiences sometimes produce good memory (flashbulbs) and sometimes no memory (repression)? In E. Winograd & U. Neisser (Eds.), *Affect and accuracy in recall: Studies of "flashbulb" memories* (pp. 212–223). Cambridge University Press.

Lysaker, P. H., & Lysaker, J. T. (2010). Schizophrenia and alterations in self-experience: A comparison of 6 perspectives. *Schizophrenia Bulletin, 36*(2), 331–340.

MacIntyre, A. (2007). *After virtue: A study in moral theory* (3d ed.). University of Notre Dame Press.

Maiese, M. (2011). *Embodiment, emotion, and cognition.* Palgrave Macmillan.

Maiese, M. (2016). *Embodied selves and divided minds.* OUP.

Marosán, B. P. (2020). *Kontextus és fenomén II. Az igazság problémája az időtudat és a keletkezés husserli filozófiájában.* L'Harmattan.

Marosán, B. P. (2023). The genesis of the minimal mind: Elements of a phenomenological and functional account. *Phenomenology and the Cognitive Sciences.* https://doi.org/10.1007/s11097-023-09946-7

McCarroll, C. J. (2018). Looking at the self: Perspectival memory and personal identity. *Philosophical Explorations, 22*(3), 259–279.

McGinn, C. (1989). Can we solve the mind-body problem? *Mind, 98*, 349–366.

Menary, R. (2008). Embodied narratives. *Journal of Consciousness Studies, 15*(6), 63–84.

Merleau-Ponty. (2012). *The phenomenology of perception.* (D. A. Landes, Trans.). Routledge.

Metzinger, T. (2003). *Being no one: The self-model theory of subjectivity.* MIT Press.

Metzinger, T. (2006). Reply to Zahavi: The value of historical scholarship. *Psyche, 12*(2), 1–7.

Metzinger, T. (2009). *The ego tunnel: The science of the mind and the myth of the self.* Basic Books.

Michie, P. T., Badrock, J. C., Waters, F. A. V., & Maybery, M. T. (2005). Auditory hallucinations: Failure to inhibit irrelevant memories. *Cognitive Neuropsychiatry, 10*(2), 125–136.

Mink, L. O. (1970). History and fiction as modes of comprehension. *New Literary History, 1*(3), 541–558.

Mishara, A. (1990). Husserl and Freud: Time, memory and the unconscious. *Husserl Studies, 7*, 29–58.

Mishara, A. (2010). Klaus Conrad (1905-1961): Delusional mood, psychosis, and beginning of schizophrenia. *Schizophrenia Bulletin, 36*(1), 9–13.

Moran, D. (2011). Edmund Husserl's phenomenology of habituality and habitus. *Journal of British Society for Phenomenology, 42*(1), 53–77.

Moran, D. (2013a). The phenomenology of embodiment: Intertwining and reflexivity. In R. T. Jensen & D. Moran (Eds.), *The phenomenology of embodied subjectivity* (pp. 285–303). Springer.

Moran, D. (2013b). 'Let's look at it objectively': Why phenomenology cannot be naturalized. *Royal Institute of Philosophy Supplement, 72*, 89–115.

Moran, D. (2017). Husserl's layered concept of the human person: Conscious and unconscious. In D. Legrand & D. Trigg (Eds.), *Unconsciousness between phenomenology and psychoanalysis* (pp. 3–23). Springer.

Morley, J. (2003). Phenomenological and biological psychiatry: Complementary or mutual. *Philosophy, Psychiatry, and Psychology, 9*(1), 87–90.

Nagel, T. (1974). What is it like to be a bat? *The Philosophical Review, 83*(4), 435–450. Retrieved from: https://www.jstor.org/stable/2183914

Neisser, U. (2014 [1967]). *Cognitive psychology* (Classic ed.). Psychology Press.

Nelson, K. (2003a). Narrative and the emergence of a consciousness of self. In G. D. Fireman, T. E. McVay, & O. J. Flanagan (Eds.), *Narrative and consciousness* (pp. 17–36). OUP.

Nelson, K. (2003b). Narrative and self, myth and memory: Emergence of the cultural self. In R. Fivush & C. A. Haden (Eds.), *Autobiographical memory and the construction of a narrative self: Developmental and cultural perspectives* (pp. 3–28). Lawrence Erlbaum Associates Publishers.

Nelson, B., Fornito, A., Harrison, B. J., Yücel, M., Sass, L. A., Yung, A. R., Thompson, A., Wood, S. J., Pantelis, C., & McGorry, P. D. (2009). A disturbed sense of self in the psychosis prodrome: Linking phenomenology and neurobiology. *Neuroscience and Biobehavioral Reviews, 33*(6), 807–817.

Nelson, B., & Sass, L. S. (2009). Medusa's stare: A case study of working with self-disturbance in early phase of schizophrenia. *Clinical Case Studies, 8*(6), 489–504.

Nichols, S. (2017). Memory and personal identity. In S. Bernecker & K. Michaelian (Eds.), *The Routledge handbook of philosophy of memory* (pp. 169–179). Routledge.

Noel-Smith, K. (2016). *Freud on time and timelessness*. Palgrave Macmillan.

Olsen, E. T. (1998). There is no problem of the self. *Journal of Consciousness Studies, 5*(5–6), 645–657.

Overgaard, S. (2007). What, if anything, can phenomenology teach psychopathology (and vice versa)? *Husserl Studies in Japan, 4*(5), 32–42.

Pardiñas, A., & Holmans, P. (2018). Common schizophrenia alleles are enriched in mutation-intolerant genes and in regions under strong background selection. *Nature Genetics, 50*, 381–389.

Parnas, J., & Sass, L. A. (2011). The structure of self-consciousness in schizophrenia. In S. Gallagher (Ed.), *The Oxford handbook of the self* (pp. 524–546). OUP.

Pienkos, E., Silverstein, S., & Sass, L. S. (2017). The phenomenology of anomalous world experience in schizophrenia: A qualitative study. *Journal of Phenomenological Study, 48*(2), 188–213.

Rabelhofer, B. (2019). Inhabiting a time before time. Freud's concept of trauma as a psychoanalytical figure of thought. *Studia Theodisca, 26*, 57–71.

Ratcliffe, M. (2008). *Feelings of being: Phenomenology, psychiatry and the sense of reality*. OUP.

Ratcliffe, M. (2011). Phenomenology is not a servant of science. *Philosophy, Psychiatry, & Psychology, 18*(1), 33–36.

Ratcliffe, M. (2013). Delusional atmosphere and the sense of unreality. In G. Stanghellini & T. Fuchs (Eds.), *One century of Karl Jaspers' general psychopathology* (pp. 229–244). OUP.

Ratcliffe, M. (2015). *Experience of depression: A study in phenomenology*. OUP.

Ratcliffe, M. (2016). Existential feeling and narrative. In T. Breyer & O. Müller (Eds.), *Funktionen des Lebendigen* (pp. 169–192). De Gruyter.

Ratcliffe, M. (2017a). *Real hallucinations: Psychiatric illness, intentionality, and the interpersonal world*. MIT Press.

Ratcliffe, M. (2017b). Selfhood, schizophrenia, and the interpersonal regulation of experience. In C. Durt, T. Fuchs, & C. Tewes (Eds.), *Embodiment, Enaction, and culture: Investigation the constitution of the shared world* (pp. 149–172). The MIT Press.

Revonsuo, A. (2018). *Foundations of consciousness*. Routledge.

Ricoeur, P. (1970). *Freud and philosophy: An essay on interpretation*. YUP.

Ricoeur, P. (1992). *Oneself and another*. (K. Blamey, Trans.). The University of Chicago Press.

Roediger, H. L. (1990). Implicit memory. *Retention without remembering. American Psychologist, 45*(9), 1043–1056.

Rogozinski, J. (2010). *The ego and the flesh*. (R. Vallier, Trans.). Standford University Press.

Saks, E. R. (2015). *The center cannot hold: My journey through madness*. Hachette Books.

Sass, L. A. (2004). Affectivity and schizophrenia: A phenomenological view. In D. Zahavi (Ed.), *Hidden resources: Classical resources on subjectivity* (pp. 127–147). Imprint Academic.

Sass, L. A. (2017). *Madness and modernism: Insanity in the light of modern art, literature, and thought*. (Revised edition). Oxford University Press.

Sass, L. A., & Parnas, J. (2003). Schizophrenia, consciousness, and the self. *Schizophrenia Bulletin, 29*(3), 427–444.

Sass, L. A., & Ratcliffe, M. (2017). Atmosphere: On the phenomenology of "atmospheric" alterations in schizophrenia – Overall sense of reality, familiarity, vitality, meaning, or relevance (ancillary article to EAWE domain 5). *Psychopathology, 50*(1), 90–97.

Saunders, G. (2014). *Acts of consciousness: A social psychology standpoint*. CUP.

Schacter, D. L. (1996). *Searching for memory: The brain, the mind, and the past*. Basic Books.

Schacter, D. L. (2001). *The seven sins of memory: How the mind forgets and remembers*. Houghton Mifflin Harcout.

Schacter, D. L., Chiu, C. Y., & Ochsner, K. N. (1993). Implicit memory: A selective review. *Annual Review of Neuroscience, 16*, 159–182.

Schechtman, M. (1994). The truth about memory. *Philosophical Psychology, 7*(1), 3–18.

Schechtman, M. (1996). *The constitution of selves*. Cornell University Press.

Schechtman, M. (2001). Empathic access: The missing ingredient in personal identity. *Philosophical Explorations, 4*(2), 95–111.

Schechtman, M. (2007). Stories, lives, and basic survival: A refinement and defense of the narrative view. *Royal Institute of Philosophy Supplement, 60*, 155–178.

Schechtman, M. (2009). Personal identity. In J. Symons & P. Calvo (Eds.), *The Routledge companion to philosophy of psychology* (pp. 634–646). Routledge.

Schechtman, M. (2011). The narrative self. In S. Gallagher (Ed.), *The Oxford handbook of the self* (pp. 394–418). OUP.

Schechtman, M. (2014). *Staying alive: Personal identity, practical concerns, and the Unity of a life*. OUP.

Schechtman, M. (2016). A mess inside: Emphatic access, narrative, and identity. In J. Dodd (Ed.), *Art, mind, and narrative: Themes from the work of Peter Goldie* (pp. 17–34). OUP.

Schmitz, H., Müllan, R. O., & Slaby, J. (2011). Emotions outside the box – The new phenomenology of feeling and corporeality. *Phenomenology and the Cognitive Sciences, 10*(2), 241–259.

Sechehaye, M. (1970). *Autobiography of a schizophrenic girl*. American Library.

Skrbina, D. (2005). *Panpsychism in the west*. The MIT Press.

Slors, M. V. P., & Jongepier, F. (2014). Mineness without minimal selves. *Journal of Consciousness Studies, 21*(7–8), 193–219.

Smith, N. (2010). *Towards a phenomenology of repression – A Husserlian reply to the Freudian challenge*. Stockholm University Press.

Stanghellini, G. (2009). Embodiment and schizophrenia. *World Psychiatry, 8*, 56–59.

Steinbock, A. J. (2004). Affection and attention: On the phenomenology of becoming aware. *Continental Philosophy Review, 37*, 21–43.

Stephenson, H., & Parnas, J. (2018). What can self-disorders in schizophrenia tell us about the nature of subjectivity? A psychopathological investigation. *Phenomenology and the Cognitive Sciences, 17*, 629–642.

Stern, D. N. (1985). *The interpersonal world of the infant: A view from psychoanalysis and Developmental Psychology*. Basic Books.

Stevens, G. L., & Graham, G. (2007). Philosophical psychopathology and self-consciousness. In M. Velmans & S. Schneider (Eds.), *The Blackwell companion to consciousness* (pp. 195–208). Wiley-Blackwell.

Stocker, K. (2020). Mental perspectives during temporal experience in posttraumatic stress disorder. *Phenomenology and the Cognitive Sciences, 19*(2), 321–334.

Stolorow, R. D. (2008). The contextuality and existentiality of emotional trauma. *Psychoanalytic Dialogues, 18*(1), 113–123.

Stone, M. (2006). The analyst's body as tuning fork: Embodied resonance in countertransference. *Journal of Analytical Psychology, 51*(1), 109–124.

Strawson, G. (1999). The self and the SESMET. *Journal of Consciousness Studies, 6*(4), 99–135.

Strawson, G. (2000). The phenomenology and ontology of the self. In D. Zahavi (Ed.), *Exploring the self: Philosophical and psychopathological perspectives on self-experience* (pp. 39–54). John Benjamins.

Strawson, G. (2011). The minimal subject. In S. Gallagher (Ed.), *The Oxford handbook of the self* (pp. 253–278). OUP.

Strawson, G. (2015). Self-intimation. *Phenomenology and the Cognitive Sciences, 14*(1), 1–31.

Struck, K. (2020). Mental perspective during temporal experience in posttraumatic stress disorder. *Phenomenology and the Cognitive Sciences, 19*, 321–334.

Summa, M. (2014). *Spatio-temporal intertwining: Husserl's transcendental aesthetic*. Dordrecht Springer.

Summa, M. (2015). Are emotions recollected in tranquility? Phenomenological reflections on emotions, memory, and the temporal dynamics of experience. In M. Ubiali & M. Wehrle (Eds.), *Edmund Husserl on: Feeling and value, willing and action* (pp. 163–181). Springer.

Summa, M. (2016). Gefühl, Emotion und Erinnerung beim traumatisierten Selbst. Eine Frage der Kohärenz? In R. Esterbauer, P. Schmidt, A. Paletta, & D. Duncan (Eds.), *Bodytime. Leib und Zeit bei Burnout und in anderen Grenzerfahrungen* (pp. 307–329). Verlag Karl Alber.

Summa, M., Koch, S. C., Fuchs, T., & Müller, C. (2012). Body memory: An integration. In S. C. Koch (Ed.), *Body memory, metaphor and movement* (pp. 417–444). John Benjamins Pub.

Sutyák, T. (2008). *A Freud-elv – A pszichoanalízis mint lélekfilozófia*. Kalligram.

Szabó, A. (2016). *Phenomenology and psychosomatics: A Jungian and post-Jungian approach* (dissertation). Retrieved from: https://dea.lib.unideb.hu/dea/handle/2437/229953. (09. Jul. 2019).

Tengelyi, L. (2007a). *Tapasztalat és kifejezés*. Atlantisz.

Tengelyi, L. (2007b). *Erfahrung und Ausdruck: Phänomenologie im Umbruch bei Husserl und seinen Nachfolgern*. Springer.

Thiel, U. (2011). *The early modern subject: Self-consciousness and personal identity from Descartes to Hume*. OUP.

Thompson, E. (2007). *Mind in life: Biology, phenomenology, and the sciences of mind*. HUP.

Tulving, E. (2002). Episodic memory: From mind to brain. *Annual Review of Psychology, 53*(1), 1–25.

Tye, M. (2007). Philosophical problems of consciousness. In M. Velmans & S. Schneider (Eds.), *The Blackwell companion to consciousness* (pp. 23–36). Wiley-Blackwell.

Ullmann, T. (2010). *A láthatatlan forma*. L'Harmattan.
Ullmann, T. (2012a). Az idő mint a szubjektivitás alapja és horizontja. In T. Ullmann (Ed.), *Az értelem dimenziói* (pp. 91–109). Könyvpont – L'Harmattan.
Ullmann, T. (2012b). Fantázia és intencionalitás. In T. Ullmann (Ed.), *Az értelem dimenziói* (pp. 125–149). Könyvpont – L'Harmattan.
Ullmann, T. (2015). A narratív, a traumatikus és az affektív szubjektivitás. In I. Bujalos, M. Tóth, & T. Valastyán (Eds.), *Az identitás alakzatai* (pp. 21–37). Kalligram.
Ullmann, T. (2017). Phenomenology of experience and the problem of the unconscious. In M. Gabriel, C. S. Olay, & S. Ostritsch (Eds.), *Welt und Unendlichkeit: Ein deutsch-ungarischer Dialog in memoriam László Tengelyi* (pp. 141–161). Verlag Karl Alber.
Ulmman, T. (2012). Fantázia és intencionalitás. In T. Ullmann (Ed.), *Az értelem dimenziói* (pp. 125–149). Könyvpont – L'Harmattan.
Varela, F. (1996). Neurophenomenology: A methodological remedy for the hard problem. *Journal of Consciousness Studies, 3*(4), 330–349.
Varela, F., & Shear, J. (Eds.). (2002). *The view from within: First-person approaches to the study of consciousness*. Imprint Academic.
Varela, F., Thompson, E., & Rosch, E. (1991). *The embodied mind: Cognitive science and human experience*. The MIT Press.
Varga, S. (2012). Non-reflective self-awareness: Towards a 'situated' account. *Journal of Consciousness Studies, 19*(3–4), 164–193.
Velleman, J. D. (2006). *Self to self: Selected essays*. Cambridge University Press.
Waldenfels, B. (2002). *Bruchlinien der Erfahrung*. Suhrkamp.
Weiskrantz, L. (2007). The case of blindsight. In M. Velmans & S. Schneider (Eds.), *The Blackwell companion to consciousness* (pp. 175–180). Wiley-Blackwell.
Welsh, T. (2002). The retentional and the repressed: Does Freud's concept of the unconscious threaten Husserlian phenomenology? *Human Studies, 25*, 163–183.
Wilkes, K. V. (1981). Multiple personality and personal identity. *The British Journal for the Philosophy of Science, 32*(4), 331–348.
Wilkes, K. V. (1988). *Real people: Personal identity without thought experiments*. OUP.
Wilkes, K. V. (1995). Losing consciousness. In T. Metzinger (Ed.), *Conscious Experience*. Schöningh.
Williams, G. (2011). What is it like to be nonconscious? A defense of Julian Jaynes. *Phenomenology and the Cognitive Sciences, 10*(2), 217–239.
Williams, H. L., Conway, M. A., & Cohen, G. (2008). Autobiographical memory. In G. Cohen & M. Conway (Eds.), *Memory in the real world* (3rd ed., pp. 21–90). Psychology Press.
Wollheim, R. (1984). *The thread of life*. Yale University Press.
Yaw, M. (2015). Husserl and PTSD: The traumatic correlate. *Journal of Phenomenological Psychology, 46*, 206–226.
Yiassemides, A. (2014). *Time and timelessness: Temporality in the theory of Carl Jung*. Routledge.
Zahavi, D. (1998). Brentano and Husserl on self-awareness. *Études phénoménologiques, 14*(27–28), 127–168.
Zahavi, D. (1999). *Self-awareness and alterity: A phenomenological investigation*. Northwestern University Press.
Zahavi, D. (2002). First-person thoughts and embodied self-awareness: Some reflections on the relation between recent analytical philosophy and phenomenology. *Phenomenology and the Cognitive Sciences, 1*, 7–26.
Zahavi, D. (2003). *Husserl's phenomenology*. Stanford University Press.
Zahavi, D. (2005a). *Subjectivity and selfhood: Investigating the first-person perspective*. MIT Press.
Zahavi, D. (2005b). Being someone. *Psyche, 11*(5), 1–20. Retrieved from: https://cfs.ku.dk/staff/zahavi-publications/metzinger.pdf
Zahavi, D. (2006). Thinking about (self-)consciousness: Phenomenological perspectives. In U. Kriegel & K. Williford (Eds.), *Self-representational approaches to consciousness* (pp. 273–295). The MIT Press.

Zahavi, D. (2007). Self and other: The limits of narrative understanding. *Royal Institute of Philosophy Supplement, 60*, 179–202.

Zahavi, D. (2010). Minimal self and narrative self: A distinction in need of refinement. In T. Fuchs, H. C. Sattel, & P. Henningsen (Eds.), *The embodied self: Dimensions, coherence and disorders* (pp. 3–12). Schattauer.

Zahavi, D. (2012). The time of the self. *Grazer Philosophische Studien, 84*, 143–159.

Zahavi, D. (2013). The experiential self: Objections and clarifications. In M. Siderits, E. Thompson, & D. Zahavi (Eds.), *Self, no self?: Perspectives from analytical, phenomenological, and Indian traditions* (pp. 56–78). OUP.

Zahavi, D. (2014). *Self & other: Exploring subjectivity, empathy, and shame*. OUP.

Zahavi, D. (2016). Analytic and continental philosophy: From duality through plurality to (some kind of) unity. In S. Rinofner-Kreidl & H. A. Wiltsche (Eds.), *Analytic and continental philosophy: Methods and perspectives. Proceedings of the 37th International Wittgenstein Symposium* (pp. 79–94). Walter De Gruyter.

Zahavi, D. (2017). Thin, thinner, thinnest: Defining the minimal self. In C. Durt, T. Fuchs, & C. Tewes (Eds.), *Embodiment, enaction, and culture: Investigation the constitution of the shared world* (pp. 193–200). MIT Press.

Zahavi, D. (2020). Consciousness and (minimal) selfhood: Getting clearer on for-me-ness. In U. Kriegel (Ed.), *The Oxford handbook of the philosophy of consciousness* (pp. 1–19). OUP. (in press).

Zahavi, D., & Kriegel, U. (2016). For-me-ness: What it is and what it is not. In D. O. Dahlstrom, A. Elpidorou, & W. Hopp (Eds.), *Philosophy of mind and phenomenology: Conceptual and empirical approaches* (pp. 36–56). Routledge.

Zhiqiang, L., Chen, J., & Shi, Y. (2017). Genome-wide association analysis identifies 30 new susceptibility loci for schizophrenia. *Nature Genetics, 49*, 1576–1583.

Index

L. Horváth, *The Affective Core Self*, Contributions to Phenomenology 130,
https://doi.org/10.1007/978-3-031-56920-3

MIX
Papier aus verantwortungsvollen Quellen
Paper from responsible sources
FSC® C105338

If you have any concerns about our products,
you can contact us on
ProductSafety@springernature.com

In case Publisher is established outside the EU,
the EU authorized representative is:
Springer Nature Customer Service Center GmbH
Europaplatz 3, 69115 Heidelberg, Germany

Printed by Libri Plureos GmbH
in Hamburg, Germany